R. Lipowsky · D. Richter · K. Kremer (Eds.)

The Structure and Conformation of Amphiphilic Membranes

Proceedings
of the International Workshop
on Amphiphilic Membranes, Jülich, Germany,
September 16–18, 1991

With 150 Figures

Springer-Verlag
Berlin Heidelberg New York
London Paris Tokyo
Hong Kong Barcelona
Budapest

Prof. Reinhard Lipowsky
Prof. Dieter Richter
Dr. Kurt Kremer

Institut für Festkörperforschung, Forschungszentrum Jülich,
W-5170 Jülich, Fed. Rep. of Germany

ISBN-13: 978-3-642-84765-3 e-ISBN-13: 978-3-642-84763-9
DOI: 10.1007/978-3-642-84763-9

Library of Congress Cataloging-in-Publication Data. International Workshop on Amphiphilic Membranes (1991 : Jülich, Germany). The structure and conformation of amphiphilic membranes : proceedings of the International Workshop on Amphiphilic Membranes, Jülich, Germany, September 16-18, 1991 / R. Lipowsky, D. Richter, K. Kremer (eds.). p. cm. – (Springer proceedings in physics ; 66) Includes bibliographical references and index. ISBN 3-540-55452-1 (Berlin). – ISBN 0-387-55452-1 (New York) 1. Membranes (Biology) – Congresses. 2. Membranes (Technology) – Congresses. 3. Bilayer lipid membranes – Congresses. 4. Thin films, Multilayered – Congresses. I. Lipowsky, R. (Reinhard), 1953- . II. Richter, D. (Dieter), 1947- . III. Kremer, K. (Kurt), 1956- . IV. Title. V. Series: Springer proceedings in physics ; v.66. QH601.I555 1991 574.87'5–dc20 92-14296

Softcover reprint of the hardcover 1st edition 1992

Typesetting: Camera ready by authors/editors
54/3140 - 5 4 3 2 1 0 - Printed on acid-free paper

Preface

This volume contains the proceedings of the International Workshop on Amphiphilic Membranes which took place at the Forschungszentrum Jülich, Germany, September 16–18, 1991.

About 150 participants from all over Europe, from the United States, and Canada came together in order to discuss the latest developments in this rapidly evolving field of research. Unfortunately, there were too many submitted papers and we could not accept all of them. In the end, the program consisted of five oral sessions with 31 talks and three poster sessions with 48 posters. Some participants felt that this was still too much excitement for three days but there was also a general consensus that the whole event was very successful and should be repeated.

The workshop was primarily concerned with three aspects of membranes: (I) Molecular structure of monolayers and bilayers, (II) Conformation of membranes and vesicles, and (III) Membranes in complex fluids. Consequently, these proceedings contain three parts corresponding to these three topics.

This workshop was made possible by a grant from the Forschungszentrum KFA Jülich and by support from the workshop secretariat at the Forschungszentrum. We thank Wolfgang Helfrich, Heinz Hoffmann, and Erich Sackmann for their work in the Scientific Committee. We also acknowledge the help of Liselotte Moll and Magdalena Kötter with the organization of this workshop, and thank Magdalena Kötter for her help in the editorial process.

Jülich
November 1991

R. Lipowsky
D. Richter
K. Kremer

Contents

Part II Conformation of Membranes

Part III Membranes in Complex Fluids

The Structure and Conformation of Amphiphilic Membranes: Overview

R. Lipowsky, D. Richter, and K. Kremer

Institut für Festkörperforschung, Forschungszentrum Jülich, Postfach 1913, W-5170 Jülich, Fed. Rep. of Germany

Amphiphilic molecules such as lipids or surfactants have a hydrophilic (polar) 'head' and a hydrophobic (non–polar) 'tail' consisting of one or two hydrocarbon chains. In aqueous solution or in oil–water mixtures, these molecules aggregate in such a way that the tails have essentially no contact with the aqueous medium. This 'hydrophobic effect' which is mainly entropic in origin [1] leads to a variety of different aggregates. In many cases, the molecules form membranes, i.e., monolayers or bilayers.

The workshop was primarily concerned with three aspects of membranes; (I) Molecular structure of monolayers and bilayers; (II) Conformation of membranes and vesicles; and (III) Membranes in complex fluids. Therefore, these proceedings contain three parts corresponding to these three topics. As one can see from the table of contents, each part has been further divided into several sections. In this overview, we will provide an introduction to and a brief summary of these sections.

Part I: Molecular structure of membranes

Monolayers or Langmuir–Blogdett films — At the interface between water and air (or between water and oil), lipids and surfactants form monolayers or Langmuir–Blogdett films in which the head groups point towards the aqueous medium. These layers or films exhibit several 2–dimensional phases which can be studied by a variety of experimental techniques. Three relatively new methods are described below: x–ray scattering [Möhwald et al], neutron reflectivity [Penfold et al], and atomic force microscopy [Radmacher et al].

Structure and dynamics of bilayers — If one 'glues' two monolayers of lipids together, one arrives at a bilayer which represents the universal building block of all biological membranes. In aqueous solution, lipid bilayers usually form multi–lamellar vesicles. Alternative model systems are supported bilayers, i.e., bilayers which are bound to solid substrates [Dolainsky et al].

Lipid bilayers exhibit a fluid (L_α) phase at high temperatures and one or several gel phases at low temperatures. Within the coexistence region of the fluid and one of the gel phases, relatively stable domains are formed. In such a situation, diffusion along the bilayer is fast and slow when the fluid domains are connected and disconnected, respectively. It is argued by [Vaz et al] that this can be viewed as a percolation problem.

Ripple phase — For some lipids, the bilayer exhibits a corrugated gel phase, the so–called ripple or P_β phase. It then aquires a superstructure of line defects. This phase intervenes between the fluid phase at high temperatures and another (planar) gel phase at low temperatures. It is interesting to note that this ripple phase is suppressed in supported bilayers.

Bilayers and biologically relevant molecules — Biomembranes contain a large number of macromolecules such as proteins. At the workshop, Jähnig discussed some recent attempts to calculate the structure of a membrane protein. [2] Another example is provided by local anaesthetics which strongly affect the properties of cell membranes [Böttner et al].

Molecular forces between bilayers — Two adjacent bilayers interact via a variety of molecular forces. At short distances, this interaction is dominated by the so–called hydration force which is (usually) repulsive. Several theories for the basic mechanism underlying this force have been proposed [Kornyshev and Leikin]. This topic is still very controversial; e.g., such a force could also arise from the molecular roughness of the bilayer surfaces. [3,4]

Part II: Conformation of membranes

Shape of lipid vesicles — Lipid bilayers in water usually form closed surfaces or vesicles which can exhibit a large variety of different

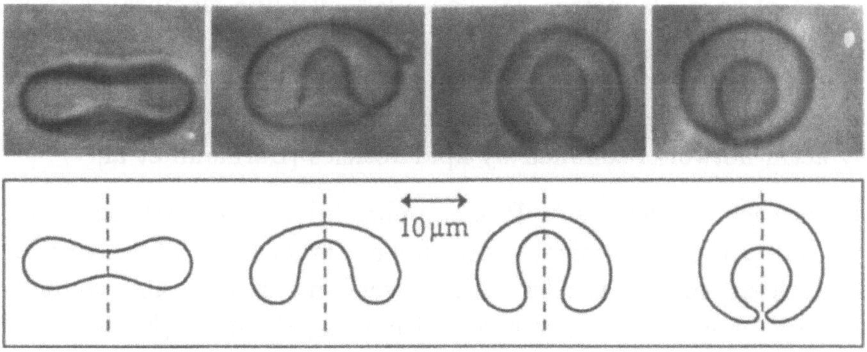

Figure 1: Shape transformations of a lipid vesicle as induced by a change in temperature. [5,6] The shapes are axisymmetric with respect to the broken line.

shapes. In fact, the shape of a single vesicle may simply be changed by changing the temperature as shown in Fig. 1. In this figure, the experimentally observed shape is compared with calculations based on the concept of bending elasticity and curvature. [5,6] More recent theoretical work on curvature models is presented below by [Seifert et al, Mosseri et al, and Sevsek et al].

It has been postulated [7] that lipid membranes exhibit a microroughness which cannot be observed by optical microscopy. So far, some evidence for such a roughness has been obtained by electron microscopy [Klösgen and Helfrich]. More experiments are certainly needed in order to clarify this issue.

Shape of cells and cell models — The discocyte shape displayed in Fig. 1 resembles the shape of a red blood cell. The plasma membrane of this cell consists of a lipid–protein bilayer and an essentially 2–dimensional network of rod–like spectrin molecules. It is argued by [Zilker et al] that thermally–excited shape fluctuations of this membrane are *not* affected by the presence of the network.

The network can be detached from the bilayer and then be studied directly [Schmidt et al]. So far, the experiments are not able to distinguish a crumpled state from a rough but uncrumpled one. The dynamics of rough polymerized membranes is theoretically discussed by [Frey and Nelson]. At the workshop, Baumgärtner presented results on computer simulations for tethered surfaces consisting of hard plaquettes. [8] These self–avoiding membranes seem to attain a crumpled state.

The spectrin network represents an example for the cytoskeleton of cells. A simple model system for the latter structure is provided by actin networks confined by lipid vesicles [Bärmann et al].

Microtubes and tethers — If one applies a local external force, one can pull a thin cylindrical microtube or tether from a lipid vesicle. In such a process, the two monolayers are dynamically coupled and are then governed by nonlocal binding elasticity [Evans et al, Svetina et al].

Adhesion and fusion of vesicles — A vesicle can adhere to another surface. The shape of such a bound vesicle state can be observed by optical methods [Rädler and Sackmann]. Bound vesicles may fuse and then form essentially planar bilayers. Such a process has recently been discovered at the surface of a vesicle dispersion and was found to be very sensitive to changes in temperature [Cevc et al]. At the workshop, Helfrich reviewed the process of tension–induced adhesion as observed in multi-lamellar systems. [9]

Electric field effects — External electric fields can be used to deform lipid vesicles and to induce fusion between them.

Part III : Membranes in complex fluids

In parts I and II, the structure and conformation of a *single* monolayer or bilayer has been considered. In part III, we focus on systems which contain a *large* number of membranes.

Aggregation of micelles and vesicles — At relatively low concentrations, amphiphilic molecules can form micelles and vesicles. This is studied here for gangliosides [Cantu et al], bipolar lipids [Relini et al], lipopolysaccharide [Maurer and Glatter], and single–chain surfactant mixtures [Szönyi et al]. Spontaneous formation of vesicles is found in several cases.

Monolayers and microemulsions — In oil–water mixtures, the surfactant molecules form monolayers which separate water and oil domains or droplets. These mixtures can exhibit a variety of different thermodynamic phases. This polymorphism can be theoretically studied in the framework of a very simple model [Gompper and Zschocke]. Experimentally, the static and dynamic properties of the domains or droplest can be investigated by neutron scattering [Farago]. A brief review on the applications of microemulsions is also given [Stickdorn and Schwuger]. At the workshop, Kahlweit presented new experimental results on the wetting properties of microemulsions. [10]

Hexagonal to lamellar transition — In many systems, there is a region in the temperature–concentration diagram in which the amphiphilic molecules are aggregated into cylindrical micelles which are arranged in a hexagonal manner. As the concentration is increased, this phase typically undergoes a transition to a lamellar phase [Lerche et al, Templer et al].

Cubic phases — Another phase which has recently attracted a lot of interest is the bicontinuous cubic phase which consists of two interwoven labyrinths or subvolumes. These two subvolumes are separated by a membrane the shape of which resembles a triply–periodic minimal surface [Charvolin and Sadoc, Clerc and Sadoç]. In lipid–water systems, several cubic phases have now been identified [Seddon and Bartle, Templer et al].

Bilayers and sponge phases — The ordered cubic phase built up from bilayers can melt into a sponge phase which is still bicontinuous but has no long–range translational order. Such phases have been observed in several surfactant systems [Thunig et al, Chen et al, Olsson et al]. From the theoretical point of view, the sponge phase is particularly interesting because it exhibits an Ising–like symmetry which corresponds to the interchange of the two solvent domains [Cates].

Hydrodynamic effects — The last two contributions address the flow properties of thin fluid films.

Acknowledgements

This workshop was made possible by a grant from the Forschungszentrum KFA Jülich and by support from the workshop secretariat at the Forschungszentrum. We thank Wolfgang Helfrich, Heinz Hoffmann and Erich Sackmann for their work in the Scientific Committee. We acknowledge the professional help of Bernhard Krahl–Urban, Liselotte Moll and Magdalena Kötter with the organization of this workshop, and thank Magdalena Kötter for her help in the editorial process.

[1] C. Tanford, The hydrophobic effect, (Wiley, New York 1980)

[2] F. Jähnig and O. Edholm, Z. Phys. B**78**, 137 (1990)

[3] J.N. Israelachvili and H. Wennerström, Langmuir **6**, 873 (1990)

[4] V.A. Parsegian and R.P. Rand, Langmuir (in press)

[5] K. Berndl, J. Käs, R. Lipowsky, E. Sackmann, and U. Seifert, Europhys. Lett. **13**, 659 (1990); J. Käs and E. Sackmann, Biophys. J. **60**, 1 (1991); U. Seifert, K. Berndl, and R. Lipowsky, Phys. Rev. A**44**, 1182 (1991)

[6] For a review, see R. Lipowsky, Nature **349**, 475 (1991)

[7] W. Helfrich, Liqu. Cryst. **5**, 1647 (1989)

[8] A. Baumgärtner, J. Phys. I France **1**, 1549 (1991)

[9] R.M. Servuss and W. Helfrich, J. Phys. France **50**, 809 (1989)

[10] M. Aratono and M. Kahlweit, J. Chem. Phys. (Dec. 1991)

Part I

Molecular Structure of Membranes

Monolayers of Amphiphilic Molecules

H. Möhwald[1], *R.M. Kenn*[1], *K. Kjaer*[2], *and J. Als-Nielsen*[2]

[1]Department of Physical Chemistry, Universität Mainz, W-6500 Mainz, Fed. Rep. of Germany

[2]Risø National Laboratory, DK-4000 Roskilde, Denmark

1 Introduction

There are good reasons why the organizers have placed this lecture at the beginning of a conference on the multiple aspects of membranes. In many respects the monolayer may be considered half of a membrane and it obviously is the most simple and best-defined model system. Yet we will show that the system is more complex than many have anticipated. On the other hand, there are many general features which can be discussed without looking into chemical details.

This work will concentrate on studies of the microstructure of Langmuir monolayers predominantly by X-ray scattering techniques. We will demonstrate the analogy with smectic liquid crystals as well as with alcane crystals. There are additionally interesting questions concerning ordering in two dimensions and the structural mismatch between head groups and tails. The monolayer is ideally suited to study these aspects because it is possible to control density and other variables over a broad range. However, it requires rather sophisticated characterization techniques, and in the development of the latter there has been considerable progress. This will become apparent below.

2 General Phase Diagrams

Typical pressure/area isotherms of a Langmuir monolayer are shown in Fig. 1, left. In analogy with three-dimensional systems increasing the two–dimensional density, one traverses from a gaseous via a fluid to a solid state, sometimes, i. e. at low temperatures, the system can also be sublimed. Generally, these transitions are of first order and then the isotherm slope should be horizontal. Experimentally, this is nearly but not exactly fulfilled, and we do not want to discuss this discrepancy here, which is probably due to residual impurities. Instead, we will first concentrate on that part of the isotherm, where a sharp increase of pressure can be observed, indicative of many ordered phases. Judged from the isotherms, there do not appear to be gross differences between phospholipids, fatty acids, alcohols or esters, and so we will first discuss ordered phases of the most systematically studied class of surfactants, fatty acids.

From isotherm measurments at different temperatures, phase boundaries have been obtained, whereas the structure of the phases was subject to spec-

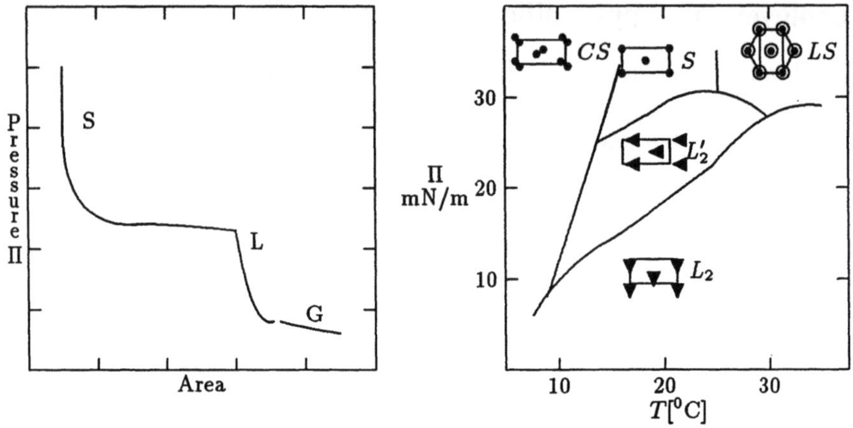

Figure 1: Left: Typical isotherm (lateral pressure Π vs. area) of a Langmuir monolayer. Right: (Π, T) phase diagram for behenic (docosamic) acid as deduced from (Π, A) isotherm measurments [1]. The phases are labeled in accordance with Ref. [2]. The schematics show the structures as deduced from X-ray diffraction [3].

ulation [1, 2, 4]. The system depicted in Fig. 1, right, behenic acid, is well suited for structural studies, since different ordered states are accessible near room temperature. On the other hand, the structural data presented here are not unique for this system. In fact, it has been shown that varying the chain length merely shifts the phase boundaries by 8 - 10 K per CH_2 unit and also diffraction data of fatty alcohols are comparable to those presented below [5].

3 Behenic Acid

The local structure sketched in Fig. 1 has recently been obtained measuring X-ray diffraction with monolayers at the air/water interface [3]. For these studies, a film balance is placed at the sample stage of the D4 diffractometer at HASYLAB, Hamburg. A monochromatized Synchrotron beam ($\lambda = 0.138$ nm) traverses the surface at glancing incidence (angle$\approx 0.1^0$). For a periodic two-dimensional density distribution, the X-ray beam is Bragg diffracted and from the peak positions a lattice spacing d_{hk} is assigned according to Bragg's law:

$$d_{hk} = \frac{2\pi}{Q_{\text{hor}}^{hk}} \qquad \text{with} \qquad Q_{\text{hor}}^{hk} = \frac{4\pi}{\lambda}\sin(\theta_{hk}) \tag{1}$$

being the horizontal wave vector transfer at the Bragg angle θ_{hk}. Measuring the intensity distribution of a Bragg peak along the plane normal z (Bragg rods) we obtain the perpendicular electron distribution [6]. Of special interest for us is the following information: Assume the aliphatic tails are uniformly tilted with a tilt angle t with respect to the plane normal and the tilt projected on the surface forms an angle Ψ_{hk} with the normal to the lattice plane $\{hk\}$. Then

$$\tan(t)\cos(\Psi_{hk}) = \frac{Q_z^{hk}}{Q_{\text{hor}}^{hk}} \tag{2}$$

where Q_z^{hk} is the peak positon along Q_z. It is this relation which eventualy allows distinction between many different phases. It also enables determination of tilt angles to accuracies near 1^0 even if they are close to 0^0 (due to the tangential dependence).

The distribution $I(Q_z)$ is the Fourier transform of the electron density along the tails. Thus, if we model these as rods of length L the full width at half maximum equals $2\pi/L$. From the data we deduce that L is within 10% equal to the length of the fully extended aliphatic tails proving that the entire chain contributes to coherent scattering [6, 7]. This indicates a low density of kink defects at least along 90% of the tails. Fig. 2, top, shows the scattered intensity as a function of in–plane wave vector transfer Q_{hor} for a behenic acid monolayer increasing the pressure at low temperature and at room temperature. At 7^0C and above 10mN/m the peaks are sharp and their width resolution limited, proving crystalline order. In all the other cases the peaks are broader and from their widths one can deduce positional coherence lengths between 10 and 70 lattice spacings [3]. On the other hand, polarization micrographs [8] and electron diffraction of transferred monolayers [9] reveal a correlation of the directions of tilt and local crystallographic axes over tens of μm. This altogether indicates a liquid crystalline structure.

The diffraction data in Fig. 2 top, show two peaks, either resolved in Q_{hor} or in Q_z, wich can be ascribed to an orthorhombic lattice which basically results from a unidirectional expansion or compression of a nearly hexagonal lattice of the aliphatic tails. As expected, the lattice spacings do not depend on pressure for the crystalline phase, but they do so for a liquid crystal. The data presented in Fig. 2 were taken from (at least) four different phases which can be shown from the measurement of the diffraction intensity along Q_z. This is given for representatives of these phases in Fig. 2, bottom, considering both diffraction peaks. At room temperature and low pressures (L_2) one peak has a maximum for $Q_z \approx O$, the other one for $Q_z > O$. This can be understood with a model on the tilt as shown in the insert: For a nearest neighbour tilt $\Psi_{\{2,0\}} = 90^0$ and $Q_z^{\{2,0\}} = O$ whereas $Q_z^{\{1,1\}} > 0$ as observed. In contrast for a next nearest neighbour tilt $\Psi_{2,0} = 0^0$ and $\Psi_{1,1} \approx 60^0$, hence one expects $Q_z^{\{2,0\}} \approx 2Q_z^{\{1,1\}} > 0$. This obviously holds for the phase L_2'.

For the other two phases there is clearly no chain tilt. In addition to these orthorhombic phases shown in Fig. 2, there exists a hexagonal phase (LS) at high pressures and temperatures without chain tilt and with liquid crystalline order [3].

Knowing the two-dimensional lattice, one can easily calculate A_t, the area per tail projected on the surface and having determined t we can derive A_c, the cross section per tail: $A_c = A_t\cos(t)$. We obtain $A_c = 0.197, 0.192, 0.192,$

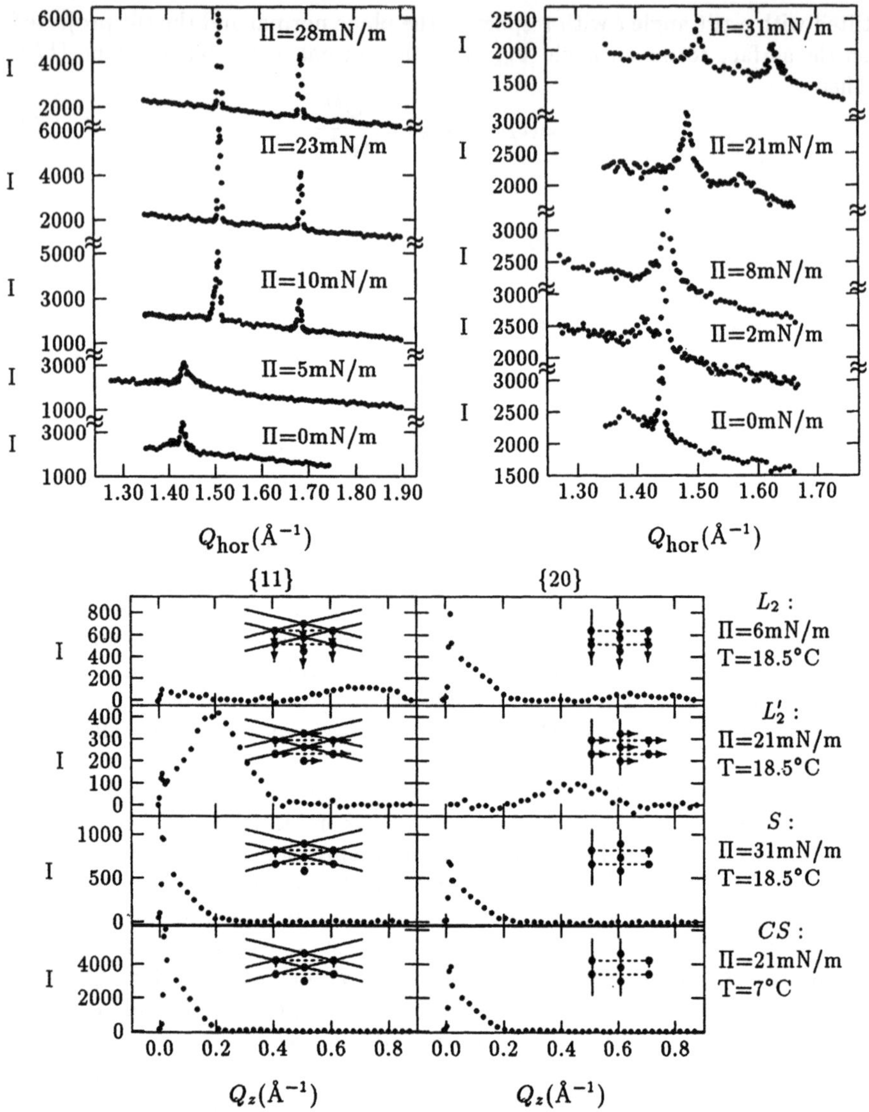

Figure 2: Top: Horizontal X-ray diffraction scans (intensity I vs. horizontal scattering vector Q_{hor}) for 7^0C (left) and 18.5^0C (right). Bottom: Bragg rod intensity (intensity I vs. vertical scattering vector Q_z) for behenic acid monolayers on water.

0.192 and 0.187 (±0.1)nm^2 for the LS, L_2, L_2', S and CS phases, respectively. Comparing this with data on alkane crystals [10], one finds, that LS is a phase where the tails are allowed to rotate freely about their long axes (rotator I). For the L_2, L_2' and S–phase this rotation is still possible but with restricted amplitude (rotator II). In the *CS* phase, the cross section is close to that observed for crystals with the tails in a herringbone arrangement. Although a direct proof of this structure is not possible – an observation of the reflections which directly prove the translational inequivalence of the molecules at the corner and the center of the cell fails due to the low contrast in the formfactors – our diffraction data support this picture.

4 Phospholipid Monolayers

Basically, a phospholipid consists of two hydrocarbon chains linked by a hydrophilic head group. Hence we may expect structures similiary to those of fatty acids, and in addition see influences of the head group like head group ordering, symmetry breaking due to linkage of two tails in nonsymmetric directions, head/tail incompatibility and head/head repulsion. For the fatty acids at molecular areas between 0.19 and 0.24nm^2 we have shown that for the larger area there is a uniform tilt of about 30^0 and the tilt angle can be reduced towards zero by compression [11, 12]. This reduction occurs continuously, but there are also discontinuous transitions involving tilt azimuth and density [3]. This is also valid for phospholipids. Specifically we have found that even if the head group is small compared to the tail cross section, the tails are not aligned vertically for an average tail area of, say 0.23nm^2[13]. Instead there is a uniform tilt which is continuously reduced upon compression. This is again derived measuring diffraction resolved in $Q_{\rm hor}$ as well as in Q_z. These measurements also reveal one of the peculiarities of phospholipids (see Fig. 3). The in plane diffraction with L-α-dymyristoylphosphatidylethanolamine (DMPE) shows, at Π = 12mN/m, 3 different low order reflections with maxima at different Q_z (Fig. 3, left, scan labeled α). This suggests a distortion of the rectangular arrangement, resulting in a triclinic lattice. This behaviour is also found in Cd arachidate monolayers, where a counter ion is attached to two acids, forming the link between them. Additionally, in this system a commensurate head group lattice, a crystalline Cd layer, could be found below the aliphatic lattice [14]. The structure is also observed for multilayers of DMPC varying the water content [15]. In the phase diagram of this system, it is located between the phases with nearest and with next nearest neighbour tilt. In other words, the tilt azimuth passes through an intermediate position upon hydration (or expansion). These similarities suggest that the head group linking of two tails reduces the symmetry of the tail lattice. This suggests positional and orientational head group order. There are other indirect indications that this may be true, but a direct proof by observing e. g. the Bragg scattering due to the different periodicty of the head groups or superlattice peaks is missing. Maybe the best indication results from atomic force microscopy data, but these are taken from bilayers on support [16].

Expanding surfactant layers one enters fluid phases and these phases and transitions are better studied with phospholipids because of their higher stability compared to single chain compounds. For these studies, however, diffraction

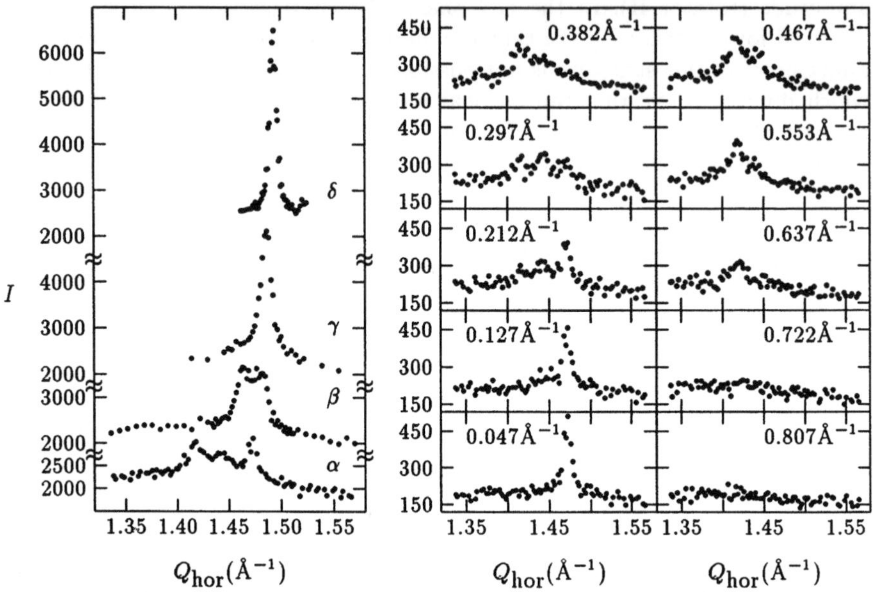

Figure 3: Left: Q_z integrated intensity vs. horizontal scattering vector for DMPE monolayers at 20^0C and different lateral pressures: α : 12mN/m, β : 20mN/m, γ : 27mN/m and δ : 35mN/m. Right: Scattered intensity vs. horizontal scattering vector for Q_z values indicated in the figure, (DMPE, 20^0C, Π =12mN/m, peak "α" of left Fig.)

is inapplicable and one has to resort to reflection techniques. With the same set-up as for diffraction one can measure the specular X-ray reflectivity as a function of incidence angle, hence wave vector transfer Q_z. For simplicity the data can be analyzed assuming slab models for the electron density distribution and the reflectivity R divided by the one expected for a flat and stepwise interface R_F is given by the equation [17, 18]

$$\frac{R}{R_F} = \frac{1}{\rho_\infty^2} \left| \int_0^\infty \frac{\mathrm{d} < \rho >}{\mathrm{d}z} e^{\imath Q_z z} dz \right|^2 , \tag{3}$$

where ρ_∞ is the electron density of the substrate. It is fortunate that there is structure in the reflectivity curves (Fig. 6). This facilitates the analysis and is easily understood. As seen from eq. 3 the dominant contributions to R result from regions with large gradients of the electron density. Assuming the liquid to exist of a tail slab with density ρ_t and thickness l_t and a head slab (ρ_h, l_h) waves reflected at these interfaces interfere and the minima in Fig. 6 result from a destructive interference of a wave reflected at the air/tail interface with the sum of waves coming from the hydrocarbon/head and head/water interfaces. In the case $\rho_t \approx \rho_\infty$ the minimum in Q_z, $Q_z^{\min}$ is related to the layer thickness, defined as $l_t + \frac{1}{2} l_h$ according to [6]

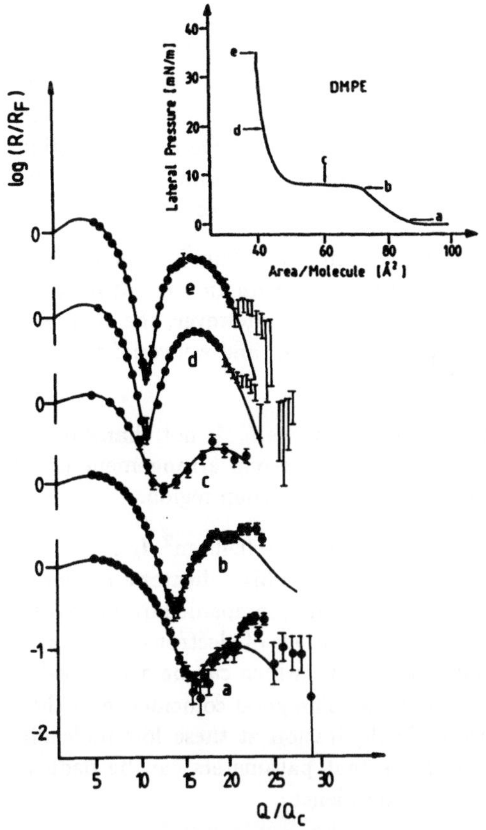

Figure 4: Normalized X-Ray reflectivity vs. Q/Q_c of DMPE monlayers at pressures indicated in the insert. $Q_c = 0.02764\text{Å}^{-1}$

$$l_t + \frac{1}{2} l_h = \frac{1}{Q_z^{\min}} \frac{3}{2} \pi. \tag{4}$$

Accordingly, one can qualitatively understand the shift of $Q_z^{\min}$ to lower values on compression as an increase in film thickness. One can also understand the shallower minimum in the flat part of the isotherm, if there are large domains of coexisting phases of different thickness. In that case, however, there are too many free parameters to obtain a meaningful fit and, therefore, we present in Fig. 4 only parameters derived from measurments in single phase regions. These data were obtained for the two phospholipids DMPE and L-α-dilauroylphosphatidylethanolamine (DLPE) [19]. They exhibit the same head group and hence the consistency of the analysis is established from the finding of nearly equal head group parameters. It is also established from a comparison of the tail parameters. ρ_t should be identical whereas l_t should be 2.5Å larger for the most condensed film and somewhat less for higher molecular areas. The data as a function of molecular area reveal four interesting aspects

1. The volume decreases by more than 25% on going from the fluid to the ordered phases, this corrersponds to an increase of the electron density of 10%. The comparable large volume reduction is due to dehydration. This is much more than observed for the main transition of bilayers [20] and must be ascribed to the different opposing surfaces.

2. Decreasing the molecular area from 0.43nm^2 to 0.40nm^2, l_t increases by about 0.1nm. This can be understood due to a reduction in tilt in agreement with the diffraction data presented above. However, the change of about 0.1nm is about the accuracy of the data fit and we would not have dared to conclude on it from reflectivity alone.

3. Compressing the film throughout the fluid phase l_h, is not changed, but ρ_h is increased. This indicates that the head group arrangement is not affected, but water is squeezed out of the head group region.

4. Decreasing the molecular area A from 0.43nm^2 to 0.40nm^2, l_h is reduced to about 0.8nm and decreased slightly further. This indicates an ordering of the head groups at least in normal direction accompanied by a dehydration. In fact, calculating $N = \rho_h A l_h$ the number of electrons in the head group slab for $A = 0.40\text{nm}^2$ one obtains 155, which can be attributed to the head group plus 1.5 water molecules. The good coincidence of these values with the postulated state of dehydration at these low molecular areas may be fortuitous, but it shows that parameters can be deduced from these measurements with high confidence.

5 Conclusions

This report should show that there is a richness of ordered phases for insoluble monolayers. We did not stress the point here, but all these phases can be assigned smectic categories [21]. Varying chain lengths or head group interactions, these phases may be shifted in the pressure/temperature space, but their existence is not affected. The coupling of two tails via the head group introduced still more complexity. This has led to old and new ideas on head group ordering and frustration, but more and reliable data will be needed to arrive at further conclusions on this interesting topic.

6 Acknowledgement

This work partly presents new data, partly we collected information obtained in collaboration with C. A. Helm, C. Böhm, P. Tippmann–Krayer and D. Degenhardt. We also appreciate technical support and beam time at HASYLAB and financial support by the Bundesministerium für Forschung und Technologie (BMFT).

References

[1] A. M. Bibo and I. R. Peterson, *Adv. Mat.*, **2**, 309 (1990)

[2] S. Ställberg-Stenhagen and E. Stenhagen, *Nature*, **156**, 239 (1945)

[3] R. M. Kenn, C. Böhm, A. M. Bibo, I. R. Petersen, H. Möhwald, K. Kjaer and J. Als-Nielsen, *J. Phys. Chem.*, **95**, 2092 (1991)

[4] M. Lundquist, *Chem. Scripta*, **1**, 5 and 207 (1971)

[5] B. Lin, M. C. Shih, T. M. Bohanon, G. E. Ice and P. Dutta, *Phys. Rev. Lett.*, **65**, 191 (1990)

[6] J. Als-Nielsen and K. Kjaer, In: *Phase Transitions in Soft Condensed Matter*, T. Riste and D. Sherrington, editors. Plenum Press, New York (1989)

[7] D. Möbius and H. Möhwald, *Adv. Mat.*, **3**, 19 (1991)

[8] I. R. Peterson, *J. Mol. Elec.*, **2**, 95 (1986)

[9] C. Böhm, R. Steitz and H. Riegler, *Thin Solid Films*, **178**, 511 (1989)

[10] B. Ewen, G. R. Strobl and D. Richter, *Faraday Discussion*, **69**, 19 (1980)

[11] K. Kjaer, J. Als-Nielsen, C. A. Helm, P. Tippmann-Krayer and H. Möhwald, *J. Phys. Chem.*, **93**, 3200 (1989)

[12] P. Tippmann-Krayer and H. Möhwald, *Langmuir, in press*, (1991)

[13] H. Möhwald, R. M. Kenn, D. Degenardt, K. Kjaer and J. Als-Nielsen, *Physica A*, **168**, 127 (1990)

[14] F. Leveiller, D. Jacquiemain, M. Lahav, L. Leiserowitz, M. Deutsch, K. Kjaer and J. Als-Nielsen, *Subm. to Science*, (1991)

[15] G. S. Smith, E. B. Sirota, C. R. Safinya and N. A. Clark, *Phys. Rev. Lett.*, **60**, 813 (1988)

[16] J. A. N. Zsadzinski, C. A. Helm, M. L. Longo, A. L. Weisenhorn, S. A. C. Gould and P. K. Hansma, *Biophys. J.*, **59**, 755 (1991)

[17] J. Als-Nielsen and H. Möhwald, to appear in: *Handbook of Synchrotron Radiation*, S. Ebashi, E. Rubinstein and K. Koch, editors, vol. 4. North-Holland, Amsterdam

[18] C. A. Helm, H. Möhwald, K. Kjacr and J. Als-Nielsen, *Europhys. Lett.*, **4**, 697 (1987)

[19] C. A. Helm, P. Tippmann-Krayer, H. Möhwald, J. Als-Nielsen and K. Kjaer, *Biophys. J.*, in press, (1991)

[20] G. Schmidt and W. Knoll, *Ber. Bunsenges. Phys. Chem.*, **89**, 36 (1985)

[21] I. R. Peterson, *Mol. Cryst. Liq. Cryst.*, **147**, 1 (1987)

The Determination of the Structure of a Mixed Surfactant Monolayer by Specular Neutron Reflection

J. Penfold[1], *R.K. Thomas*[2], *E.M. Lee*[2], *E.A. Simister*[2], *J.R. Lu*[2], and *A.R. Rennie*[3]

[1]ISIS Science Division, Rutherford Appleton Laboratory, Chilton, Didcot, Oxon. OX11 0RA, UK
[2]Physical Chemistry Laboratory, Oxford University, South Parks Road, Oxford OX1 3QZ, UK
[3]School of Chemistry, Bristol University, Cantocks Close, Bristol BS8 1TS, UK

Abstract. Neutron reflection in combination with isotopic substitution has been used to determine the surface structure of a mixed monolayer of sodium dodecyl sulphate (SDS) and dodecanol at the air/solution interface. Two different methods of analysis have been applied to the data. The first fits a single structural model to different reflectivity profiles arising from manipulaton of the refractive index distribution by isotopic substitution. The second, based on the kinematic approximation, is more direct, and in particular allows the distances between the centres of the distributions of the two surfactants to that of the solvent to be obtained without recourse to modelling. At a SDS concentration of 0.0067M and at a SDS/dodecanol ratio of 0.005 the dodecanol/solvent separation is 9.0 ± 0.5Å and the SDS/solvent separation is 6.5 ± 0.5Å. The two methods of analysis provide a consistent structural model, which for concentrations less than the critical micellar concentration (cmc) the SDS is more immersed in the solvent than the dodecanol, and that the SDS molecules are less fully extended than the dodecanol molecules.

Introduction

In a neutron reflection experiment the specular reflection is measured as a function of the wavevector transfer, Q, (where $Q = 4\pi \sin \theta/\lambda$, θ is the angle of incidence, and λ the neutron wavelength) perpendicular to the reflecting surface. By analogy with light optics (1) the neutron refractive index can be written as,

$$n = 1 - (\lambda^2/2\pi)\,\varrho \,, \qquad (1)$$

where ρ is the scattering length density given by,

$$\varrho = \Sigma_i\, n_i b_i \qquad (2)$$

and,
n_i is the number density of the ith nucleus, b_i its scattering length.

A reflectivity profile is related by an optical transform (2) to the refractive index profile normal to the surface, and so for neutrons is related to the scattering length density profile normal to the surface.

For neutrons isotopic substitution can be used to produce large contrasts in scattering densities. Of particular importance in surface chemistry is the large difference in scattering powers between hydrogen and deuterium, where selective deuteration can be used to manipulate refractive index profiles. This has been exploited extensively in the study of surfactant adsorption at the air/solution interface (3,4), where not only can adsorption isotherms be determined unambiguously(5) but the detailed surface structure can be obtained(6). A requirement of this approach is that isotopic substitution does not change the interfacial properties, and this has been clearly demonstrated in recent studies(5).

The use of isotopic substitution is especially important in the study of multi-component surfactant systems, and by selective deuteration of each surfactant we have already determined the amount adsorbed at the interface for each component in the binary SDS/dodecanol system(7). We present here an extension of that work, where we have now obtained in detail the structure of the SDS/dodecanol mixed monolayer at the air/solution interface at a concentration below cmc. Particular emphasis is placed here on the use of the kinematic approximation, in combination with isotopic substitution, to determine directly the structure.

Theory

The most commonly used method to analyse neutron reflectivity data has been to assume a structural model for the interface, and to calculate the reflectivity exactly using the optical matrix methods(2). The question of uniqueness always arises with such a procedure, and this can be partially alleviated by constraining that the same model predicts the reflectivity from other refractive index profiles (obtained from isotopic substitution) for the same system. This approach has already been successfully applied to a number of single component surfactant systems.

For a one-component system a simple model has been derived which is consistent with the range of systems measured so far. It is a two layer model where the first layer adjacent to the vapour phase contains some fraction of hydrocarbon chains (l–f), with a thickness given by

$$d_1 = l_c (l-f)\,\epsilon\,, \tag{3}$$

where l_c is the length of a fully extended chain, f is the fraction of chain in the headgroup region, and ϵ is a factor which is related to the degree to which the chain is fully extended. The second layer is adjacent to the solvent phase, and contains the headgroups, some fraction of the chains, f, and bound counterions, and has a thickness d_2. The model is then characterised by three structural parameters, the area per molecule, A, f and ϵ. From known molecular volumes, sizes and scattering lengths it is then possible to calculate the extent and scattering densities of the two regions in the model. This model has been extended to include both components in a binary system and applied here to the SDS/dodecanol mixture, where there is now three refinable parameters for each surfactant component.

Some of the difficulties associated with the model fitting approach can be overcome by using the recent developments of the kinematic approximation (6,8) for such systems. In the kinematic approximation the reflectivity, (R(Q), is written as,

$$R(Q) = \frac{16\pi^2}{Q2}\,|\,\varrho\,(Q)\,|^2\,, \tag{4}$$

where $\rho(Q)$ is the one dimensional Fourier transform of $\rho(z)$, the z dependence of the scattering length density given by,

$$\varrho(z) = \Sigma b_i n_i(z)\,. \tag{5}$$

An alternative expression to equation 4 can be written in terms of $\rho' = d\rho/d_z$ and is often used.

The Fourier transform of $|\rho(Q)|^2$ is the "Patterson" function of the scattering length density,

$$P(z) = \int \varrho(z)\,\varrho\,(u-z)du \tag{6}$$

and a knowledge of P(z) is often sufficient to determine ρ (z) (4). This approach has been extended(6) by considering the different components in the interfacial region.

In the case of SDS/dodecanol we can write,

$$\varrho(z) = b_s n_s(z) + b_d n_d(z) + b_w n_w(z)\,, \tag{7}$$

where s,d and w refer to the SDS, dodecanol and solvent.

In combination with (4) this gives,

$$R(Q) = \frac{16\pi^2}{Q^2}[b_s{}^2 h_{ss} + b_d{}^2 h_{dd}\;\; b_w{}^2 h_{ww} + 2b_s b_d h_{sd} + 2b_s b_w h_{sw} + 2b_d b_w h_{dw}] \tag{8}$$

An equivalent expression in terms of $h'_{ij}(Q)$ can be obtained such that,

$$h'_{ij}(Q) = Q^2 h_{ij}(Q) \tag{9}$$

If at least six reflectivity profiles are measured, using isotopic substitution to obtain the different values of b_s, b_d and b_w, then the different h_{ij} functions, which are referred to as partial structure factors (6). can be determined. The Fourier transform of these partial structure factors give a Patterson function of the form

$$P_{ii}(z) = \int_{-\infty}^{\infty} n_i(u) n_i(u-z)du \tag{10}$$

This approach has now been applied to a number of surfactant systems, and in most detail recently to the cationic surfactant tetradecyltrimethyl ammonium bromide ($C_{14}TAB$) (6), where the self partial structure factors for the chain, head and solvent, and their cross-terms have been obtained and interpreted.

The cross partial structure factors, h_{ij}, are particularly important in this study as they contain information about the relative positions of the distribution of the labelled components. In the example described here the surfactant/solvent cross terms result from the combination of a even and odd distribution and in this case it has been shown that (6),

$$h_{iw} = (h_{ii}h_{ww})^{1/2} \sin Q\delta \tag{11}$$

where δ is the distance between the centres of the two distributions. Whereas the surfactant/surfactant (that is, the dodecanol/SDS) cross term results from the combination of two even functions, and can be written as (6),

$$h_{ij} = (h_{ii}h_{jj})^{1/2} \cos Q\,\delta \tag{12}$$

For the interpretation of the self partial structure factors it is necessary to resort to modelling, as there is insufficient Q range for a reliable Fourier inversion, whereas the cross terms provide important structural information which is model independent.

Unlike the cationic surfactants (6) we do not here have an ideal combination of isotopic labels; that is, we cannot separate the chain and headgroup combinations and a correction to account for this have been devised (9). Assuming that the dodecanol contribution is dominated by its chain then the cross terms involving the SDS in equation (8) can be written explicitly in terms of the headgroup and chain contributions (9), such that,

$$2b_d b_s h_{ds} = 2b_d b_{sc} h_{dsc} + 2b_d b_{sh} h_{dsh} \tag{13}$$

where,
b_{sc} and b_{sh} refer to the SDS chain and headgroups
and similarly,

$$2b_d b_s h_{ds} + 2b_s b_w h_{sw} = 2b_d b_{sc} h_{dsc} + 2b_d b_{sh} h_{dsh} + 2b_{sc} b_w h_{scw} + b_{sh} b_w h_{shw} \tag{14}$$

The SDS/dodecanol cross terms in equations (13) and (14), and the SDS/solvent cross terms in equation (14) have the form of equations (11) and (12).

If we assume that the head/solvent separation is much smaller than the chain/solvent then the sin Q δ_{shw} term will be much smaller than sin Q δ_{scw}, and similarly if the dodecanol/SDS chain separation is much smaller than the dodecanol/SDS head separation then cos Q δ_{dsc} will be much larger than cos Q δ_{dsc} will be much larger than cos Q δ_{dsh}. Therefore in equation (13) and (14) we can ignore the terms in h_{dsh} and h_{shw} such that equations (13) and (14) become,

$$2b_d b_s h_{ds} = 2b_d b_{sc} h_{dsc} \tag{15}$$

and

$$2b_d b_s h_{ds} + 2b_s b_w h_{sw} = 2b_d b_{sc} h_{dsc} + 2b_{sc} b_w h_{scw} \tag{16}$$

This correction provides a small but important adjustment to the partial structure factors, and the factors described later have been obtained using it. The partial structure factors involving SDS now strictly refer to the SDS chain and not the whole amphiphile. In the evaluation of the cross terms using equations 11 and 12 then strictly h_{scsc} and not h_{ss} should be used. However, in the context of this study we can assume that the scattering from the SDS headgroup is negligible, and h_{scsc} can be taken as equivalent to h_{ss} without any significant loss of accuracy.

Results and Discussion

The reflectivity measurements have been made on the reflectometer CRISP (10) at the ISIS Pulsed Neutron Source, where the measurements have been made using the fixed geometry white beam time of flight method in the Q range 0.05 to 0.65Å^{-1}. The detailed experimental procedures are now well established and are described in detail elsewhere.

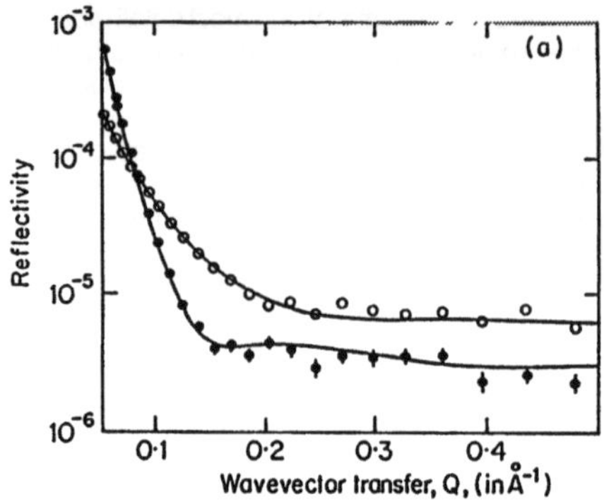

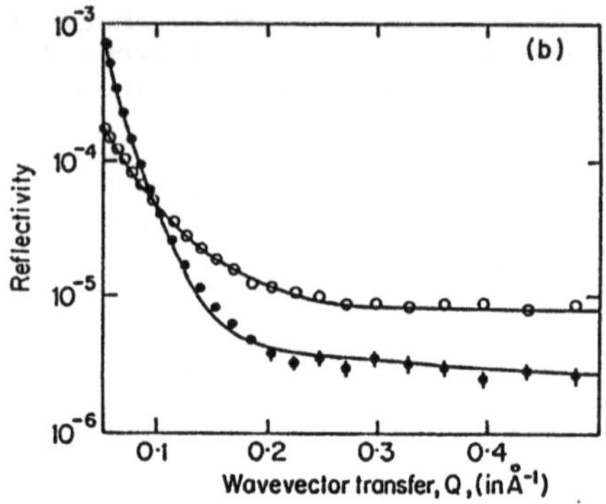

Figure 1. Neutron reflectivity profiles for SDS/dodecanol in (O) nrw and (●) D_2O for (a) deuterated dodecanol/protonated SDS and (b) protonated dodecanol/deuterated SDS. The solid lines are model fits for the parameters described in table 1.

Table 1 Model parameters for 0.0067M SDS and a dodecanol/SDS ratio of 0.005.

	A	f	ε	d_1	d_2
dodecanol	42±2	0.25±0.05	0.85±0.05	10.5±1	2.0±1
SDS	48±2	0.4±0.05	0.6±0.05	-	-

The surface structure of the mixed SDS/dodecanol monolayer at the air/solution interface has been evaluated at an SDS concentration of 0.0067M and a dodecanol/SDS ratio of 0.005. Six different reflectivity profiles have been measured for either surfactant deuterated, both or neither in D_2O and water index matched to air (nrw). Shown in figure 1 are the profile for either surfactant deuterated in nrw and D_2O. The solid lines are model fits using the model described earlier for the parameters summarised in table 1.

The model fitting implies that compared to a pure SDS monolayer (7) the presence of 0.5% dodecanol increases the fraction of chains in the headgroup region, and results in the chains being less extended, whereas the surface excess is rather similar. In contrast the dodecanol molecules are closer to being fully extended and have a smaller fraction of their chains in the solvated headgroup region.

The six partial structure factors h_{dd}, h_{ss}, h_{ww}, h_{ds}, h_{dw}, and h_{sw} were determined from the six isotopic measurements. In the Q range available it is not possible to distinguish between a layer of uniform composition and a distribution such as a Gaussian. For direct comparison with the earlier model fitting the h_{ss} and h_{dd} partial structure factors have hence been fitted to uniform layers and give thicknesses of 22.5±0.5Å and 19.0±0.5Å for the dodecanol and SDS respectively.

The cross terrm h_{sd} has been evaluated from equation 12 (see figure 2) and we obtain a δ (the distance between the centres of the two surfactant distributions) of 4.0Å.

These results are entirely consistent with the assumptions of the model fitting; suggesting that the dodecanol molecules are more extended in the surface layer than the SDS, and that the two surfactants are staggered with respect to one another.

The most striking confirmation of this observation comes from the h_{sw} and h_{dw} partial structure factors (see figure 3), where the difference between those two partial structure factors immediately indicate that the dodecanol/solvent separation is larger than that of SDS/solvent.

Using equation 11 we obtain a δ, the distance from the centre of the surfactant distribution to that of the solvent, of 9.0Å for dodecanol and 6.5Å for SDS. This provides a direct confirmation of the earlier assumption that the SDS molecules are more immersed in the solvent than the dodecanol molecules.

The two methods of analysis provide a consistent and detailed description of the structure of the mixed SDS/dodecanol monolayer at the air/solution interface, in which the SDS molecules are more immersed in the solution subphase than the dodecanol molecules, and that the SDS molecules are less fully extended than the dodecanol molecules.

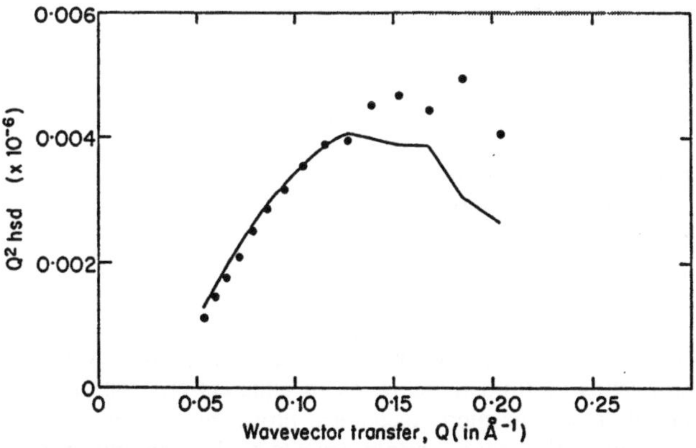

Figure 2 Partial structure factor for SDS/dodecanol. The solid line is the geometric mean $Q^2 (h_{ss}h_{dd})^{1/2}$ multiplied by cos Q δ, with δ = 4.0Å.

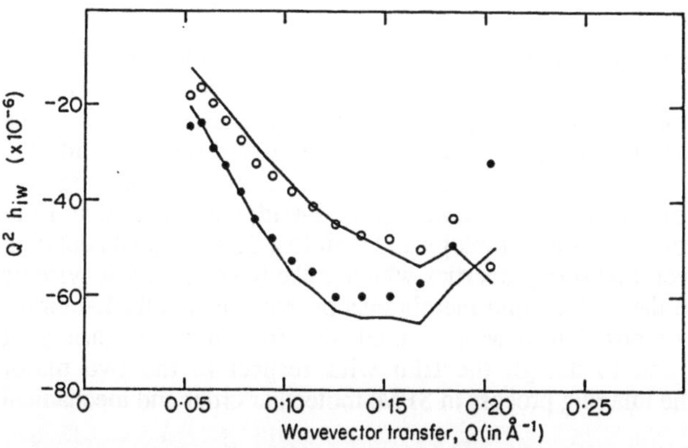

Figure 3 Partial structure factors for SDS/solvent (O) and dodecanol/solvent (●). The solid lines are for $Q^2 (h_{ii}h_{ww})^{1/2}$ sin Q δ with δ = 6.5Å. for SDS/solvent and δ = 9.0Å for dodecanol/solvent.

References

[1] J Lekner (1987) Theory of reflection (Dordrecht:Nijhoff)

[2] M Born and E Wolf (1975) Principles of optics. 5th Edn (Oxford:Pergamon)

[3] J Penfold and R K Thomas, J Phys:Condens Matter 2 (1990) 1369

[4] T L Crowley, E M Lee, E A Simister, R K Thomas, J Penfold and A R Rennie, Colloids and Surfaces (1990) (in press)

[5] E A Simister, R K Thomas, J Penfold, R Aveyard, B P Binks, P Cooper, P D I Fletcher, J R Lu and A Sokolowski, J Phys Chem (1991) (in press)

[6] E A Simister, E M Lee, R K Thomas and J Penfold, J Phys Chem (1991) (in press)

[7] J Penfold, R K Thomas, E A Simister, E M Lee, A R Rennie, J Phys:Condens Matter 2 (1990) SA411

[8] T L Crowley, E M Lee, E A Simister, R K Thomas, Physica (1991) (in press)

[9] J R Lu, E A Simister, E M Lee, R K Thomas, J Penfold (to be published)

[10] J Penfold, R C Ward and W G Williams, J Phys E : Sci Instrum (1987) 20 1411

Does the Scanning Force Microsope Resolve Individual Lipid Molecules?

M. Radmacher[1], *R.M. Zimmermann*[1], *and H.E. Gaub*[2]

[1]Physik Department, Technische Universität München, W-8046 Garching, Fed. Rep. of Germany

[2]Chemistry Department, Stanford University, Stanford, CA 94305, USA

The invention of the Scanning Force Microscope has opened new ways of addressing the unique properties of lipid membranes in aqueous environment. In addition to structural information certain micromechanical properties can be accessed on the molecular level. Here we will discuss certain aspects of the imaging process and the forces involved.

1. Introduction

Recent progress in the area of scanning probe microscopes has led to designs which allow the imaging of nonconducting surfaces with very high resolution (for an overview see [1]). Scanning Force Microscopy has drawn special attention as it seems to be applicable to the widest range of samples [2]. Imaging hard samples with this technique at even atomic resolution has become routine. However, imaging soft samples like organic molecules is still challenging. Besides the fact that lipid membranes play a very prominent role in biology and therefore deserve great attention, lipid membranes have several unique properties which make them ideal test systems for a novel technique like the SFM. Lipid membranes or, more generally Langmuir Blodgett films, can be composed in a well defined way from a broad variety of molecules. This allows one to design the film with respect to the two major properties that dominate the imaging process in SFM: molecular order and mechanical susceptibility.

2. Imaging Lipids

We [3-5] and recently also others [6, 7] have shown that molecular resolution on LB films can be achieved by SFM. Yet some key questions concerning details of the very nature of the imaging process are not well understood. In this paper, we address these questions.

2.1 How localized is the imaging process?

All high resolution images obtained with the SFM so far were reported from periodic structures. This resulted in the concern that the images obtained with the SFM might result from a coherent superposition of a delocalized tip-sample interaction rather than a single atomically localized interaction. Images of defects in periodic structures or of non-periodic structures at high resolution can answer this question. To our knowledge there was no point defect reported in the literature until recently; Binnig et al now showed a step in a layered crystal at about 5 Å resolution [8]. We have taken the alternative approach and have imaged the hydrated surface of amorphous silicon oxide with an SFM. Fig 1 shows a sequence of high resolution images taken from the same spot at time intervals of 22 sec. Two features are important: i) The surface of the sample appears to be amorphous and the smallest

Springer Proceedings in Physics, Vol. 66 The Structure and Conformation of Amphiphilic Membranes
Editors: R. Lipowsky · D. Richter · K. Kremer

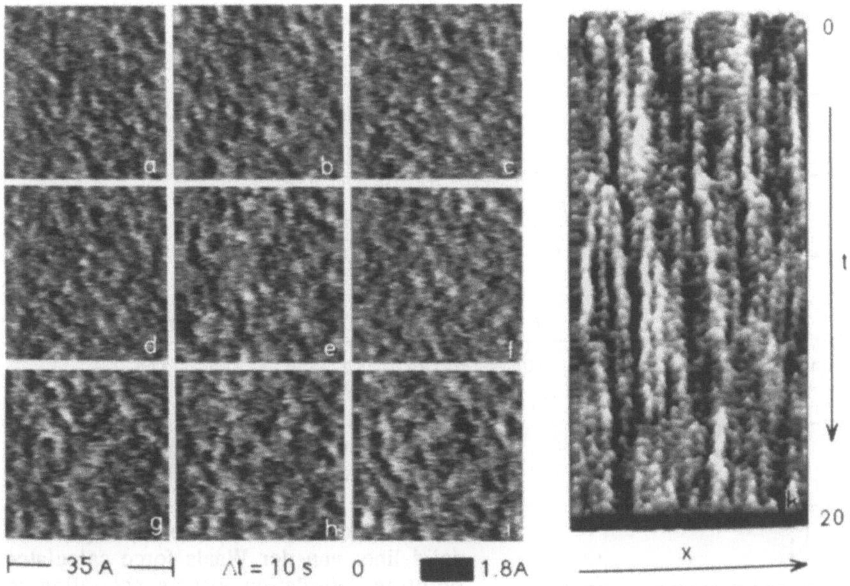

Fig 1 Surface topology of silicon oxide in water as seen by SFM. a)-i) series of images taken from the same spot at time intervals of 10 sec. k) x-scan plottet versus time. All SFM images were taken at room temperature with a home built instrument [12]. Details of the experimental conditions will be given in [9].

structures that can be resolved are in the size of ≈ 3 Å. ii) The image of the surface is changing slowly in time, which in a detailed study given elsewhere [9], we can attribute to a dynamic reorganization of the sample surface. As the timescale of this reorganization is too fast to be resolved in a series of images, this process was visualized in fig 1k, where only the x coordinate was scanned and successively plotted, line for line, in the y direction. Again, clearly resolved rows representing surface structures on the sample can be distinguished with a resolution of ≈3Å. This clearly shows that under suitable conditions the interaction of the tip with a flat surface is localized to an area with the characteristic size of 3Å. So the question raised above can be answered in the sense that depending on the substrate and the tip, the resolution of the SFM is such that indeed individual lipid molecules can be touched with this technique.

2.2 Which forces dominate image formation?

The forces between two solid bodies at near distances are well known from measurements with the surface force apparatus. Under our experimental conditions the most important forces are the Van der Waals attraction, the electron-electron hard core repulsion and in water or other fluids structural and electrostatic double layer forces. By recording the displacement of the tip upon approaching and retracting the sample, the sum of these forces is directly measurable with an SFM. Such a cycle (Fig 2a) shows a well pronounced hysteresis which is due to the different characteristic lengths of the attractive force and the spring stiffness. A comparison of the experimental data with a calculation from first principles however requires detailed knowledge of the tip

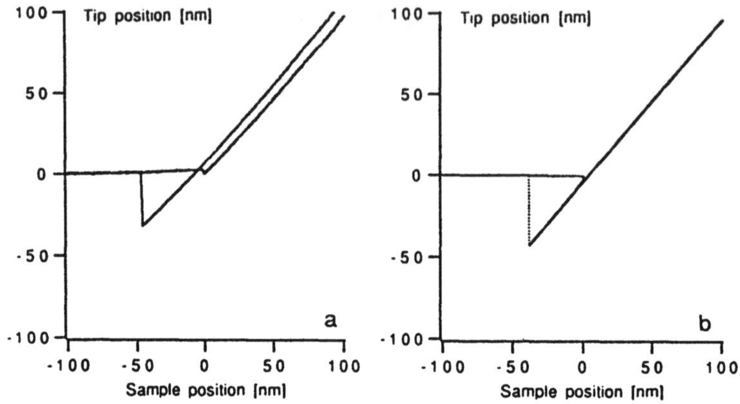

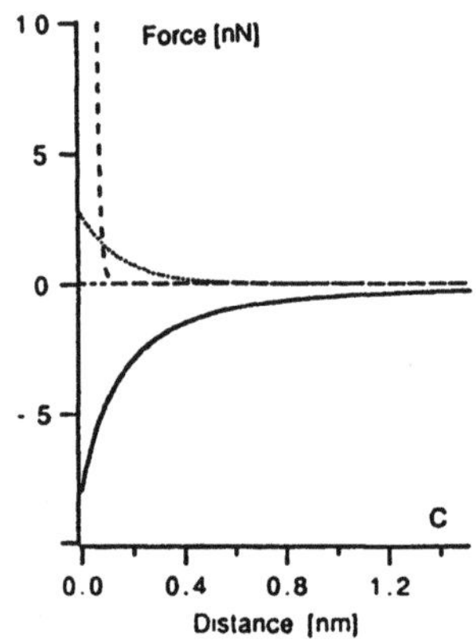

Fig 2a) Force scan of an integrated silicon nitride tip on a crystalline LB-film.b) Simulated force scan. c) Calculated contributions of the three dominating forces. solid line: van der Waals force calculated omitting retardation effects according to $F_{vdW}=H^*R_0/3(d+R_p)^2$ with $H=3^*10^{-20}$ J/m and $R_0=300$Å,. Dotted line: Hydration Force. The force between sphere and plane is related to the energy per area between two planes via the Derjaguin approximation[13]. Using this, we can take the experimental data from Rand and Parsegian for planar lipid films and calculate the hydration force to be $F_{hyd}= P_0 * \lambda * R_0/3*\exp(- (d+R_p)/\lambda)$ with $P_0= 10$ GN/m^2 and $\lambda = 1.5$Å. Dashed line: hard core repulsion. In the sphere/plane geometry the repulsive contribution in the Lenard Jones potential decays with the 8th power. The force constant was normalized so that the point of zero force is 1Å above the surface of the protrusion. All distances d were measured between the tip of the protrusion and the sample. The stiffness of the spring was 1N/m.

geometry. On the large scale the curvature radius of the tip can be determined by imaging steps of a known height. Montmorrilonit, which is a clay mineral consisting of flakes that are 20 Å high, was used here. The apparent width of the steps shown in Fig 3 indicate a typical mean curvature radius of about 300Å. This is a value that is in good agreement with measurements on proteins and other larger objects. If we combine this finding with the results from the preceding paragraph, we have to assume a relatively dull tip carrying a few small imaging protrusions. This situation is depicted in the model in Fig 4. Unfortunately, we have at present no direct information about the imaging protrusions or, in other words, about the spectral distribution of the micro-roughness of the tip. We have particularly no knowledge about the typical height of these structures. However, all high resolution images reported so far were obtained on atomically flat samples only. This indicates that the height of these imaging protrusions is only on the order of some atom diameters. Their lateral dimension must be on the order of the best lateral resolution.

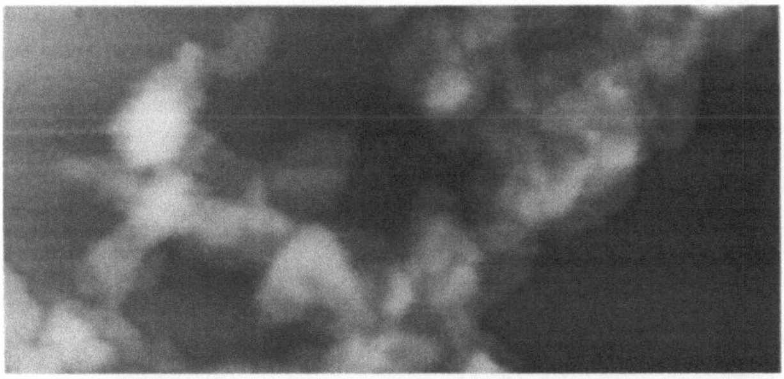

Fig 3 SFM image of Montmorillonit. The width of the image is 0.5μm.

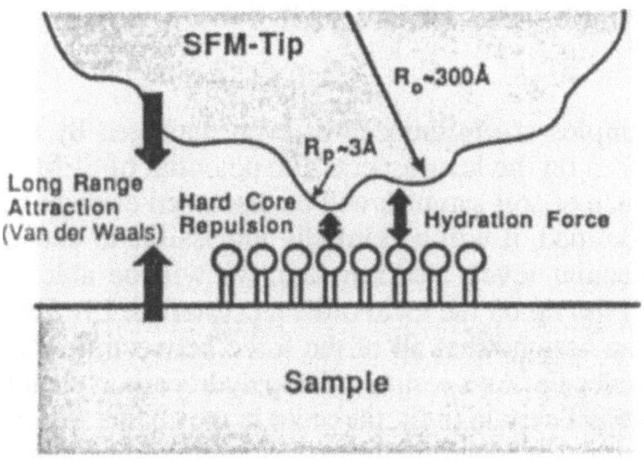

Fig 4 Sketch of the tip/sample interaction illustrating the various forces involved in the imaging process. The long range attractive interaction is integrated over the large tip radius and is counter balanced by the local repulsive interaction between the imaging protrusion and the sample. Especially when imaging under water, a considerable contribution to the repulsive interaction can be attributed to the hydration of both the sample and the tip.

Based on these ideas we can in the simplest way represent our tip by a large sphere (R_o=300 Å) with a little hemisphere (R_p= 3Å) pointing towards the sample. Following this model we have simulated the force distance curves (Fig 2c)based on van der Waals attraction, repulsive hydration force between the large sphere and a flat sample and hard core repulsion between the protrusion and the flat sample. The resulting displacement curve where the external spring was taken into account is given in Fig 2b. This curve can be directly compared with the experimental result in Fig 2a. An examination of the individual contributions (Fig 2c) shows that in the minimum, that is close to working point of the SFM, the major part of the van der Waals force (<3nN) is counter balanced by the delocalized structural force of the water. Only a small fraction (<1nN) of the total force acts via the protrusion and contributes to high resolution image formation. It should be pointed out again that the above calculations are based on a very crude model which does not take into account the local surface structure and chemistry.

2.3 Can lipid films withstand the applied forces?

In the preceding paragraphs we have tried to develop some ideas concerning tip geometry and the forces which dominate image formation. When the SFM technique is applied to organic samples rather than minerals or other hard surfaces the question arises how the micromechanical properties of the sample contributes to the apparent

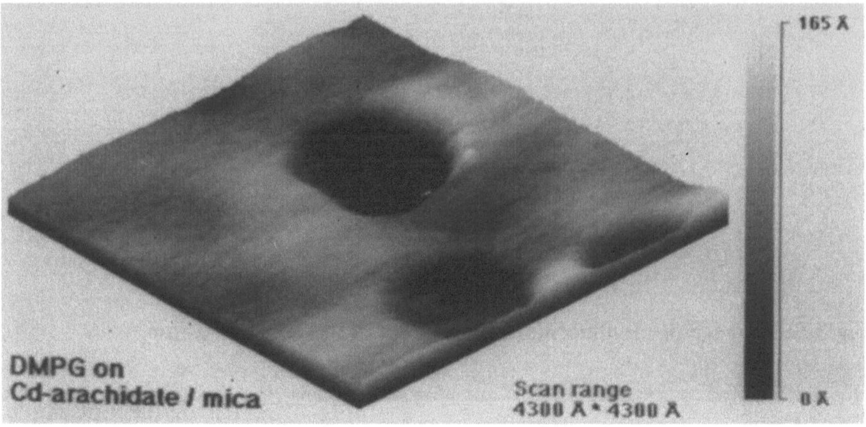

Fig 5 SFM image of a hole in a lipid bilayer. For details see [4].

topology, and, in turn, how these samples are influenced or even damaged by the applied force. In this paragraph we focus on the later aspect. The potential of SFM to locally probe the mechanical compliance of soft samples will be discussed elsewhere. As long as the tip geometry is so ill-defined, it will be virtually impossible to derive quantitative information on the molecular level. Nevertheless, we will be able to derive a good estimate on the effect of the tip on the local order a crystalline LB film. As a worst case approximation we can assume that all of the force between tip and sample is loaded on one single lipid molecule. As a result, ordered hydrocarbon chains will be squeezed, and if the force exceeds a certain limit, the order in the chains will be broken up locally. If we assume this to occur when the lipid is compressed by about 3Å, which roughly corresponds to the difference in the thickness of a solid and a fluid film, and if we take the energy which is necessary to break up the order from calorimetric measurements [10],we can estimate the required force to be $F \approx \Delta H / \Delta s \approx 30kJ / (6*10^{23} *2 Å) \approx 0.3$ nN. In comparison with the previous paragraph this approximation shows that with the present restrictions of the tip geometry we are just barely capable of non-destructively imaging lipid membranes. It tells us also that it will be extremely difficult to image packing defects in the hydrocarbon chains, as the cooperativity of the chain melting which additionally stabilizes molecules inside the crystal, will not stabilize the molecules close to the defect. The mechanical damage that might occur to lipid films is shown in Fig 5. A hole in the lipid bilayer appeared after scanning the center part of the sample with increased force. This local damage may in turn be utilized for nano-lithography if one can write pattern into monomolecular films with lateral dimensions of some nanometers. For a detailed description of this potential application see ref [11].

3. Concluding remarks

We have shown in the preceding paragraphs that i) the resolution of the SFM can be in the order of 3Å and that the image is formed by a localized tip sample interaction. ii) the minimum forces for imaging are mainly determined by the tip size and the Hamacker constants, iii) hydration forces help to reduce the local load required for image formation and iv) lipids can withstand the applied forces when they are packed in a crystal. The answer to the title question will therefore be yes.

4. Acknowledgements

This work was supported by the Deutsche Forschungsgemeinschaft.

5. Literature References

(1) Wickramashing, K. *Scientific American* Okt.1989,
(2) Hansma, P.; Elings, V. B.; Marti, O.; Bracker, C. E. *Science* **1988**, *242*, 209.
(3) Egger, M.; Ohnesorge, F.; Weisenhorn, A.; Heyn, S. P.; Drake, B.; Prater, C.B.; Gould, S.; Hansma, P.; Gaub, H. E. *Journal of Struct.Biol.* **1990**,*103*, 89-94.
(4) Radmacher, M.; Goettgens, B. M.; Tillmann, R. W.; Hansama, H. G.; Hansma, P. K.; Gaub, H. E. In *Scanned Probe Microscopies*; McDonald, Ed.; 1991.
(5) Weisenhorn, A.; Egger, M.; Ohnesorge, F.; Gould, S.; Heyn, S. P.; Hansma, H. G.; Sinsheimer, R. L.; Gaub, H. E.; Hansma, P. *Langmuir* **1991**, *7*, 8-12.
(6) Fujii, T.; Okane, T.; Arisawa, S.; Yamamoto, R. *Proceedings STM 91* **1991**,
(7) Akamine, S. *Proceedings STM 91* **1991**,
(8) Binnig, G. *Proceedings STM 91* **1991**,
(9) Tillmann, R.; Radmacher, M.; Gaub, H. E. *Subm. to Phys. Rev.Letters* **1991**,
(10) Jürgens,E.; Höhne,G.; Sackmann,E. *Ber.Bunsenges.Phys. Chem.* **1983**, *87*, 95
(11) Hansma, H. G.; Gaub, H. E.; Zazadinski, J. A. N.; Longo, M.; Gould, S. A. C.; Hansma, P. K. *Langmuir* **1991**,
(12) Radmacher, M.; Eberle, K.; Gaub, H. E. *Ultramicroscopy* **InPress May1992**,
(13) Derjaguin, B. V. *Kolloid* Z **1934**, *69*, 155-164.

Domain Formation in a Lipid Monolayer

A. Gliozzi[1], *A.C. Levi*[2], *M. Menessini*[1], and *E. Scalas*[1]

[1]Dipartimento di Fisica, Università di Genova,
via Dodecaneso 33, I-16146 Genova, Italy
[2]SISSA, via Beirut 4, I-34014 Trieste, Italy

Abstract. We introduce a growth model, which may be useful for the understanding of solid domain growth in a lipid monolayer, where the morphology of the grown cluster is dependent on temperature, disequilibrium, and impurity concentration.

1. Introduction

Out-of-equilibrium domain formation in lipid monolayers at the air-water interface, during the liquid-to-solid phase transition, has been the object of both experimental and theoretical studies [1,2,3,4,5].

In lipid monolayers the solidification process is controlled by the lateral diffusion of both lipid molecules and impurities, since the latent heat of the transition can diffuse into the aqueous substrate [2,4].

The competition between the diffusion field, leading to an instability of the Mullins-Sekerka type [6], and the effect of line tension, tending to keep smooth the interface between the two coexisting phases, results in a characteristic dense-branching morphology [7].

The diffusive instability can be amplified either by a stronger disequilibrium or by a lower interfacial tension, which can be lowered both by increasing the temperature and by adding impurities to the monolayer. In these cases more ramified solid domains are obtained.

In this paper, we introduce a DLA-like model for the growth of two-dimensional aggregates, including the effects of temperature, disequilibrium, and impurity concentration; this model may be useful for the understanding of domain growth in lipid monolayers.

2. Domain Growth in a Monolayer without Impurities

We use a modified version of the pedestrian model introduced by Kadanoff in order to simulate the Hele-Shaw cell experiments [8].

On a finite square lattice the central site is filled with a nucleation seed. Then a random walker is released at great distance from the seed and it moves until either it gets lost or it hits the cluster. In the latter case a pseudo-random number $\xi \in [0,1]$ is extracted with uniform probability, and is compared with the probability of "condensation" C, which is given by the Glauber formula:

Springer Proceedings in Physics, Vol. 66 The Structure and Conformation of Amphiphilic Membranes
Editors: R. Lipowsky · D. Richter · K. Kremer

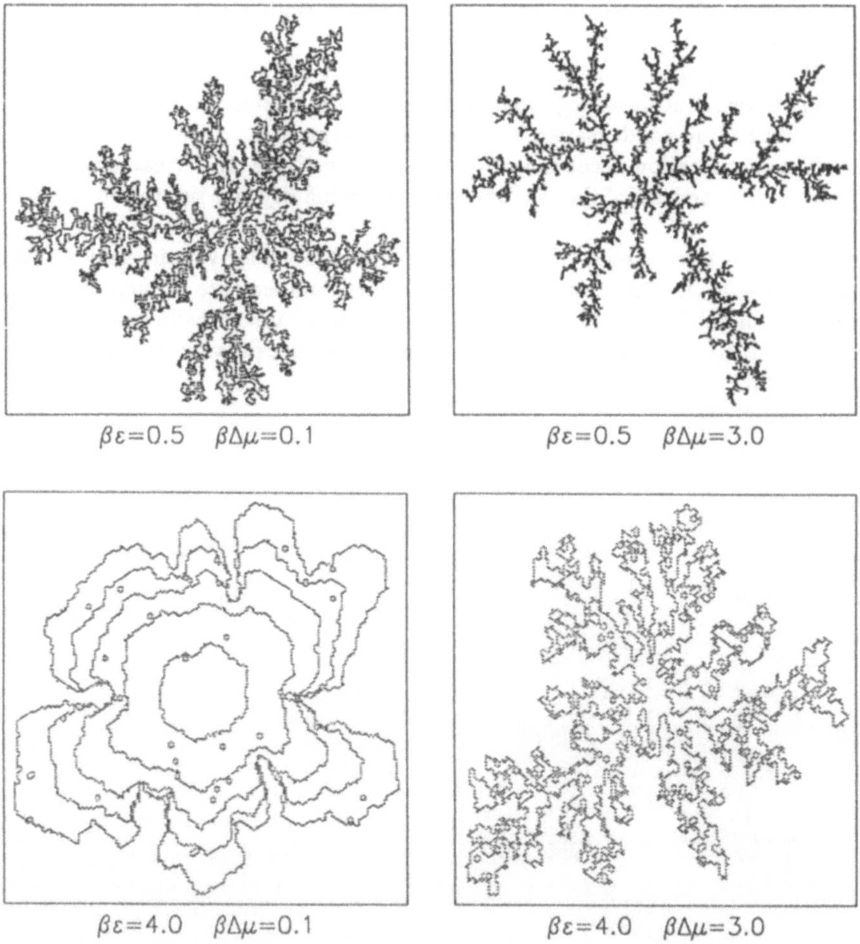

Figure 1: Typical cluster shapes for various values of $\beta\varepsilon$ and $\beta(\mu - \mu_{eq})$. (Lines refer to succesive times).

$$C = \frac{1}{1 + \exp[\beta(\Delta E - \mu)]}, \tag{1}$$

where β is the inverse temperature, ΔE is the energy difference due to the condensation of a single particle and μ is the chemical potential. If $\xi < C$ a new site is added to the cluster where it is hit by the random walker, otherwise the walker keeps on moving.

As soon as the cluster includes two sites, one of them is chosen for "evaporation". Again a uniform pseudo-random number ξ' is extracted and compared with the probability of evaporation E:

$$E = \frac{1}{1 + \exp[\beta(\Delta E + \mu)]}. \tag{2}$$

If $\xi' < E$ the site evaporates and starts a random walk until it either again the cluster hits or it gets lost.

If μ is greater than the equilibrium value $\mu_{eq} = -2\varepsilon$ (where $-\varepsilon$ is the binding energy between nearest-neighbour cluster sites) the cluster starts growing, while evaporation and condensation movements are repeated.

The evaporation–recondensation moves constitute the "rearrangement" process. The arrivals of the random walker from infinity give rise to the diffusive instability.

In figure 1 are shown typical results of simulations for various values of $\beta\varepsilon$ and $\beta(\mu - \mu_{eq})$. One can observe that, when temperature is increased, the diffusive instability tends to overcome rearrangement eventually giving rise to fractal DLA-like clusters. The same phenomenon happens by increasing disequilibrium.

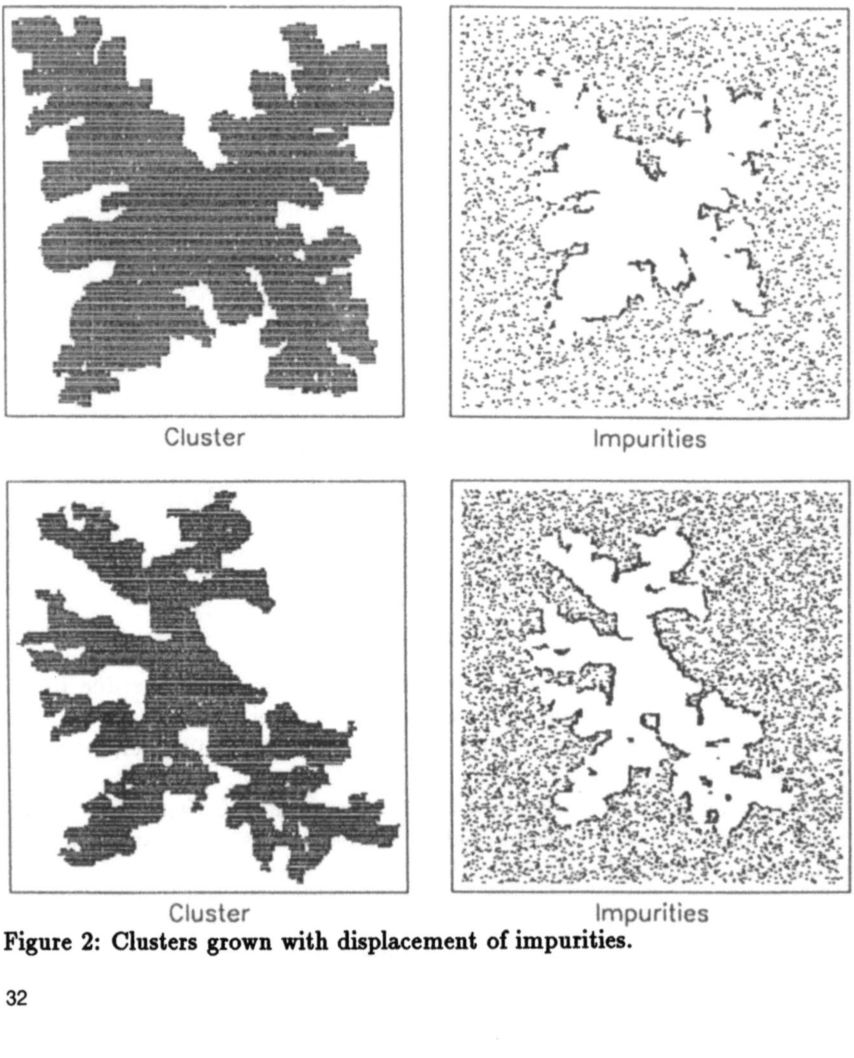

Figure 2: Clusters grown with displacement of impurities.

3. The Inclusion of Impurities

In order to take into account the effect of impurities (introduced to represent e.g. the fluorescent dye used in the experiments), the algorithm described previously has to be slightly modified. Now the nucleation seed is placed in the centre of a matrix which has been randomly (but uniformly) filled with impurities with a probability p related to the impurity concentration. As the solid cluster grows, the impurities are displaced and accumulate at its boundary. They are allowed to diffuse away from the cluster in order to avoid stopping the growth process. In figure 2 the accumulation of impurities at the growing cluster boundaries is shown.

4. Conclusion and Outlook

We have introduced a growth model that presents some qualitative features of domain growth in lipid monolayers, namely the dependence of domain shapes on temperature, disequilibrium, and impurity concentration. We are currently quantitatively studying the morphology diagram of the model and we hope to be able to compare it with the results of fluorescence microscopy experiments.

5. Acknowledgements

The authors wish to thank Dr. R. Rolandi for many interesting discussions on the issue. One of the authors (E. Scalas) is indebted to Professor W. Helfrich for bringing the works of H. Möhwald to his attention.

This work has been funded by ESPRIT II, BASIC RESEARCH ACTION 3200-OLDS, MPI 40% and 60% grants and by the CNR grant "Progetto Finalizzato Chimica Fine II ".

References

[1] A. Miller, W Knoll and H. Möhwald, Phys. Rev. Lett. **56**, 2633 (1986).

[2] A. Miller and H. Möhwald, J. Chem Phys. **86**, 4253 (1987).

[3] H.C. Fogedby, E.S. Sørensen and O.G. Mouritsen, J. Chem. Phys. **87**, 6706 (1987).

[4] K.A. Suresh, J. Nittmann and F. Rondelez, Europhys. Lett. **6**, 437 (1988).

[5] B. Berge, L. Faucheux, K. Schwab and A. Libchaber, Nature **350**, 322 (1991).

[6] W.W. Mullins and R.F. Sekerka, J. Appl. Phys. **34**, 323 (1963).

[7] E. Ben-Jacob and P. Garik, Nature **343**, 523 (1990).

[8] L.P. Kadanoff, J. Stat. Phys. **39**, 267 (1985).

Structure and Dynamics of Planar and Spherical Supported Phospholipid Bilayers

C. Dolainsky, T. Köchy, C. Naumann, T. Brumm, S.J. Johnson, and T.M. Bayerl

Physik Department, Biophysics Group E22, Technische Universität München, James Franck-Str., W-8046 Garching, Fed. Rep. of Germany

Abstract. The structure and dynamics of planar and spherical supported phospholipid bilayers is studied by NMR methods, specular reflection of neutrons (SRN), DSC and computer simulations. The combination of these methods allows the assessment of a number of basic physical properties of these new membrane model systems. Spherical supported vesicles (SSV), which offer the advantage over multilamellar vesicles (MLV) of being geometrically well defined and adjustable in their size, are shown to exhibit a structure similar to the MLV. The existence of an ultrathin water layer of ≈15Å thickness between the bilayer and the solid support is confirmed by both NMR and SRN. The main phase transition temperature of SSV's of DPPC is reduced by 2°C as compared to MLV's and there is no pretransition. A new method is applied to determine the lateral diffusion coefficients of SSV's at various temperatures. Moreover, the new model system is used to study the effect of slow motions (on the NMR timescale) on the anisotropy of the transverse relaxation time T_{2e}. The results suggest that the major relaxation contribution comes from the lateral diffusion of the phospholipids.

1. Introduction

Recently, we proposed spherical supported vesicles as a new model system for spectroscopic studies of membranes [1,2]. Owing to their known shape and adjustable size, SSV's exhibit a variety of features which give them potential in particular for the study of membrane dynamics. Two types of spherical supported

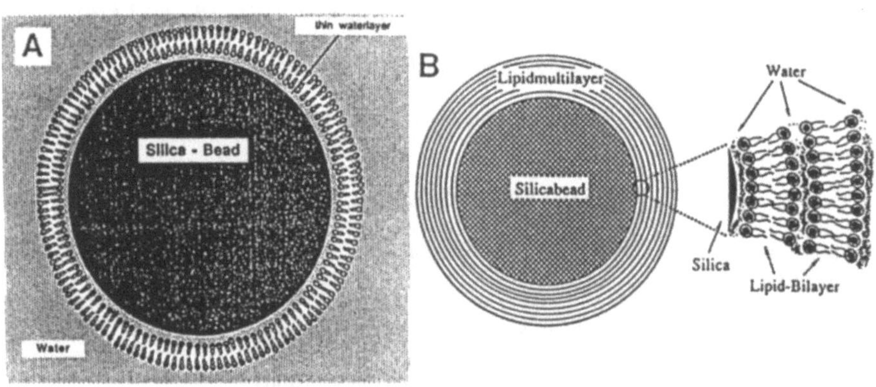

Fig.1: Schematic depiction of (A) Spherical Supported Vesicles (SSV) and of (B) Spherical Supported Multilayer Vesicles (SSMV).

Springer Proceedings in Physics, Vol. 66 The Structure and Conformation of Amphiphilic Membranes
Editors: R. Lipowsky · D. Richter · K. Kremer

systems have been developed so far: spherical supported vesicles (SSV) and spherical supported multilayer vesicles (SSMV) as depicted in Fig. 1A,B. The aim of this paper is to discuss their basic physical properties as well as to show the usefulness of some of their unique features in the study of membrane dynamics by NMR relaxation methods.

2. Results and Discussion

2.1 Thermodynamics of Spherical Supported Vesicles (SSV's)

We studied the phase behaviour of SSV's by differential scanning calorimetry (DSC). In Fig. 2A typical DSC thermograms of DPPC-SSV's and of a DPPC multilamellar vesicle dispersion (MLV) are shown. On the first view, three salient features are obvious: 1) The main phase transition temperature T_m of the SSV's is 2°C below that of the MLV's. 2) SSV's exhibit no so-called pretransition, i.e. a transition from the L_β to the $P_{\beta'}$ ("rippple") phase. 3) The cooperativity of the SSV main phase transition is a factor of 5 lower than that of MLV's, as assessed from the FWHM values of both thermograms. The DSC results prove that all phospholipids in single bilayer SSV systems are affected in the same way by the interaction of the single bilayer with the silica surface, in particular, there is no decoupling between the inner and the outer bilayer leaflet as regards the main phase transition.

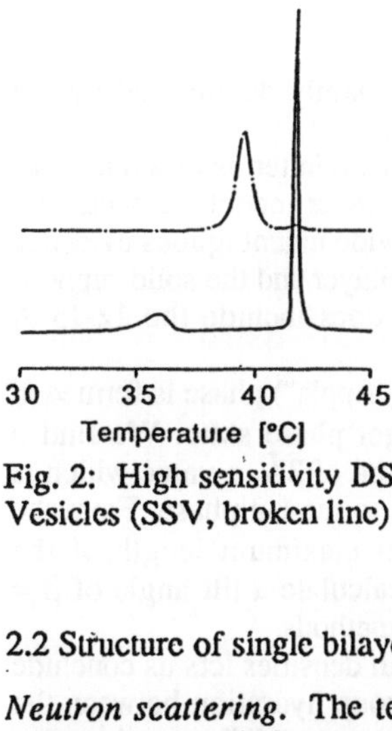

Fig. 2: High sensitivity DSC thermograms (heating scans) of Spherical Supported Vesicles (SSV, broken line) and of multilamellar vesicles (full line) of DPPC.

2.2 Structure of single bilayers on a solid support

Neutron scattering. The technique of neutron specular reflection in combination with contrast variation was used in order to obtain information about the structure of the single supported bilayer and about the dimensions of an anticipated ultrathin water layer between the bilayer and the silica surface [3]. For this study, the single bilayer was adsorbed to a planar quartz plate and three different cases of contrast were studied: a) the bulk water was contrast matched to the quartz. This allows a sensitive measurement of changes of the bilayer thickness. b) Heavy water (D_2O) was used as bulk water, providing a strong contrast between the quartz surface and the single bilayer. This arrangement is sensitive to changes of the lipid headgroup hydration and the thickness of the confined water layer. c) The bulk water and the lipid chains

Table I: Results for the analysis of the neutron reflectivity data of a single DMPC supported bilayer in the fluid state and in the gel state for three cases of scattering contrast. d_L, d_W, d_H, d_T are reported in Å and represent the respective thicknesses of the total bilayer, the water layer between bilayer and quartz, the head group and the tail region. The scattering densities ρ for the corresponding regions are reported in 10^{-6} $Å^{-2}$.

Case	T(°C)	d_L	ρ_L	d_W	ρ_W	d_H	ρ_H	d_T	ρ_T
a)	31	43	0.7	-	-	-	-	-	-
b)	31	43	2.0	30	3.7	-	-	-	-
c)	31	43	-	-	-	8	1.6	13.5	3.5
a)	20	46	0.7	-	-	-	-	-	-
b)	20	46	2.1	30	3.7	-	-	-	-

are matched to the quartz such that the reflectivity is mainly determined by the headgroups. This provides the thickness of the headgroups.

The analysis of the reflectivity data for case a) was done in terms of a one layer model, for cases b) and c) a three layer model and two layer model, respectively, were applied. The results are shown in Table I. They provide unambiguous evidence for the existence of an ultrathin water layer between the bilayer and the solid support. The thickness of 30Å given for this layer in Table 1 does include the 12-15 Å surface roughness of the quartz plate.

A further interesting result is the finding that no $P_{\beta'}$ ("ripple") phase is formed at the transition of the bilayer from the fluid into the gel phase state. We find a thickness change of the bilayer due to the phase transition of 3Å, a value which is normally observed for the transition from the L_α phase to the $L_{\beta'}$ phase. From the measured tail length in the gel phase and the known maximum length of the hydrocarbon tail in an all-trans conformation we can calculate a tilt angle of $\beta = 26 \pm 7°$, in good agreement with values obtained by other methods.

Finally, the analysis of the obtained scattering length densities lets us conclude that there is no significant change of the DMPC headgroup hydration between the two phase states studied. This finding indicates a major structural difference between supported bilayers and phospholipid monolayers. For the latter we found a significant change of the hydration between the extended and the condensed phase state [4].

NMR methods. A further proof for the existence of a water layer between the single bilayer and the supporting solid surface comes from the application of proton NMR on SSV systems. We obtained a water layer thickness of $d_w = 17 \pm 5$Å using this method [1]. The surface roughness of the silica beads is less than 3 Å (as seen by atomic force microscopy) and can be neglected.

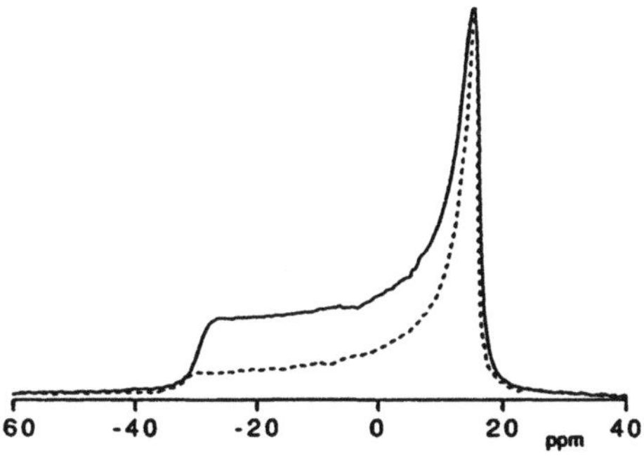

Fig. 3: 161MHz ^{31}P-NMR spectra of spherical supported vesicles (SSV, full line) and of multilamellar vesicles (MLV, broken line) of DPPC in the liquid crystalline phase state.

A wealth of information about the bilayer structure in SSV's and SSMV's can be obtained by wide line deuterium and phosphorus NMR. Figure 3 shows representative ^{31}P-NMR spectra of a SSV sample with a silica bead diameter of 640±40 nm and, for comparison, the corresponding MLV spectra. Since the ^{2}H-NMR spectra (data not shown) were obtained using acyl chain perdeuterated DPPC-d_{62}, the combination of both NMR methods allows the assessment of the molecular order in both head and tail region of the bilayer. The flexibility along the fatty acyl chain, which can be expressed as the orientational order parameter profile, is not significantly affected by the bilayer-silica surface interaction as compared to the MLV [1]. Similarly, the effective chemical shift anisotropy obtained from the ^{31}P-NMR spectra exhibits no significant differences for the two model systems. A salient feature of the MLV spectra (Fig. 3) is a lineshape obstruction due to a macroscopic orientation in the field of the NMR magnet. Recently, we proved that MLV's in the fluid phase state adopt an elliptical shape with the long axis parallel to H_0 at a magnetic field stregth of 9.6 Tesla (400 MHz proton resonance frequency) [5]. Thus, one important virtue of the SSV's is that such systems prevent completely the lineshape obstructions caused by macroscopic orientation effects (cf. Fig. 3).

2.3 Dynamics of spherical supported vesicles

Lateral diffusion of phospholipids. So far, we have focused on the study of the processes slow on the NMR time scale, in particular the lateral diffusion of phospholipids along the membrane surface. The well defined geometry of the SSV's enables the application of the so called Carr-Purcell-Meiboom-Gill pulse sequence, modified for spin-1 nuclei (quadrupolar CPMG), for the measurement of the lateral diffusion coefficient D of the lipids [6,7]. The transverse relaxation time T_2^{CPMG} obtained by this pulse sequence is increasingly affected by slow motions such as lateral diffusion for increasing values of the echo spacing τ. For a SSMV sample of POPC-d_{31} (Fig. 4) we obtained the following values of D at three different temperatures: D= (2±0.6) 10^{-12} m^2/s (10°C), D= (3.8±0.9) 10^{-12} m^2/s (30°C), D= (9.8±2.3) 10^{-12} m^2/s (50°C). SSV's of DPPC-d_6 gave D= (5.5±0.8) 10^{-12} m^2/s at 30°C. These values and their temperature dependence are in good agreement with

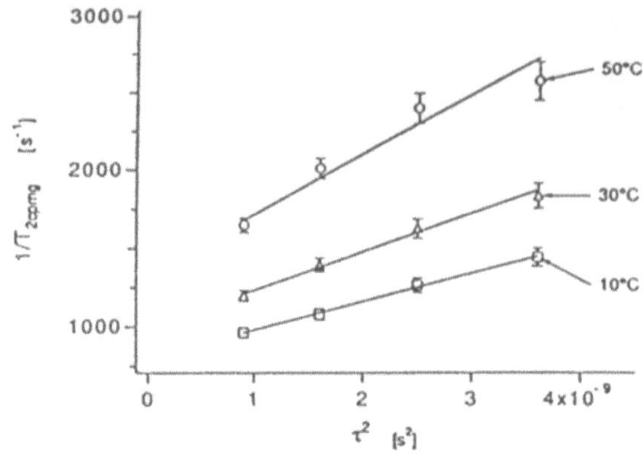

Fig. 4: Plot of $1/T_2^{CPMG}$ versus the squared echo spacing τ^2 obtained for a SSMV system of POPC-d_{31} by ^{2}H-NMR at three different temperatures.

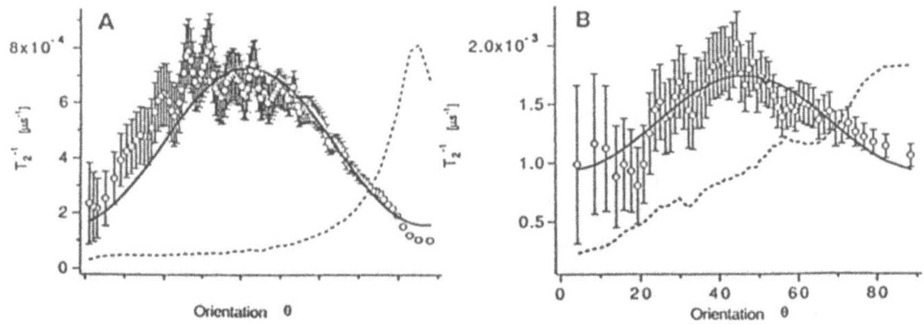

Fig. 5: ^{31}P-NMR spectra (dotted lines) and their T_{2e} relaxation spectra (o).
A: POPC multilamellar vesicles at 30°C, B: POPC spherical supported vesicles of 640nm diameter at 30°C. The full lines represent fits of the relaxation spectra to equation (1).

those obtained by other methods such as fluorescence methods and dynamic neutron scattering in phospholipid multilayers.
Modulation of transverse relaxation by slow motions. The calculation of D from the transverse relaxation time is a consequence of a more general formalism of the effect of slow motions on the orientation dependence of the relaxation times [7,8]. The theory predicts that all slow motions mentioned above can modulate the transverse relaxation behaviour in such a way that T_{2e} obtained by a quadrupolar or spin echo experiment scales with the angle θ between the molecular director axis and H_0 as [7]

$$1/T_{2e} = A + B \sin^2\theta \cos^2\theta . \qquad (1)$$

where A and B are appropriate constants. Spherical supported bilayers represent a model system which is, by virtue of its well defined geometry, particularly well suited for an experimental verification of this theoretical prediction. Figure 5A,B shows ^{31}P-NMR spectra of MLV and SSV, respectively, obtained by the Hahn spin echo sequence as well as their relaxation spectra. It is obvious that the T_{2e} orientation dependence of both model systems obeys the theoretical prediction. However, the slope of d T_{2e} / dθ is significantly lower for the SSV's. This indicates that relaxation mechanisms other than lateral diffusion provide a major contribution to the MLV relaxation. A further proof for the effect of diffusion on the T_{2e} orientation dependence of SSV's comes from a simulated relaxation spectrum. This simulation solely considers the averaged squared phase of the spin system acquired in the echo time 2τ by a random walk of the individual spins on the surface of a sphere with the same diameter as the SSV's (640 nm), using the diffusion coefficient $D = 3.8\ 10^{-12}\ m^2/s$ as determined by the CPMG method [10] and gives an T_{2e} orientation dependance according to (1).

3. References

[1] Bayerl, T. M. & M. Bloom. 1990. Biophys. J. 58: 357-362

[2] Bayerl, T. M. & M. Bloom. 1990. J. Magn. Reson. 93: 279-290

[3] Johnson, S. J., T. M. Bayerl, D. C. McDermott, G. W. Adam, A. R. Rennie, R. K. Thomas & E. Sackmann. 1990. Biophys. J. 59: 289-294

[4] Bayerl, T. M., R. K. Thomas, J. Penfold, A. Rennie & E. Sackmann. 1990. Biophys. J. 57: 1095-1098

[5] Brumm, T., A. Möps, C. Dolainsky, S. Brückner & T.M. Bayerl. 1991. Biophys. J. (submitted)

[6] Koechy, T. & T. M. Bayerl. 1991 (in preparation)

[7] Boom, M. & E. Sternin. 1987. Biochemistry 26, 2101-2105

[8] Bloom, M., E. Evans & O.G. Mouritsen. 1990. Quart. Rev. Biophys. (in press)

[9] Nezil, F. A., C. Morrison, K. P. Whittall & M. Bloom 1991

[10] Dolainsky, C. & T. M. Bayerl. 1991. (in preparation)

Translational Diffusion and Fluid Phase Connectivity in Multi-Component, Multi-Phase Lipid Bilayer Membranes

W.L.C. Vaz[1], *T.E. Thompson*[2], *P.F. Almeida*[2], *T. Bultmann*[1], and *E.C.C. Melo*[3]

[1]Max-Planck-Institut für Biophysikalische Chemie, Postfach 2841, W-3400 Göttingen, Fed. Rep. of Germany
[2]Department of Biochemistry, University of Virginia, Charlottesville, VA 22908, USA
[3]Centro de Tecnologia Química e Biológica – IST, R. da Quinta Grande, 6, Apartado 127, P-2780 Oeiras, Portugal

Abstract. We have used the fluorescence recovery after photobleaching technique (FRAP) to investigate the long range translational diffusion of a fluid-phase soluble, gel-phase insoluble fluorescent lipid derivative (NBD-dilauroylphosphatidylethanolamine or NBD-1-palmitoyl-2-oleoylphosphatidylethanolamine) in lipid bilayers made up of two or more lipid components as a function of temperature and chemical composition of the bilayers. The FRAP experiment gives information on two aspects of the diffusivity of the test particle (fluorescent probe): 1) the diffusion coefficient is related to the characteristic recovery time for the fluorescence after photobleaching, and 2) the freedom of diffusion in the membrane plane over long distances ($\approx 10\mu m$) is related to the fraction of initial fluorescence recovered at "infinite" time after photobleaching. For the case where solid and fluid phases co-exist in the plane of the lipid bilayer the two parameters taken together provide information on the connectivity of the fluid (or, reciprocally, the gel) phase domains and, when applicable, to what extent an archipelago of gel phase domains slows diffusion in the fluid phase. The results indicate that the gel phase in some cases forms highly elliptical domains, which at a very low mass fraction ($\leq 20\%$) impede percolation of the fluid phase, whereas in others the gel phase domains are more centro-symmetric so that the fluid phase is continuous at a very low mass fraction ($\leq 10\%$). A comparison of the experimental results with existing theoretical descriptions of diffusion in an archipelago becomes possible in systems of the latter type and was undertaken.

1. Introduction.

Evidence that lateral phase separations in the lipid bilayer of biological membranes occur under physiologically relevant conditions has been accumulating rapidly in the recent literature [1]. These phase separations manifest themselves in the formation of chemically and physically distinct and co-existent domains in the

Springer Proceedings in Physics, Vol. 66 The Structure and Conformation of Amphiphilic Membranes
Editors: R. Lipowsky · D. Richter · K. Kremer

plane of the membrane. Since these domains act as solubilizing media for the integral proteins as well as adsorption surfaces for the extrinsic proteins in a membrane, they could play a significant role in the physiology and biochemical dynamics of membranes [2].

It has been a biophysical tradition of long standing that in order to understand processes that occur in complex multi-component biological systems researchers first try to understand similar physical processes in much less complex "model systems". Over the last three decades this approach has been an excellent source of insights into the structure and function of biological membranes. On this basis we have undertaken the study of translational diffusion in artificial lipid bilayer membranes of known chemical composition using the FRAP [3] technique. Several binary mixtures and one ternary mixture have been examined.

2. Results and Discussion.

A lipid bilayer membrane, viewed from above, is a two dimensional system in which lateral phase separations resulting in domain formation present some interesting problems with respect to domain shape, size and distribution. When only two types of domains co-exist within the membrane the point at which one of the domain types forms a continuous cluster, completely disconnecting the other domain type, is a singular point of the system. The long range (relative to average domain size) translational diffusion of a probe, chosen such that it is soluble in only one of the two co-existing domain types, will show this singularity. The problem is one of two-dimensional percolation and the singularity is the percolation threshold. A knowledge of the percolation threshold in a region of co-existing domains combined with a knowledge of the mass fraction of the phases which form the domains (obtained from a temperature-composition phase diagram, at constant pressure) allows some conclusions to be made with respect to domain shape. If the probe is fluorescent, e.g. a fluorescent lipid derivative, FRAP allows the examination of the translational diffusion of the probe over distances on the order of about 10μm which are very much larger than the lipid domains. In these experiments the fluorescence does not recover to the initial (pre-bleach) intensities at "infinite" times after photobleaching if the domains into which the probe partitions are not connected. Thus in going from a non-connected state (<100% recovery) to a fully connected state (100% recovery) the point in temperature and composition at which complete fluorescence recovery is first obtained has been arbitrarily identified as the percolation threshold. In addition the characteristic recovery time in the FRAP experiment, which is related to the translational diffusion coefficient, also provides information on the percolation, size and shape of the domains, though this relationship is somewhat more complex than the fluorescence recovery.

We have examined several binary mixtures of phosphatidylcholines for which detailed temperature-composition phase diagrams are available in the literature. Connectivity of the fluid phase was examined using FRAP under conditions (temperature and composition) where gel and fluid phases were known to be co-existent from the phase diagrams. The fluorescent probe used was either NBD-dilauroylphosphatidylcholine (NBD-DLPE) or NBD-1-palmitoyl-2-oleoylphosphatidylcholine (NBD-POPE), both of which were shown to be soluble only in the fluid phase. The recovery of fluorescence (at times which were much longer than the characteristic recovery times) compared to the pre-bleach fluorescence intensity was used as a measure of fluid phase connectivity.

In the isomorphous mixture of dimyristoylphosphatidylcholine and dipalmitoylphosphatidylcholine the percolation threshold in the region of co-existing solid and fluid phases was obtained at a 0.5 mass fraction of solid phase in the system [4]. Since the gel phase lipid molecular area in the plane of the bilayer is about 75% of the fluid phase lipid molecular area, this corresponds to an area fraction of 0.375 for the gel phase. If the gel phase domains were circular in shape and uniform in size, continuum percolation arguments would need a gel phase area fraction of 0.67 [5]. Using the same arguments we deduce an elliptical shape with an axial ratio of about 0.15 for the solid phase domains in this case.

In the peritectic mixture of dimyristoylphosphatidylcholine and distearoylphosphatidylcholine we have obtained percolation thresholds between mass fractions of 0.14 and 0.67 gel phase [6]. This corresponds to elliptical gel domains with axial ratios between 0.03 and 1.0, assuming uniform size and random orientations. The ellipticity of the gel phase domains was related to the chemical composition of this phase: the broader the range of possible chemical compositions for the gel phase on cooling the mixture at a given composition through the liquidus to the solidus, the higher the mass fraction of gel phase required to attain the percolation threshold. This suggests that compositionally purer gel phases form more asymmetric domains.

A eutectic mixture of dihexadecanoylphosphatidylcholine and 1-docosanoyl-2-lauroylphosphatidylcholine was also examined [7]. In this mixture liquid and gel phases co-exist in two distinct areas of the temperature-composition phase diagram. In one case the liquid phase co-exists with a P_{β}' gel phase (as was the case in the other two studies described above) and the percolation threshold in temperature and composition was very similar to that observed in the peritectic mixture. In the other co-existence region the same liquid co-exists with a mixed-interdigitated L_c phase and the percolation threshold occurs at ≥0.9 mass fraction of the gel phase. This can be understood only if either some polygonal structure is assumed for the L_c gel phase or this phase forms roughly circular domains of non-uniform size.

The archipelago effect on diffusion in a continuous liquid phase in which the "islands" are gel phase domains was examined in a mixture of tetracosanoyl-galactosylceramide and dipalmitoylphosphatidylcholine [8]. The percolation threshold in this mixture was found to lie close to the solidus (≥0.7 gel phase area fraction). As may be expected, the diffusion coefficient of the liquid-phase-soluble, solid-phase-insoluble probe was found to be reduced due to the presence of the solid phase domains relative to the diffusion coefficient in their absence. However, the reduction was found to be considerably larger than predicted by Monte Carlo simulations [9]. It was shown that a model which attributed a fluid phase boundary (or annulus) around the gel phase domains with a coherence length of about 3-4 lipid radii in which the probe diffusion was slower than in the bulk fluid phase adequately fit the experimental result. In the same work [8] it was shown that a membrane-spanning fluorescent lipid derivative [10] enabled an estimation of trans-bilayer superposition of gel phase domains in the two apposed monolayers of the bilayer. The gel phases were shown to be superposed in the dimyristoylphosphatidylcholine / distearoylphosphatidylcholine (eutectic) mixture discussed earlier, but not superposed in the galactosylceramide / dipalmitoylphosphatidylcholine mixture.

Further work [11] has examined domain connectivities in binary mixtures of cholesterol and phosphatidylcholines as well as ternary mixtures of cholesterol with two phosphatidylcholines. In these systems liquid/liquid domain co-existence was also examined primarily through the effects of such phase co-existence upon the diffusion coefficient for the probe and its temperature-dependence.

3. Acknowledgements.

This work was supported in part through grants GM-14628 and GM-23573 from the National Institutes of Health, U.S.A.

4. References.

[1] W.L.C. Vaz (1991) Comments Mol. Cell. Biophys., in press.
[2] T.E. Thompson, M.B. Sankaram and R.L. Biltonen (1991) Comments Mol. Cell. Biophys., in press.
[3] T.M. Jovin and W.L.C. Vaz (1989) Meth. Enzymol. **172**, 471-513.
[4] W.L.C. Vaz, E.C.C. Melo and T.E. Thompson (1990) Biophys. J. **58**, 273-275.
[5] W. Xia and M.F. Thorpe (1988) Phys. Rev. **A38**, 2650-2656.
[6] W.L.C. Vaz, E.C.C. Melo and T.E. Thompson (1989) Biophys. J. **56**, 869-876.
[7] T. Bultmann, W.L.C. Vaz, E.C.C. Melo, R.B. Sisk and T.E. Thompson (1991) Biochemistry **30**, 5573-5579.
[8] P.F. Almeida, T.E. Thompson and W.L.C. Vaz (1991) Biophys. J. **59**, 613a.
[9] M. Saxton (1987) Biophys. J. **52**, 989-997; and (1989) Biophys. J. **56**, 615-622.
[10] W.L.C. Vaz, D. Hallmann, R.M. Clegg, A. Gambacorta and M. De Rosa (1985) Eur. Biophys. J. **12**, 19-24.
[11] P.F. Almeida (1991) unpublished results.

Estimation of Gel and Fluid Domain Sizes in Two-Component Lipid Bilayers

M.B. Sankaram[1], *D. Marsh*[2], *and T.E. Thompson*[1]

[1]Department of Biochemistry, University of Virginia, Charlottesville, VA 22908, USA

[2]Max-Planck-Institut für Biophysikalische Chemie, Postfach 2841, W-3400 Göttingen, Fed. Rep. of Germany

Abstract. The widths and shapes of electron spin resonance (ESR) spectroscopic lines were analyzed to determine the average size of gel and fluid domains in a two-component, two-phase lipid bilayer system. The ratio of the maximum peak-to-peak amplitude to the amplitude in the wings of electron spin resonance spectral lines from a spin labelled lipid was found to be diagnostic of the connectedness of the system. Analysis of the ESR lineshapes suggested that for a statistical distribution of the spin labels in disconnected domains, the intensity ratio is always greater than the value for connected domains. The intensity ratio was calculated as a function of the average number of spin labels per domain. This method was applied to the binary mixtures of dimyristoyl phosphatidylcholine (DMPC) with distearoyl phosphatidylcholine (DSPC) probed with a spin labelled derivative of DMPC. The results suggest that the size of the fluid (or gel) domains increases with increasing fluid (or gel) mole fraction.

1. Introduction

Phospholipid bilayers are capable of undergoing a variety of thermotropic and lyotropic phase transitions which are affected by the bulk aqueous phase and the chemical composition of the lipid bilayer [1]. Lateral in-plane phase separation [1-3] is a common phenomenon observed in many biological membranes at physiological temperatures [2] and in one- and multi-component lipid bilayers under appropriate conditions. This differentiation of two-dimensional spatial regions leads inevitably to the formation of isolated domains of one of the phases [4]. The existence of domain structure is particularly interesting because of its possible regulatory role in biological systems [5].

2. The DMPC/DSPC System

The phase diagram [6] and the phase connectivity properties [4] of the binary system formed from DMPC and DSPC are well characterized. In an attempt to obtain a quantitative determination of the average size of fluid

Springer Proceedings in Physics, Vol. 66 The Structure and Conformation of Amphiphilic Membranes
Editors: R. Lipowsky · D. Richter · K. Kremer

and gel domains in this system, we incorporated a derivative of DMPC spin labelled at the 13-position in the *sn*-2 acyl chain (13-DMPCSL) into mixed bilayers of DMPC and DSPC. The ESR spectra of the spin label in the gel and fluid phases were always found to be apparently one-component in nature. Surprisingly, the spectra were still apparently single component-like even in the gel-fluid coexistence region. The absence of two spectral components in the coexistence region may be indicative of a) an exclusive partitioning of the spin label into one of the two phases, b) the rates of spin label exchange between the two phases being faster than the ESR time scale or c) a slow exchange situation wherein the spectral components from the two phases are similar in hyperfine anisotropy. The first possibility was ruled out by independent experiments where intervesicular spontaneous phospholipid transfer was monitored by means of spin label induced fluorescence quenching (Wimley & Thompson, unpublished observations).

3. Linewidth Analysis

The peak-to-peak linewidths (ΔH^{pp}) of the three hyperfine lines were measured as a function of temperature and lipid composition to test the applicability of the fast exchange approximation in this system. The linewidths were found to decrease with increasing temperature in the gel and fluid phases. Discontinuities were observed at three temperatures corresponding to the solidus, the point of connectivity and the fluidus. At the point of connectivity (or the percolation point), the system goes from a disconnected fluid phase in a continuous gel phase to the complementary situation where the fluid phase is connected and the gel phase is disconnected [4].

The linewidths in the coexistence region were calculated as weighted averages of the gel and fluid contributions. The mole fractions of the two phases were determined from the phase diagram [6]. The linewidths of the two phases were obtained by extrapolating linewidth data from the all-gel and all- fluid phases into the coexistence region. The calculated linewidths were found to agree with the observed linewidths at temperatures below the point of connectivity but not above it. Thus, the fast exchange model did not provide a satisfactory fit to the experimental data.

4. Lineshape Analysis

If the inter-phase spin label exchange were slow, then the linewidth of the experimentally observed single resonance is no longer given by a simple weighted-average formalism. We thus sought an alternative measurable parameter with which one can circumvent the intrinsic temperature

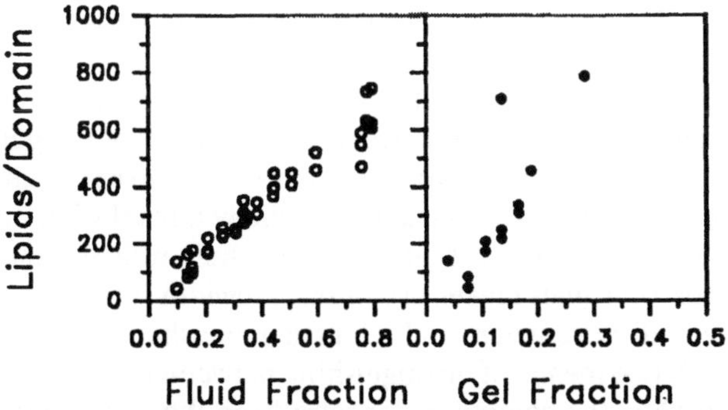

Figure 1: Average number of lipids per fluid or gel domain as a function of fluid (left panel) or gel (right panel) mole fraction calculation from intensity ratio data for the DMPC/DSPC system.

dependence of the gel and fluid phase contributions to the linewidth in the mixed phase region.

The single component spectroscopic lines frequently have either a Lorentzian, a Gaussian or a Voigt lineshape [7]. An interesting property of these spectroscopic lines is that the ratio of the maximum peak-to-peak line height to that in the wings away from the center separated by a constant multiple of ΔH^{pp} is always a constant which is independent of ΔH^{pp} [8]. We calculated the intensity ratio for a Lorentzian line using a model wherein the 13-DMPCSL spin label was distributed between the gel and fluid phases in the same way as the host DMPC, but is statistically distributed in disconnected domains of either phase. This calculation showed that the intensity ratio of the composite resonance summed over domains was always greater than that for single Lorentzian, Gaussian or Voigt lines and therefore is diagnostic of the presence of disconnected domains. The intensity ratios were calculated as a function of average number of spin labels per domain.

Figure 1 shows the results obtained by the lineshape analysis applied to the DMPC/DSPC system. The experimentally determined intensity ratios were assumed to be weighted-average values of the component gel and fluid phase intensity ratios. The observation that the intensity ratio for the connected phase is a constant allowed the calculation of the intensity ratio of the disconnected phase. This ratio was found to be always greater than the value expected from single resonance lines. The left panel shows the results obtained below the point of connectivity where the fluid domains are disconnected to form isolated domains. The right panel contains the results above the point of connectivity where the gel domains are disconnected. Both the fluid and gel domains are seen to grow in size with increasing mole fraction of that phase. This result is consistent with a model in which the domains are first formed and then grow as more fluid or gel is formed and argues against a model where the domains increase in number with no growth.

While analyzing the linewidths using the fast exchange model and modifications of it, the temperature dependence of the linewidth of the gel and the fluid phases was found to be an unknown parameter. The apparent single- component nature of the ESR spectra did not allow a resolution of the two phases. On the other hand, the lineshape analysis utilizing the intensity ratios of spectroscopic lines makes use of the linewidth (and hence temperature) invariance of Lorentzian and Gaussian lineshapes. The intensity ratio method offers a means of distinguishing between the slow or fast exchange situations when the component spectra from the two phases are very similar.

Acknowledgements. This work was supported by USPHS, NIH grants GM-14628, GM-23573 and by the Alexander von Humboldt Foundation.

References

1. G. Cevc,D. Marsh: *Phospholipid Bilayers* (John Wiley, New York 1987)
2. D.L. Melchior,J.M. Steim: Prog. Surf. Membr. Sci. **13**, 211 (1979)
3. D. Marsh: *Handbook of Lipid Bilayers* (CRC Press, Boca Raton 1990)
4. W.L.C. Vaz,E.C.C. Melo,T.E. Thompson: Biophys. J. **56**, 869 (1989)
5. T.E. Thompson,M.B. Sankaram,R.L. Biltonen: Comm. Mol. Cell Biol. 1991 (in press).
6. W. Knoll,K. Ibel,E. Sackmann: Biochemistry **20**, 1379 (1981)
7. C.P. Slichter: *Principles of Magnetic Resonance*, 3rd Edition (Springer-Verlag, Berlin 1990)
8. B.L. Bales: J. Magn. Res. **48**, 418 (1982)

Phosphatidyl-Glycerol in Mixtures with Positively Charged Amphiphiles: A ^{2}H- and ^{31}P-NMR Study of the Phase Behaviour and Headgroup Structure

R.G.K. Habiger and *J. Seelig*

Biocenter of the University of Basel, Department of Biophysics, Klingelbergstr. 70, CH-4056 Basel, Switzerland

The thermotropic behaviour of membranes composed of 1-palmitoyl-2-oleoyl-sn-3-phosphatidyl-glycerol (POPG) and dimethyl-dihexadecyl-ammonium bromide (DHAB) was investigated with differential scanning calorimetry and ^{2}H- and ^{31}P-nmr spectroscopy. The main transition temperature, T_m, from the gel state to the liquid-crystalline state (l.c.) showed a local maximum around 40 mole % DHAB in the membranes. With nmr spectroscopy we also detected non-bilayer properties of the system. Both the DSC scans and the phase diagram of the system are presented.

The relation between the PG-headgroup conformation and the DHAB concentration was investigated with ^{2}H and ^{31}P-nmr spectroscopy. The PG-headgroup undergoes distinct conformational changes when the amphiphile concentration is varied. These changes are more pronounced below 40 mole %. Above 40 mole % only minor changes can be detected; the two α-deuterons become motionally inequivalent.

1. Thermotropic studies

The thermotropic studies were performed on a Microcal MC II differential scanning calorimeter.The scans are shown in figure 1. The most obvious result is that the main transition temperature from the gel state to the liquid crystalline state, T_m, is related to the DHAB concentration. T_m increases in the range from 0 to 40 mole % DHAB and has a local maximum around 40 mole %. Connected to this effect is an increase in the cooperativity of the transition, since the enthalpy of the transition is minimal at 40 mole % DHAB. Above this concentration, T_m is shifted back to lower temperatures again. These observations are in good agreement to the recently published thermotropic studies of Inoue et. al. [1] on the system phosphatidyl glycerol in mixtures with positively charged amphiphiles. The biphasic behaviour of T_m can be explained as follows. Between 0 and 40 mole % DHAB the strong interactions, both van der Waals and electrostatic, between the POPG molecules and DHAB are responsible for the increase of T_m. These interactions also increase the cooperativity. Above 40 mole % the POPG-DHAB interaction cannot be increased further. The repulsive DHAB-DHAB interaction becomes more and more important and chain melting occurs again at lower temperatures. Measuring α-POPC/DHAB mixtures with nmr (see below) we also detected non-bilayer phases. NMR spectra were measured at intervals of

Figure 1: DSC-scans of the System POPG/DHAB

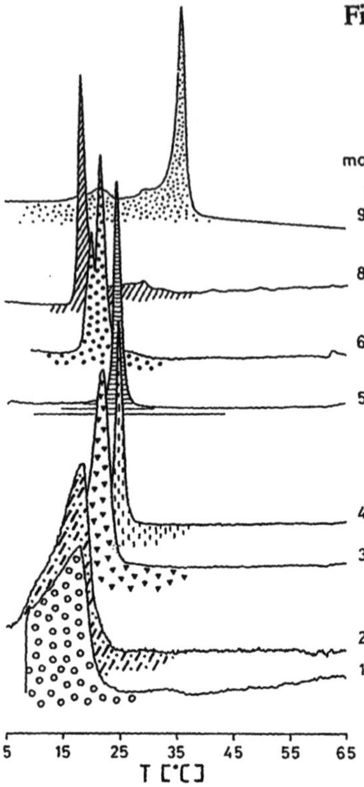

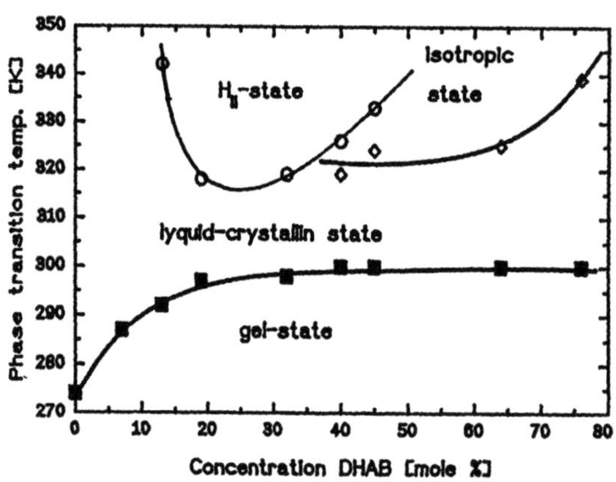

Figure 2: Phase-diagram of the system POPG/DHAB

1 K of POPG/DHAB mixtures in the range up to 80 mole % amphiphile. The shape of the ^{2}H-nmr spectra reports on the phase state. The resulting phase diagram is shown in figure 2. The H_{II}- and the isotropic phase showed coexistence with the liquid crystalline state in certain temperture ranges. In these cases we chose the beginning of the coexistence region as the phase boundary. Four diffe-

rent states can be distinguished by NMR: gel state, liquid crystalline state, hexagonal state (presumable invoted H_{II}) and an isotropic state. The latter is known to occur sometimes as an intermediate between the l.c. and the H_{II} state [2]

2. Headgroup studies

For the headgroup studies we deuterated POPG at three different headgroup segments, the α-, β-, and α-γ-segment according to the following definition. The deuteration procedure is described elsewhere [3]. The measurements were performed on a Bruker-Spectrospin MSL 400 spectrometer at 61.4 MHz and 162

O
O-P-O-CH_2-CHOH-CH_2OH
O α β γ

MHz for ^{2}H- and ^{31}P-nmr, respectively. The quadrupole splitting of all three segments and the phosphorous chemical shielding anisotropy showed a different behaviour below and above 40 mole % DHAB. The results are summarized in table 1. In figure 3 we show the α-splitting as a function of the DHAB concentration. The slope at DHAB concentrations below 40 mole % is steeper than that above this concentration. In addition the two deuterons at the α-position become motionally inequivalent above 40 mole % amphiphile. This gives rise to two quadrupole splittings. The β-splitting remains constant up to 40 mole % and decreases slightly at higher concentrations. The α-γ-data are difficult to interprete due to impurities in the samples which caused a large isotropic peak. But nevertheless, the main effects again occurred below 40 mole % amphiphile. All the nmr parameters suggest a model where the headgroup of POPG becomes

Table 1: Summary of all quadrupole splittings and chemical shielding anisotropies

α-splitting:									
mole % DHAB	0.0	07.0	13.0	19.0	32.0	40.0	45.0	64.0	76.0
$\Delta\nu$ [kHz]	10.7	10.4	9.7	9.3	7.8	7.4	7.5	6.9	6.1
$\Delta\nu_2$ [kHz]						6.4	6.3	5.2	4.4
β-splitting:									
mole % DHAB	0.0.	10.6	21.9	33.8	40.6	54.8	64.3	81.1	
$\Delta\nu$ [kHz]	3.3	3.3	3.5	3.3	3.2	3.1	3.2	3.0	
γ-splitting:									
mole % DHAB	0.0	9.7	20.5	31.8	39.6	50.3	60.0	69.9	
$\Delta\nu$ [kHz]	1.3	1.4	4.5	1.6	1.7	1.7	1.7	1.8	
Chemical shielding anisotropy:									
mole % DHAB	0.0	8.0	14.0	19.0	40.0	45.0	64.0	76.0	
σ [ppm]	-40	-42	-44	-47	-53	-51	-52	-53	

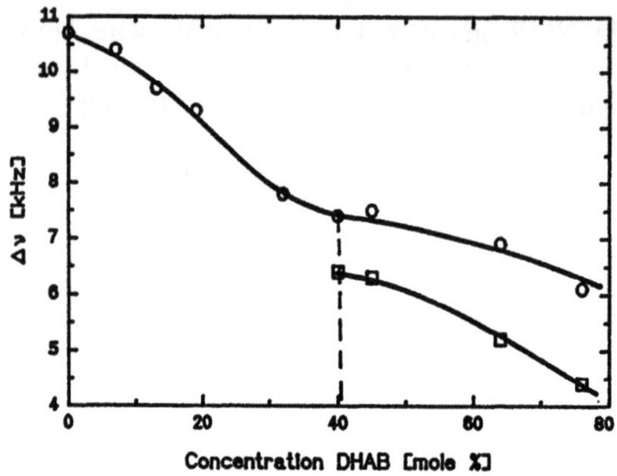

Figure 3: Quadrupole splitting of α-POPG

compressed when the positively charged amphiphiles are introduced into the membrane. Hence the well-known "molecular voltmeter concept" [4] for zwitterionic headgroups such as phosphatidyl choline, does not work for the PG headgroup. Rather we detect two different mechanisms leading to conformational changes in the PG-headgroup. The stronger changes below 40 mole % DHAB are due to the POPG-DHAB interaction. The weaker changes above 40 mole % are driven by the increase of the positive surface potential. Here the individual dipole moments of the α-, β-, and γ-group, respectively, react to the changed electrostatical environment.

References

[1] T. Inoue et. al., Chem. Phys. Lip. 55 (1990) pp. 145 - 154
[2] John M. Seddon, BBA 1031 (1990) pp. 1 - 69
[3] R. Wohlgemuth et. al., Biochem. 19 (1980) pp. 3315 - 3321
[4] J. Seelig et. al., Biochem. 26 (1987) pp. 7535 - 7541

Diffusion Controlled Reactions in Two-Dimensional Space. The Pyrene Excimer Example

J.M. Martins and E.C.C. Melo

Centro de Tecnologia Química e Biológica – IST,
R. da Quinta Grande, 6, Apartado 127, P-2780 Oeiras, Portugal

Abstract. Diffusion-controlled reactions in two dimensional or quasi two dimensional space are important in both, surface, and membrane reaction kinetics. A model reaction that have been extensively used to study the dynamics and structural properties of membranes is that of the formation of the excimer of pyrene. We present experimental evidence that in multilamelar liposomes of dimyristoylphosphatidylcholine, at temperatures above phase transition, and for probe:phospholipid ratio of 1:100, or lower, the fluorescence decay of both, pyrene, and 3-palmitoyl-2-(1-pyrenodecanoyl)-L-α-phosphatidylcholine follows a kinetic law specific of 2D diffusion-controlled reactions. The parameters obtained for the diffusion coefficient, 1.6×10^{-7} cm^2 s^{-1} for the free pyrene and 1.0×10^{-7} cm^2 s^{-1} for the pyrene-labeled phospholipid, and the activation energy for diffusion, are in perfect accordance with the figures that came out from independent techniques.

1. Introduction

Reactions that take place in a two dimensional or a nearly two dimensional space are very common either in surface chemistry or in biochemical systems. The specificity of the kinetics of diffusion-controlled reactions in such systems was first recognized by Emeis and Fehder [1], and the formalism that accounts for the reduced dimensionality have been independently presented by Razi-Naqvi [2] and Owen [3]. Despite of its importance, a complete experimental validation of these models have not yet been done.

The reaction of a pyrene molecule in the singlet excited state, **Py***, with a ground state pyrene molecule, **Py**, yielding an excimer, is a very convenient model reaction for the study of diffusion-controlled processes in phospholipid bilayers [4].

In phospholipid bilayers, and at temperatures under ca. 55° C, the excimer formation reaction is non-reversible [5]. It is, therefore, accurately described by the following kinetic scheme

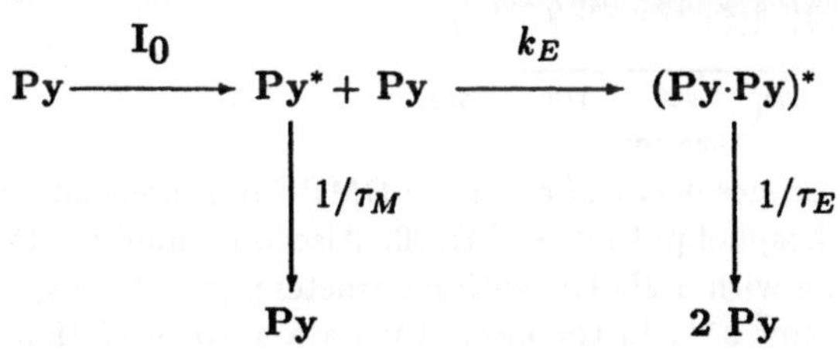

where k_E is the rate constant for excimer formation, and τ_M and τ_E are the unquenched lifetimes, into the bilayer, of the pyrene monomer and excimer, respectively.

Accordingly to the classical Birks formalism for the 3D case, a single exponential decay of the monomer fluorescence is expected. In fact, experimental results present a large deviation from the single exponential law [6]. In the literature, these deviations have been repeatedly attributed to ground state proto-aggregates of pyrene molecules [6,7]. The same arguments have been used in the case of pyrene functionalized diacyl phosphatidylcholine, **Py-PC** (3-palmitoyl-2-(1-pyrenodecanoyl)-L-α-phosphatidylcholine) [7].

2. Results and Discussion

Experiments are performed using free and functionalized pyrene in pure dimyristoylphosphatidylcholine, DMPC, multilamelar liposomes at temperatures above phase transition ($T > 23°$ C). Decay of monomer and excimer are obtained at 395 and 485 nm respectively, using the Time Correlated Single Photon Counting technique. Data was deconvoluted from lamp pulse by using the exact decay law corresponding to the above kinetic scheme with a time dependent rate coefficient [2]

$$k_E(t) = \frac{16DN}{\pi} \int_0^\infty e^{-\alpha\beta^2} \frac{d\beta}{\beta\,[\mathcal{J}_0^2(\beta) + \mathcal{Y}_0^2(\beta)]} \,,$$

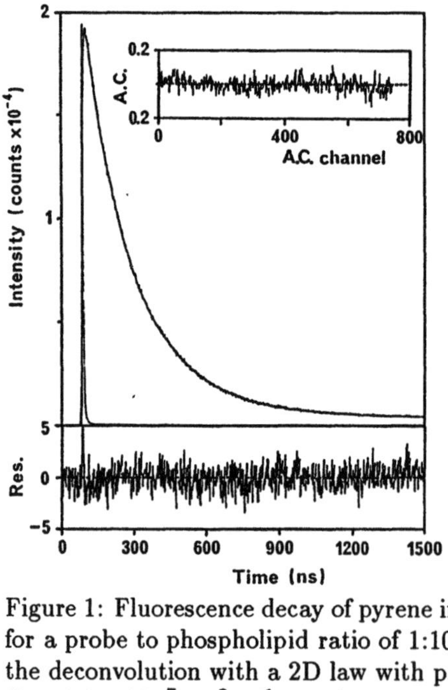

Figure 1: Fluorescence decay of pyrene in DMPC multilamelar vesicles at 30° C for a probe to phospholipid ratio of 1:100. Also presented are the residuals for the deconvolution with a 2D law with parameters: $\tau = 260$ ns, $R = 7.1$ Å, and $D = 1.6 \times 10^{-7}$ cm^2 s^{-1}. In the insert the data autocorrelation is shown.

where $\alpha = Dt/R^2$, $\mathcal{J}_0$ and $\mathcal{Y}_0$ are Bessel functions, and D and R are the diffusion coefficient and encounter distance, respectively.

Despite of the intrinsic difficulty of the analysis of the experimental data (e.g. due to light scatter and poor lamp profile definition), an excellent fit to the model is obtained, as exemplified in Figure 1 for the case of free pyrene. For the given reactional distance, $R = 7.1$ Å, we obtain for the fitting parameter, D, 1.6×10^{-7} cm^2 s^{-1} in the case of free **Py**, and 1.0×10^{-7} cm^2 s^{-1} for **Py-PC**. These values are in perfect accordance with the figures that came out from independent techniques, e.g. FRAP.

The reversibility of the excimer is negligible until 60° C (range of low temperature limit). From the linear dependence of ln D *vs.* $1/T$, a value of 34.2 KJ mol^{-1} for the activation energy for diffusion of pyrene is obtained. This energy is the one expected for a molecule that resides in the palisade near to the head-groups [8].

3. Conclusions

As it is evident from the above results, the time dependence of the rate coefficient cannot be neglected in the analysis of the pyrene fluorescence decay in bilayers. Hence, a 3D formalism is not applicable. Furthermore, for the range of probe:phospholipid ratios used, there is no evidence of probe association. For that reason, we sustain that, at least for this concentration range, there is no need to postulate that the pyrene molecules aggregate to account for its characteristic fluorescence decay.

In this context, a true steady-state cannot be defined. For the interpretation of the steady-state fluorescence data a suitable integration of the decay should be performed.

The value of the activation energy for the diffusion of pyrene molecules, 34.2 KJ mol^{-1}, is, within the experimental error, equal to that obtained for DMPE-NBD molecules with the FRAP technique [8]. From this, we may conclude that, on molecular grounds, the controlling step for both diffusion mechanisms must have the same physical origin.

4. Acknowledgement

The research described herein was partially supported by J.N.I.C.T.-Portugal, under contract no. PMCT/CEN 658/90.

5. References

[1] C.A. Emeis, P.L. Fehder (1970) J. Am. Chem. Soc. **92**, 2246-2252.
[2] K. Razi-Naqvi (1974) Chem. Phys. Lett. **28**, 280-284.
[3] C.S. Owen (1975) J. Chem. Phys. **62**, 3204-3207.
[4] H.-J. Galla, E. Sackmann (1975) Biochim. Biophys. Acta **339**,103-115.

[5] K.A. Zachariasse, G. Duveneck (1984) R. Busse, J. Am. Chem. Soc. **106**, 1045-1051.
[6] H. Lemmetyinen, M. Yliperttula, J. Mikkola, P. Kinnunen (1989) Biophys. J. **55**, 885-895.
[7] M. Sassaroli, M. Vauhkonen, D. Perry, J. Eisinger (1990) Biophys. J. **57**, 281-290.
[8] F. Podo, J.K. Blasie (1977) Proc. Natl. Acad. Sci. USA **74**, 1032-1036.
[9] W.L.C. Vaz, R.M. Clegg, D. Hallman (1985) Biochemistry **24**, 781-786.

Ripple Phase in Mixed Model Membrane

C. Cametti[1], *F. De Luca*[1], *A. D'Ilario*[1], *G. Briganti*[1], *M.A. Macri*[2], *and B. Maraviglia*[1]

[1]Dipartimento di Fisica, Università degli Studi "La Sapienza", Piazzale A. Moro, 2, I-00185 Rome, Italy

[2]Istituto di Fisica Medica, Università di Chieti, I-66100 Chieti, Italy

Abstract. We have investigated by differential scanning calorimetry and radiowave dielectric spectroscopy, in the frequency range between 1kHz and 1MHz, the pre transition or ripple $P_{\beta'}$ phase of aqueous suspensions of mixed phospholipid bilayers of DPPC and DPPE. Our results show that the endothermic peak associated with the pre transition, moves towards the main transition on increasing DPPE contents and disappears at an extrapolated concentration of 11% by weight. The observed anomalous hump in the low frequency dielectric response, associated with the pre transition, correspondingly shrinks and decreases by adding DPPE. These data suggest that DPPE decreases the mismatch between the chain packing volume and the head group area, and reduces the correlation length of the dipolar in-phase domain that, as previously suggested, is associated with the permittivity increment.

1. Introduction

Model membrane systems, like hydrated lipid bilayers, generally undergo a peculiar first order phase transition attributed to the melting of the hydrocarbon chain from a fluid disordered L_α phase to a solid like ordered L_β or $L_{\beta'}$ phase [1]. This phase transition is characterized by a decrease of the bilayer thickness [2], and by a broad endothermic peak, whose latent heat as well as transition temperature T_m is proportional to the chain length [3].

Additionally, depending on the nature of the hydrophilic head group, a transition may occur at a temperature T_p, slightly below T_m [2,3], accompanied with a very small extra bilayer contraction [4]. The phase between T_p and T_m, called $P_{\beta'}$, is characterized by an intra-bilayer ripple modulation whose amplitude and periodicity does not depend on temperature, as shown by X-Ray diffraction [5] and freeze fracture electron microscopy [6]. The endothermic peak at T_p and the temperature interval between main and pre transition are largely independent of the chain length [3]. Besides that the latent heat at T_p is much smaller than the one at T_m. Therefore, this phase seems to be mainly related to the interfacial bilayer region and to the conformational properties as well as to the interactions of the surfactant head groups.

Recently, radiowave dielectric dispersion properties of lamellar phospholipid-water mixtures have been extensively investigated in dipalmitoylphosphatidylcholine (DPPC), dimyristoylphosphatidylcholine (DMPC) and dipalmitoylphosphatidyl-ethanolamine (DPPE) [7]. The former two molecules display at least all three mentioned phases, whereas the last one undergoes only a main phase transition.

Springer Proceedings in Physics, Vol. 66 The Structure and Conformation of Amphiphilic Membranes
Editors: R. Lipowsky · D. Richter · K. Kremer

The conspicuous and interesting feature of all the dielectric spectra we have obtained is the occurrence of a marked anomalous increment of the permittivity within the $P_{\beta'}$ phase, with a marked thermal hysteresis. Moreover, the amplitude of the permittivity peaks increases when the frequency of the applied field is lowered (the lowest frequency investigated was 1kHz) and disappears at frequency higher than about 50 ÷ 100 kHz.

According to the dielectric model proposed by Kaatze et al. [8], the dielectric dispersion observed at radiowave frequencies can be attributed to the diffusive thermal rotational motion of the cationic head group around an axis perpendicular to the permanent electric dipole moment of the zwitterion. Hence we have attributed this low frequency anomalous increase in the dielectric permittivity to the creation of large dipolar in-phase domain, induced by capillary waves, whose extension define the frequency dependence of the permittivity increment [7,9].

2. Results and discussions

The present work concerns with measurements of both differential scanning calorimetry (DSC) and low frequency dielectric relaxation on DPPC-water mixtures with various amounts of DPPE, in a temperature range, between 15 °C and 55°C, complitely containing the pre transition region of pure DPPC solution. Both these lipids have the same hydrocarbon chain length but differ on the anionic head group termination showing an $N(CH_3)_3$ the former and NH_3 the later. This difference can affect the head group conformation at the interface, changing the area per head group and the direct intra-bilayer head-head interaction [10]. Thus, under the hypothesis of uniform mixing, the delicate balance between head group area and hydrocarbon chain packing, probably at the origin of the $P_{\beta'}$ phase [11], can be continuously tuned changing the relative lipid concentration. The uniform mixing hypothesis is rather well established by the linear trend of the main transition temperature with DPPE concentration [12]. The same trend was also observed in mixtures of DMPC with N-methyl substituted dimyristoylphosphatidylethanolamine (DMPE*) where DMPE* has, like DMPC, 14 hydrocarbons per chain but the head group is characterized by $NH(CH_3)_2$ [11].

The procedures for sample preparation, the DSC apparatus and the radiowave dielectric spectroscopy measurements were described previously [7]. All our measurements in pure DPPC and in the DPPC-DPPE mixtures were performed in a solution of 75% of water and 25% of surfactants by weight, well far above the hydration threshold and in a region where the thermodynamic behavior does not depend on the water content.

The DSC enthalpy variations for 0, 1, 3, 5, 7% of DPPE/DPPC show that the absolute pre transition temperature T_p as well as the temperature difference (T_m - T_p) between the main and pre transition, have a linear trend with DPPE concentration (fig.1); the concentration at which T_p superimposes on T_m is around 11% DPPE/DPPC. Previous data on DMPC and DMPE* mixtures show the same linear trend, where the $P_{\beta'}$ phase disappears at about 15% of DMPE*/DMPC [11].

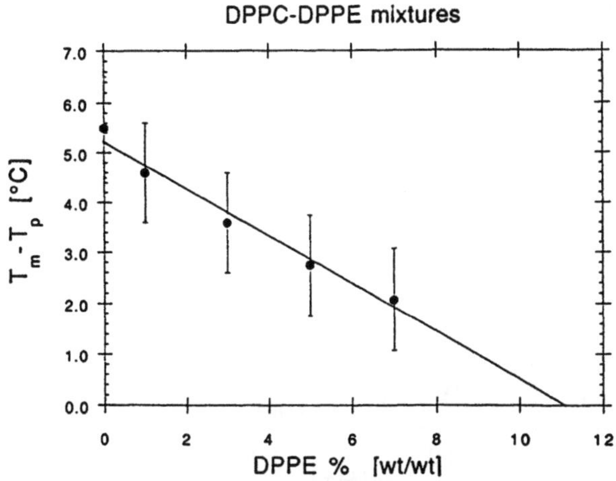

Fig.1. Concentration dependence of the difference between the main and pre transition temperature. The extrapolated value at which the $P_{\beta'}$ phase disappears is at 11% DPPE/DPPC.

These similar results suggest that the change in the volume of the anionic termination of the head group does not affect in a simple way the interfacial region. Indeed, the isomeric rotational state of the two CH_2 groups, that connect the phosphate cation with the anion, can compensate somehow the difference between DMPE* and DPPE head groups.

Coherently with DSC results, the temperature interval of the sharply defined anomaly in the dielectric permittivity reduces with DPPE concentration. Moreover the amplitude of the hump simultaneously decreases and the dielectric dispersion curve change, as shown in fig 2, mainly in the low frequency region.

Under the hypothesis of uniform mixing, the presence of DPPE molecules seems to reduce the dimension of the in-phase dipolar domain. Therefore, in these concentration range, where the head groups conformations and interactions are dominated by the DPPC characteristics, the range of dipolar correlation is limited in the long wavelength region by the presence of the DPPE molecules. This is in agreement with freeze-fracture electron microscopy data where the presence of trace amounts of DMPE* appears to disrupt the underlying hexagonal symmetry, especially on the region of high stress at the boundaries between ripples of different orientation or configuration, reducing the extent of these region [11].

The authors thank Prof. L. D'Ilario (University of Rome) for the use of the DSC apparatus.

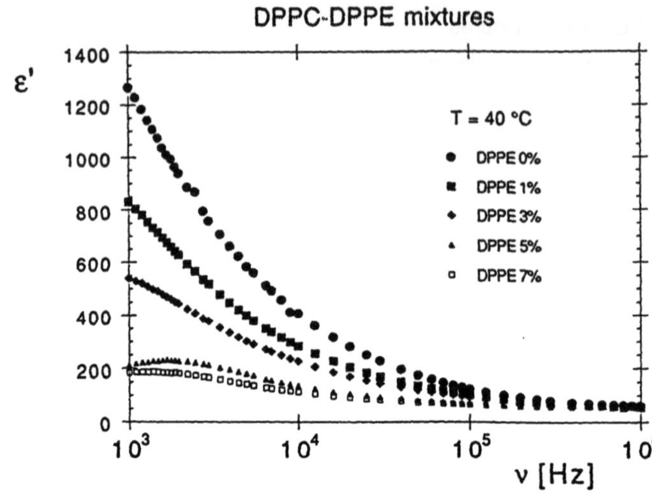

Fig.2. The dielectric permittivity as a function of frequency at 40°C. The data show that the major effect of the DPPE concentration is confined the low frequency region.

3. References

[1] D. Chapman - Biol. Membr. **1**, 125, (1968)

[2] M.J. Janiak, D.M. Small, G.G. Shipley - Biochemistry **15**, 4575, (1976)

[3] G. Cevc, D. Marsh - Phospholipid Bilayers: Physical principle and Models (1987) Wiley

[4] K. Ohki, K. Tamura, I. Hatta - Biochim. Biophys. Acta **1028**, 215, (1990)

[5] M.J. Janiak, D.M. Small, G.G.J. Shipley - J. Biol. Chem. **254**, 6068, (1979); Y. Inoko, T.Mitsui, K. Ohki, T. Sekiya, Y. Nazwa - Phys. Status Solid (a) **61**, 115 (1980)

[6] D. Ruppel, E. Sackmann - J.Physique **44**, 1025, (1983)

[7] C. Cametti, F. De Luca, M.A. Macri', B. Maraviglia, P. Sorio - Liq. Crystal **3**, 839, (1988);
C. Cametti, F. De Luca, A. D'Ilario, M.A. Macri', B. Maraviglia, P. Sorio Liq. Crystal **7**, 571, (1990)

[8] U. Kaatze, K. Gopel, R. Pottel - J. Phys. Chem. **89**, 2565, (1985)

[9] C. Cametti, F. De Luca, A. D'Ilario, M.A. Macri', G. Briganti, B.Maraviglia Progr. Colloid Polym. Sci **84**, (1991), in press.

[10] K.A. Dill, D. Stigter - Biochemistry **27**, 3446, (1988)

[11] J.A.N. Zasadzinski, M.B. Schneider J. Physique **48**, 2001, (1987)

[12] A. Blume, T. Ackermann FEBS Lett. **43**, 71, (1974).

Microscopic Theory for the Ripple Phase

R.R. Netz

Department of Physics, Massachusetts Institute of Technology,
Cambridge, MA 02139, USA

A microscopic model for the ripple phase in phospholipid bilayers is presented including translational degrees of freedom for single molecules and two types of chain excitations. Induced by low-energy chain excitations, opposing molecules form flexible units in an intermediate temperature range and can accommodate a head/tail lattice misfit by introducing a lateral shift between head and tail positions. The ripple phase involves two different head packing modi with misfits of different sign and allows the tail region to pack in the preferred hexagonal structure. Observable properties are calculated using Monte Carlo sampling.

1. Introduction

Single bilayers and fully hydrated lamellar liquid crystals of certain phospholipids have been found to be rippled in an intermediate temperature range. Below this $P_{\beta'}$ phase, the bilayers form a flat $L_{\beta'}$ phase, in which the hydrocarbon chains are fully extended and tilted with respect to the layer normal. Above the ripple phase, the chains melt at the main transition and the layer is flat again (L_α phase). The $P_{\beta'}$ phase between the main and the pretransition, exhibits a regular ripple superlattice with a wavelength of $\approx 100 - 200$ Å and amplitudes of ≈ 50 Å, depending on the specific compound. Since the superstructure forms in isolated bilayers[1,2], an interaction between membranes is not crucial for the $P_{\beta'}$ phase, although it strongly affects it. Moreover, a rippled bilayer can be pulled flat requiring mechanical work corresponding to only 1/50 of the enthalpy associated with the pretransition and without eliminating main and pretransition[2]. Thus, the superlattice is not the origin of the enthalpy at the pretransition and a water-mediated interaction between the ripples of one bilayer is also not responsible for the occurrence of the $P_{\beta'}$ phase. Raman spectra indicated that the pretransition is accompanied by an increase in population of hydrocarbon excitations toward the center of the bilayer[3]. DMR studies suggested that the pretransition does not involve a conformational change of the polar head region but has to be attributed to the hydrocarbon chain[4].

The difference in the areas occupied by the bulky head group and the fully ordered chains can in part be accounted for by a tilt of the chains, which relieves the area incompatibility in the direction of the tilt and leads to a orthorhombic packing of the hydrocarbon chains in the low-temperature $L_{\beta'}$ phase. In this work it is shown how a partial decoupling of the two opposing layers in the $P_{\beta'}$ phase , initiated by excitations of the chains toward the center of the bilayer, adjusts the area mismatch between head and tail group also in the direction perpendicular to the tilt, leading to an experimentally observed hexagonal packing of the chains in the ripple phase[5]. This is achieved by two different head conformations, which are observed in the crystalline phase of DMPC[6], a phospholipid showing the ripple phase. A working hypothesis of the present model is that the two observed head conformations have different preferred vertical displacements between nearest neighbors (nn) with different preferred lateral lattice constants, one smaller and one larger than the tail lattice constant in the intermediate temperature range. In the ripple phase, the opposing molecules form rather flexible units and the tails can bend in a lateral direction, therefore locally allowing for packing with a different lattice constant than the head lattice. This local shift propagates in a lateral direction and increases until the corresponding elastic energy becomes too large; then the other head conformation with opposite misfit is realized and the lateral shift decreases. This scenario leads to ripples with well defined wavelength, as shown by the simulation.

2. The Model

2×384 carbon chains are placed on a triangular lattice organized in four rows of 96 parallel molecules. Molecules can move on discrete lattice sites in a direction parallel to the layer normal, vertical nn displacements of maximal one lattice unit are allowed (for a similar model for thermotropic liquid crystals see [7]). By storing these relative displacements rather than the vertical positions of the molecules, periodic boundary conditions can be used without restricting the possible conformations of the membrane. For each chain the following degrees of freedom are available[8]: i) one all-trans ground state with zero excitation energy in which the chain is straight, ii) a small number (typically 5) of intermediate energy states which correspond to excitations in the center of the bilayer and slightly increase the tail area and therefore reduce the van der Waals interaction, and iii) a highly degenerate (≈ 100) high energy "melted" state with large tail area and therefore highly reduced van der Waals interaction between the tails. These chain states are now coupled to the translational degrees of freedom and to the local shift variable. In the chain ground state, the opposing molecules are tightly bound, forming a rigid symmetric unit. Therefore no lateral shift between heads and tails is possible and the bilayer is flat. The intermediate state decouples the opposing molecules, a mismatch of the head and tail lattice constants is possible by a varying lateral shift of the tails relative to the heads. This local relative shift, which is not a tilt, is a function of the head and tail areas and propagates from neighbor to neighbor. It is constantly calculated during the simulation assuming close packing of the tails. The van der Waals attraction is increased; however, the lateral shift introduces an elastic energy which grows as a power of the local shift variable. If the shift is too large, the other possible head conformation with inverted head/tail area misfit is energetically favored. One of the two possible nn configurations is of lower energy and dominates in the $L_{\beta'}$ phase. In the $P_{\beta'}$ phase, both nn configurations occur in domain-like stripes perpendicular to the direction of tilt, thus keeping the average lateral shift between heads and tails low and maximizing the tail van der Waals interaction at the same time. The Hamiltonian consists of the single chain energies, i.e. the chain excitation energies, the chain-chain van der Waals interaction which depends on the state the chains are in, the head-head interaction, which prefers zero vertical displacement for melted chains and has two nearly equivalent energy minima at non-zero vertical displacement for the other chain states, and the energy associated with the local lateral shift between heads and tails. Only one of the two layers making up the membrane is simulated; the effect of the second layer is taken into account by enforced zero lateral shift at zero temperature and the elastic shift energy at intermediate temperatures. All modulation effects discussed above are effective in a direction perpendicular to the tilt; parallel to the tilt strong forces keep the molecules at the same height. We use Monte Carlo techniques to calculate equilibrium properties of the model system as a function of temperature. Details of the Hamiltonian used will be published elsewhere. It is noted that the transitions observed are rather independent of the detailed numbers used.

3. Results and Discussion

Fig. 1a shows the calculated specific heat as a function of temperature using 600 Monte Carlo steps per data point. The data reproduce the experimentally observed two peaks corresponding to the main (T_M) and to the pretransition (T_P)[9]. The different height of the peaks simply reflects the fact that the energy cost of the intermediate excitations of the chains responsible for the ripple phase is much smaller than the cost of the melting excitation of the chain. Fig. 1b shows the average area per molecule A relative to the area of the ground state A_O. The curve is calculated assuming that the intermediate excitations increase the area by 1/10 and the melting excitation by 3/10 and agrees well with experimental findings[2]. Fig. 1c shows the Fourier transformed of the height function

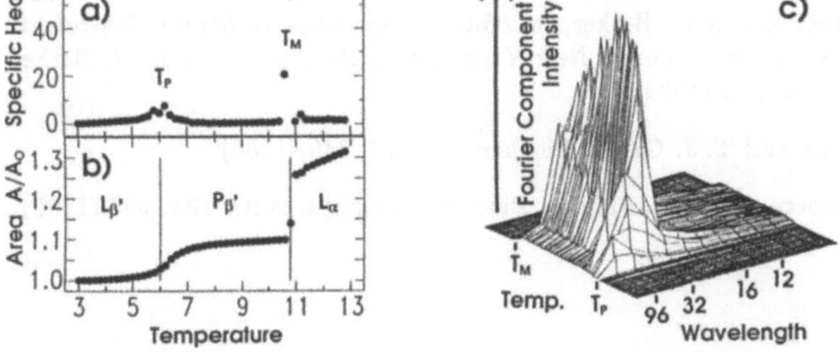

Fig. 1: Specific Heat (a) and relative area per molecule (b) for the model system using 600 Monte Carlo steps per data point. Pretransition (T_P) and main transition (T_M) are indicated by specific heat signals. The ripple phase ($P_{\beta'}$) is characterized (c) by the Fourier component signaling a modulation with a definite wavelength corresponding to 32 tail cross sections ($\approx$ 150 Å).

of the bilayer parallel to the ripple wave vector. The data show a pronounced peak for wavelengths corresponding to 32 tail cross sections (≈ 150Å) in the intermediate temperature range. The wavelength depends on the elastic energy function associated with the lateral shift. Asymmetric ripples are observed if i) the area misfit between preferred head and tail lattice constants is different for the two head configurations, ii) the vertical displacements between nearest neighbors of the two head configurations are different. Both possibilities are relevant for phospholipids.

The water mediated interaction with other bilayers introduces an additional coupling between monolayers and increases the energy associated with the bending of molecules. The wavelength therefore increases and the ripple phase finally disappears for high interaction strength corresponding to small interlayer spacing (to be published).

Thanks are due to A.N. Berker and R. Lipowsky for stimulating discussions.

This research was supported by NSF Grant No. DMR-90-22933 and by JSEP Contract No. DAAL 03-89-C0001.

4. References

[1] C. Gebhardt, H. Gruler, and E. Sackmann, Z. Naturforsch. **32c**, 581 (1977).

[2] D. Needham and E. Evans, Biochemistry **27**, 8261 (1988).

[3] I.W. Levin and S.F. Bush, Biochim. Biophys. Acta **640**, 760 (1981).

[4] H.U. Gally, W. Niederberger, and J. Seelig, Biochemistry **14**, 3647 (1975); J. Seelig, Quarterly Rev. of Biophys. **10**, 353 (1977).

[5] D.G. Cameron, H.L. Casal, E.F. Gudgin, and H.H. Mantsch, Biochim. Biophys. Acta **596**, 463 (1980).

[6] R.H. Pearson and I. Pascher, Nature **281**, 499 (1979).

[7] R.R. Netz and A.N. Berker, in *Phase Transitions in Liquid Crystals*, S. Martellucci, ed., Plenum, New York (1991); R.R. Netz and A. N. Berker, M.I.T. preprint (1991).

[8] D.A. Pink and T. J. Green, Biochemistry **19**, 349 (1980).

[9] M.J. Ruocco and G.G. Shipley, Biochim. Biophys. Acta **691**, 309 (1982).

The Influence of Local Anaesthetics on the Temperature and Pressure Dependent Phase Behaviour of Model Biomembranes

M. Böttner, M.-H. Christmann, and R. Winter

Institute of Physical Chemistry, Philipps-Universität,
Hans-Meerwein-Str., W-3550 Marburg, Fed. Rep. of Germany

Abstract. The influence of the local anaesthetic tetracaine (TTC) on the temperature and pressure dependent phase behaviour of model biomembrane systems, such as pure and cholesterol containing 1,2-dimyristoyl-phosphatidylcholine (DMPC) bilayers, has been examined by differential scanning calorimetry and neutron small-angle scattering. As tetracaine can be viewed as a model system for amphiphilic molecules, these results also provide insight into the general understanding of the physicochemical action of amphiphilic molecules on membranes.

1. Introduction

The molecular mechanisms of the anaesthetic action on nerve membranes are still poorly understood. It is still controversial whether the membrane proteins or the surrounding lipid matrix are the primary target sites of anaesthetic action [1]. An important key in understanding the molecular mechanisms of anaesthesia might be the antagonistic effect of hyperbaric pressures against anaesthetic action, which has been observed in vivo (see e.g. [2]). In order to elucidate the effect of the local anaesthetic on the properties of the pure lipid matrix, we performed experiments on phospholipid bilayers, which have been studied intensively as model systems of biological membranes [3]. Phospholipid bilayers undergo a series of phase transitions as a function of temperature, pressure and water contents (see e.g. [3,4]). Following the discovery that anaesthesia can be reversed by application of hydrostatic pressure, suggests to study also the pressure effect on the phase behaviour of the anaesthetic-lipid model membrane systems.

2. Results and Discussion

Most of our experiments on TTC-DMPC dispersions have been performed at pH=9.5, where the anaesthetic is in its uncharged form. It has been directly shown by ^{2}H-NMR studies that uncharged tetracaine intercalates into the lipid bilayer, whereas TTC at pH=5.5 is located at the phospholipid headgroup [5]. To study the thermotropic phase behaviour of the anaesthetic-phospholipid mixtures, we performed calorimetric measurements with a Perkin Elmer DSC 7 differential scanning calorimeter [6]. Fig. 1 exhibits a selection of calorimetric

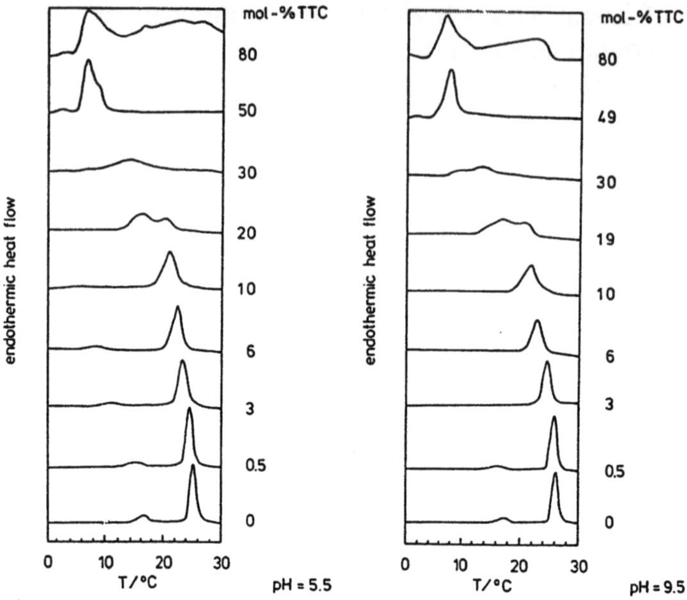

Fig. 1 DSC scans of different TTC-DMPC-H_2O mixtures at pH=5.5 and pH=9.5 (scan rate 2°C/min).

scans of DMPC dispersions at pH=9.5 and pH=5.5 as a function of TTC concentration (given in mol-%). Clearly, the well-known main transition, corresponding to the "ripple" gel ($P_{\beta'}$) to liquid-crystalline (L_{α}) bilayer conversion, of pure DMPC liposomes is seen at T=24°C, and the smaller pretransitional endotherm peak, which is due to the $L_{\beta'}$ gel to $P_{\beta'}$ gel phase transition, occures at about 15°C. The pretransition is abolished already at a concentration of about 3 mol-% TTC at pH=9.5. At pH=5.5, a higher concentration is needed to cause the same effect, which might be due to the smaller partition coefficient for the charged form of TTC in membranes [5]. The main transiton temperature T_m becomes continuously depressed with the addition of up to 10 mol-% TTC for both pH-values, with little modification of the enthalpy change for the main transition ($\Delta H_m \approx 25$kJ/mol). Parallel to the depression of T_m the peak half width of the DSC signal broadens, indicating that the temperature range over which the transition is observed is increased and the cooperativity of the transition is reduced. The cooperativity unit size of the lipid molecules is reduced by about 50% upon addition of 10 mol-% TTC. The decrease of T_m with increasing anaesthetic concentration can be explained by noting that these molecules increase the free volume of the lipid molecules. At high anaesthetic concentrations, rather complex DSC peak structures appear, which indicate a demixing process or the appearance of different phase transitions.

In order to elucidate the structural properties of the anaesthetic-lipid system, small-angle neutron scattering (SANS) experiments [4,6] have been

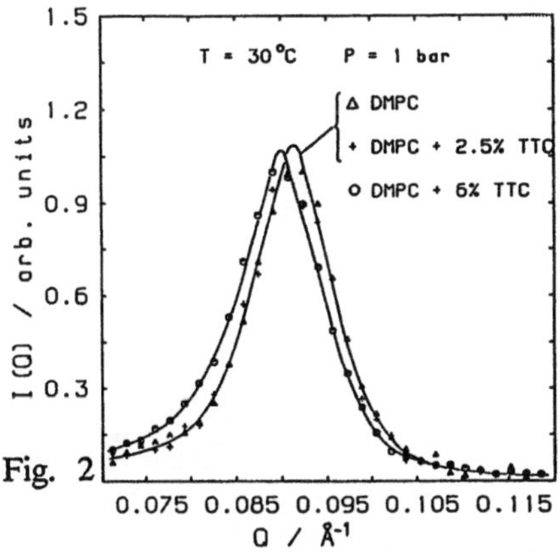

Fig. 2

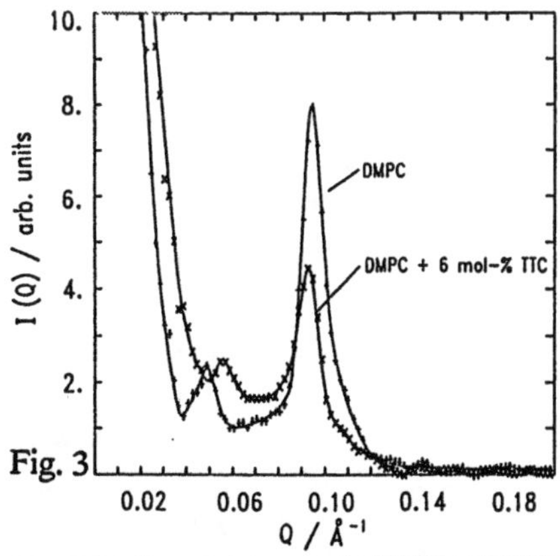

Fig. 3

Fig. 2 SANS pattern I(Q) of pure DMPC liposomes and TTC-DMPC dispersions at low anaesthetic concentrations (pH=9.5, T=30°C). Only the intermediate Q-range is shown ($Q=(4\pi/\lambda)\sin\theta$, 2θ scattering angle).

Fig. 3 SANS pattern I(Q) of pure DMPC liposomes and a 6 mol-% TTC-DMPC dispersion (pH=9.5, T=18°C).

performed as a function of temperature, pressure and concentration of TTC. Figs. 2 and 3 show the SANS patterns of pure DMPC and the corresponding curves for lipid samples containing low concentration of TTC at T=30°C (in the L_α-phase) and at T=18°C (in the $P_{\beta'}$ gel phase). The large Bragg reflections observed here result from ordered stacks of many membrane bilayers [4]. The corresponding bilayer repeat unit d, which can be calculated from the Bragg equation, consists of the bilayer thickness and the interlamellar water layer. For pure DMPC at T=30°C , the first order Bragg peak corresponds to a lamellar repeat unit d of about 69Å. Addition of 2.5 mol-% TTC to DMPC has a negligible effect on the structure of the multilamellar vesicle. The addition of 6 mol-% TTC, however, already causes a small (about 1Å) increase in the lamellar repeat period. For the diffraction profiles at T=18°C, a further small peak is observed around $Q_0=0.05Å^{-1}$, which can be assigned as the peak corresponding to the ripple repeat distance $\lambda_0=2\pi/Q_0$. For pure DMPC, $\lambda_0=128Å$ and the d-spacing is 66Å. The observed value of λ_0 is in good agreement with data reported previously [3], obtained also directly by scanning tunneling microscopy [7]. Also in the $P_{\beta'}$ gel phase, the d-value increases about 1Å upon addition of 6 mol-% TTC, the ripple wave length λ_0, however, decreases to 114Å.

The incorporation of higher concentrations of TTC leads to a drastic shift and broadening of the diffraction peaks. Fig. 4 exhibits the effect of adding 19 mol-% TTC to DMPC on the diffraction pattern I(Q) at T=30°C and increasing hydrostatic pressure. At ambient pressure, a broad peak is observed around 0.053 $Å^{-1}$, which corresponds to a d-spacing of 118 Å. The large increase in d-spacing can be explained by an increase in the interlamellar water layer around the lipid headgroups. As can be clearly seen, the diffraction peak shifts to higher Q-values,

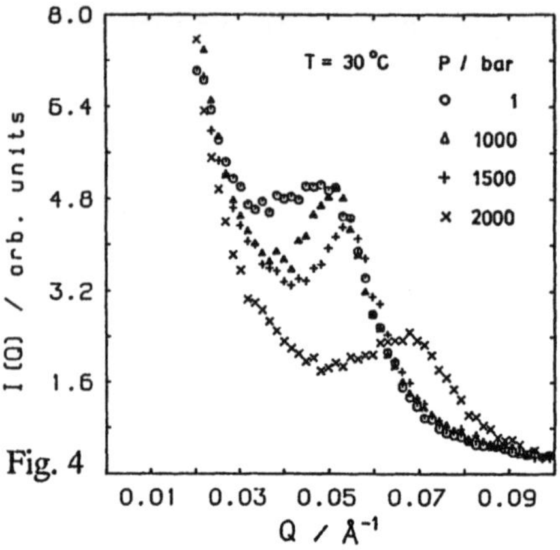

Fig. 4

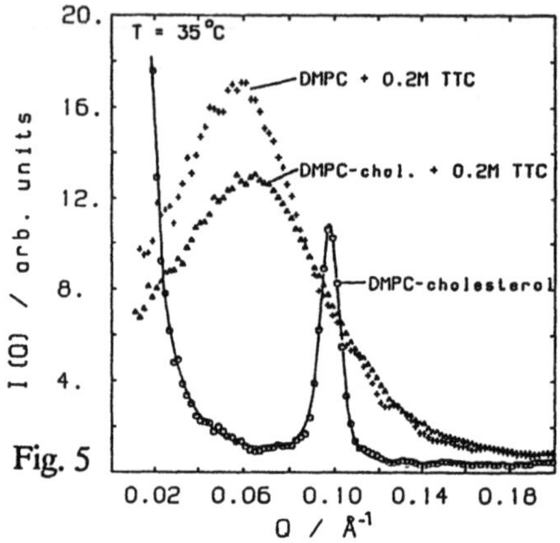

Fig. 5

Fig. 4 SANS pattern I(Q) of a 19 mol-% TTC-DMPC dispersion as a function of pressure (pH=9.5, T=30°C).

Fig. 5 SANS pattern I(Q) of a) DMPC-23mol-% cholesterol, b) DMPC in 0.2M (80 mol-%) TTC and c) DMPC-4mol-% (23 mol-% with respect to lipid) cholesterol in 0.2M (80 mol-%) TTC at T=35°C and P=1bar.

i.e. lower d-values, with increasing pressure, from 118Å at 1 bar to 89 Å at 2000 bar. Obviously, there is a partial reversibility of the anaesthetic induced effect on the structure of the multilamellar vesicle system upon pressurization.

The effect of very high anaesthetic concentrations on model membrane systems is shown in Fig. 5. The DMPC-23mol-% cholesterol system shows the typical first order Bragg reflection, corresponding to a lamellar repeat distance of 64Å at T=35°C. By addition of 0.2M TTC, the peak and the large scattering at low Q disappears, indicating a supramolecular rearrangement of the lipid assembly to a new phase consisting of small particles of radius r≈25 Å, like mixed micelles from lipid and TTC, with strong micellar interparticle interaction. The effect of increasing hydrostatic pressure on the micellar system is shown in Fig. 6. The changes of I(Q) observed around 2000 bar indicate that a swollen lamellar phase with d≈88Å has probably formed again under pressure. Pressure thus reverses the anaesthetic-induced change of the membrane structure.

Finally, the effect of diluting these samples, containing high molar ratios of TTC , is discussed. Fig. 7 exhibits the SANS curves of the 0.2M TTC-DMPC sample as a function of water concentration. The form of I(Q) indicates that upon dilution, unilamellar vesicles are formed from the micellar solution. Applying a spherical shell model, their radius r and bilayer thickness d_L can be calculated. The vesicle radius decreases from 285Å at dilution 1:5 to 169Å at dilution 1:12. The lipid bilayer thickness is about 50Å, which is about 15% larger than that of pure DMPC vesicles due to the incorporation of TTC in the headgroup region. Obviously, the particle structure is also critically dependent on the water concentration.

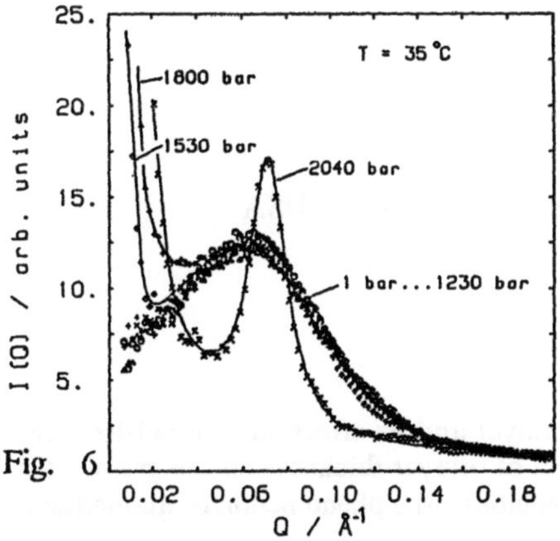

Fig. 6

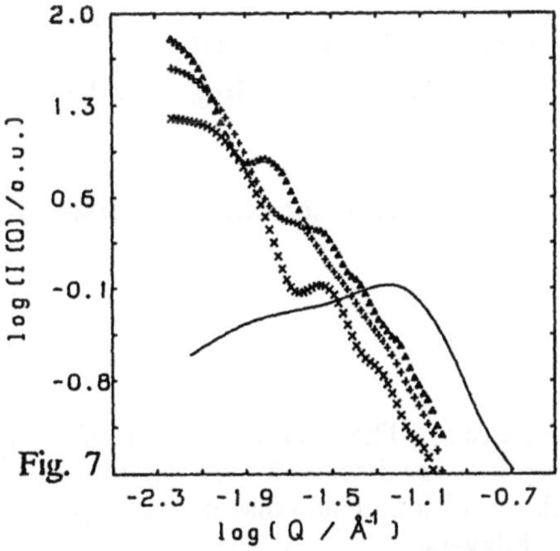

Fig. 7

Fig. 6 Pressure effect on the diffraction pattern I(Q) of the system DMPC-4 mol-% cholesterol in 0.2M TTC at T=35°C.

Fig. 7 SANS pattern I(Q) (log-log plot) of 0.2M TTC-DMPC (full line), and diluted with water by a factor a) 1:5 (Δ) , b) 1:8 (+) and c) 1:12 (×) (pH=9.5, T=30°C, P=1bar).

References

[1] I. Ueda and H. Kamaya, Anesth. Analg. 63, 929-945 (1984)

[2] F.H. Johnson and E.A. Flagler, Science 112, 91 (1950); ibid 92, 112 (1951)

[3] G. Cevc and D. Marsh, "Phospholipid Bilayers", John Wiley & Sons, New York 1987

[4] R. Winter and W.C. Pilgrim, Ber. Bunsenges. Phys. Chem. 93, 708-717 (1989)

[5] M. Auger, H.C. Jarell and I.C.P. Smith, Biochemistry 27, 4660-4667 (1988)

[6] R. Winter, M.-H. Christmann, M. Böttner, P. Thiyagarajan and R.K. Heenan, Ber. Bunsenges. Phys. Chem. 95, 811-820 (1991)

[7] J.A.N. Zasadzinski, J. Schneir, J. Garley, V. Elings, P.K. Hansma, Science 239, 1013-1015 (1988)

Bilayer Elasticity and Its Effects on Channel-Forming Peptides

H.W. Huang

Physics Department, Rice University, Houston, TX 77251, USA

Abstract. The internal elastic mode of a bilayer and its effect on channel-forming peptides are discussed. We explain the effect of bilayer thickness on the gramicidin channel lifetime and discuss a new phase-transition-like phenomenon of alamethicin in bilayers.

1. Internal Elastic Modes of Bilayer

The theory of bilayer elasticity is extended to include the internal deformation modes by regarding a bilayer as a continuous medium with two surfaces (or interfaces). The unperturbed bilayer is assumed to be in the x-y plane. Let $u_+(x,y)$ and $u_-(x,y)$ be the displacements of the top and bottom surfaces from their unperturbed positions, respectively. The deformation energy of the bilayer per unit area is then given by [1]

$$f=(B/2h)(u_+-u_-)^2+(K_c/4)[(\Delta u_+)^2+(\Delta u_-)^2]+(\gamma/2)[(\nabla u_+)^2+(\nabla u_-)^2], \tag{1}$$

where h is the bilayer thickness in equilibrium, 1/B the thickness compressibility, K_c the curvature elastic modulus and γ the surface tension. The Gaussian curvature terms are irrelevant if we do not consider topological changes of the (closed) surfaces or if we consider the boundary conditions fixed (the Gauss-Bonnet theorem). From u_+ and u_- we can construct two independent modes: $D=u_+-u_-$ and $S=(u_++u_-)/2$. The S mode describes the elastic deformation of a thin membrane [2]. The D mode is governed by the free energy

$$f_D=(hB/2)(D/h)^2+(K_c/8)(\Delta D)^2+(\gamma/4)(\nabla D)^2. \tag{2}$$

Its corresponding Euler-Lagrange equation can be solved by letting $\Delta D=\alpha D$, $\alpha=[\gamma\pm(\gamma^2-4K_cB/h)^{1/2}]/K_c$. For example, for bilayers of monoacylglycerols, h~22-29 Å, B~5.0x10^{-8} dynes/Å^2 [3], γ~1.0-2.5x10^{-8} dynes/Å. Most measurements indicate $K_c/h \gtrsim 10^{-6}$ dynes [4,5]. Thus $\alpha \sim \pm i2(B/K_ch)^{1/2}$; the surface tension term is negligible. The internal deformation of a lipid bilayer is described by a linear combination of Thomson functions [1]. These properties may be important for membrane-protein interactions.

Springer Proceedings in Physics, Vol. 66 The Structure and Conformation of Amphiphilic Membranes
Editors: R. Lipowsky · D. Richter · K. Kremer

2. Effect of Bilayer Thickness on Gramicidin Channel Lifetime

Gramicidin, a linear pentadecapeptide, spontaneously forms ion conducting pores across lipid bilayers. The cylindrical pore is formed by two monomers, each in a single stranded β-helical conformation, linked N-terminus to N-terminus by six hydrogen bonds. Most experimental observations are consistent with the assumptions that the dimeric channel loses its ion conductivity when it dissociates into monomeric units and that the dissociation can be described as a first-order reaction with a well defined dissocation constant $k_d \propto \exp(-G^*/k_BT)$, where G^* is the activation energy of dissociation [6].

It is assumed that, when a protein is embedded in a bilayer, their hydrophobic regions will tend to match each other because the hydrophobic-hydrophilic mismatch is energy costly (about 24 cal per $Å^2$ of accessible surface area of protein [7]). And since lipid bilayers are more deformable than proteins, the bilayer is likely to be deformed from its protein-free equilibrium configuration in order to accommodate an inserted protein. (For example, the thickness compressibility of a lipid bilayer is about 400 times the volume compressibility of proteins.) Thus the elastic property of a bilayer may affect the function of an embedded protein. As a concrete example of such an effect, the lifetime ($1/k_d$) of the gramicidin channel was measured as a function of bilayer thickness by Elliott et al. [8]. As expected, the lifetime decreases with increasing bilayer thickness. In the original analysis [8], it was assumed that the change of the activation energy ΔG^* equals the surface tension γ times a distance z, the distance over which the tension pulls before the dimer dissociates. Since z should be of the order of 1 Å, Elliott et al. found that the measured surface tension is at least one order of magnitude too small to account for the observed variation in k_d.

We reanalyzed the data with the more complete free energy (2) [1]. Let D_o be the unperturbed bilayer thickness minus the (hydrophobic) length of the gramicidin channel. Assume that the bilayer deforms in such a way that $D(x,y)=D_o$ where the membrane contacts the channel. We can calculate the total elastic energy for such a deformation, $F(D_o)$, but it contains one unknown constant due to the unknown contact angle between the bilayer and the channel. Then we have $\Delta G^*=-[F(D_o)-F(D_o-z)]$ (valid only for $D_o>z\sim1$ Å) and

$$\ln k_d(D_o)=[F(D_o)-F(D_o-z)]/k_BT+\text{const}, \tag{3}$$

where the constant was determined by the value of k_d at $D_o=0$ where $\Delta G^*=0$. Another data point was used to determine the unknown constant in $F(D_o)$, i.e., the contact angle mentioned above. The relation (3) is then compared with the other data points in Fig. 1. The agreement between the theory and the data is good, indicating that the effect of membrane thickness on the gramicidin channel lifetime can be accounted for by the elastic theory of membrane.

Remarkably, the contact angle as determined above is almost 90°; i.e., the hydrocarbon chains neighboring a channel are essentially parallel to the channel wall (Fig. 2). Thus the concept of membrane surface tension pulling the channel is not valid. The configuration of membrane in the vicinity of a channel may be important to the function of the channel. It is interesting to note that the deformation near the channel is in fact slightly concave rather than convex as suggested by the so-called

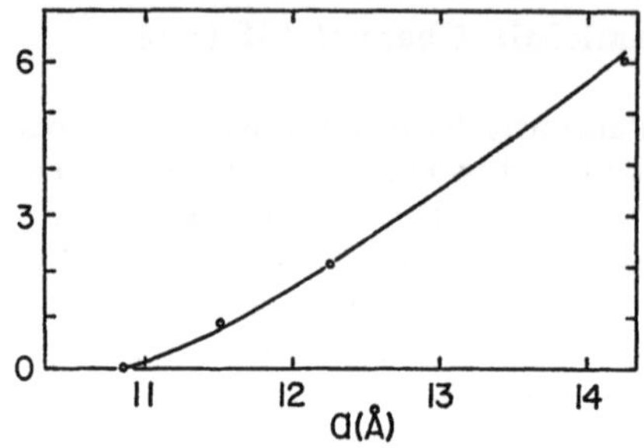

Fig. 1. The effect of bilayer thickness on gramicidin channel lifetime. The ordinate is $\ln k_d(D_o)-\ln k_d(0)$. The abscissa is the half-bilayer thickness a=h/2. D_o=0 when h=21.7 Å. The small circles are the measurements by Elliott et al. [8]. The curve is Eq. 3.

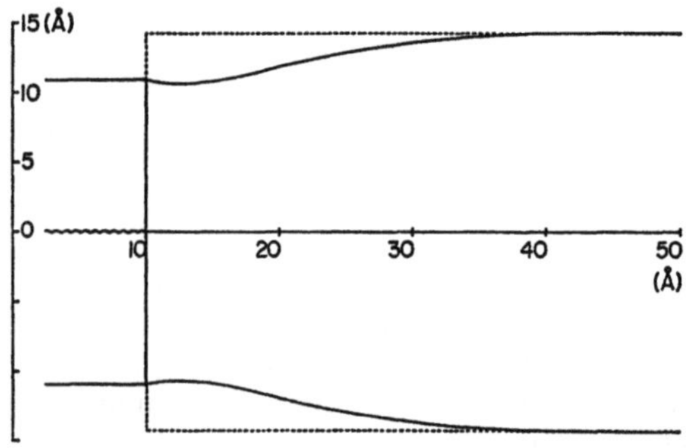

Fig. 2. The shape of bilayer deformation-the cross section of a bilayer with a gramicidin channel embedded in it. The unperturbed (dotted lines) thickness of the hydrocarbon core of the bilayer is 28.5 Å. The length of the lipophilic exterior of the channel is 21.7 Å. The outside radius of the channel is 10 Å.

dimple model [e.g., 8]. As a result, the range of deformation is substantial, of the order of $2(hK_c/B)^{1/4}$~22 Å for monoacylglycerols;~32 Å for lecithins. This may have important consequences on the membrane-mediated interactions between proteins.

3. Interaction of Alamethicin with Bilayers

Alamethicin is a linear icosapeptide which spontaneously inserts into lipid bilayers and produces a voltage-dependent ion conductance. Unlike gramicidin, the conformation of alamethicin channel is still unknown. We do know that alamethicin, in the membrane-associated state, is at least 60-70% α-helical. The helix is amphiphilic with its polar residues lying along a strip parallel to the axis [9]. In the conducting state, presumably the helices form a water-filled pore like the staves of a barrel, with their hydrophobic surfaces facing the lipid and the hydrophilic residues lining the pore. However, in order to understand how the conductance

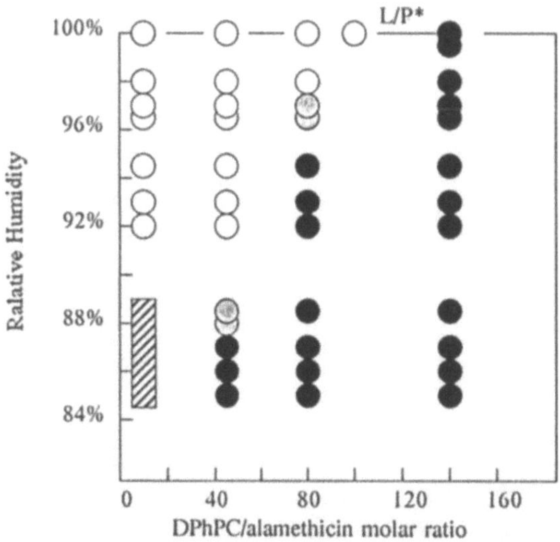

Fig. 3. The phase diagram for the orientation of alamethicin in DPhPC bilayer. For L/P>L/P*, the majority of alamethicin molecules are on the membrane surface irrespective of the value of RH. For L/P<L/P*, the majority of alamethicin molecules are inserted in the membrane at high RH; but as the RH is loweded, there is a transitional RH, below which the majority of alamethicin molecules switch to the surface state.

depends on voltage, we also need to know the configuration of alamethicin in the nonconducting state, or more generally how alamethicin interacts with lipid bilayers.

Recently we used the method of oriented circular dichroism [10] to investigate the orientation of alamethicin (its α-helical section) in bilayers and obtained rather surprising results. Lipid-alamethicin mixtures of various molar ratios (L/P) were prepared into the L_α (smectic A liquid crystalline) phase. The multilayer samples were equilibrated in humidity chambers. For example, if the lipid is diphytanoylphosphatidylcholine (DPhPC), x-ray diffraction showed that the phosphate-to-phosphate distance across the bilayer is about 34 Å while the repeat distance of the multilayers is 48.8 Å at 100% RH and about 45.5 Å at 60% RH (the length of alamethicin is about 30 Å). The orientation of alamethicin in DPhPC multilayers (in the L_α phase) is shown in Fig. 3. An open circle indicates that the great majority of the alamethicin molecules are perpendicularly inserted in the bilayer; a black circle indicates that the great majority bind parallel to the membrane surface; there is a narrow transition region (gray circles) where comparable amounts of alamethicin are in either state. This phase-transition-like behavior of alamethicin was also observed in dioleoylphosphatidylcholine (DOPC) multilayers (Fig. 4), and we suspect that it is a general property of alamethicin in most lipid membranes.

There seems to be no obvious explanation for the alamethicin-bilayer interactions observed in this experiment. For example, the insertion of alamethicin into membrane can not be explained as due to the lack of water between bilayers--the hydration dependence is contrary to this argument. Nor can it be understood as the peptide solubility to lipid--if it were so, one would expect a larger fraction of alamethicin molecules inserted at a higher L/P. Perhaps, the most unexpected result

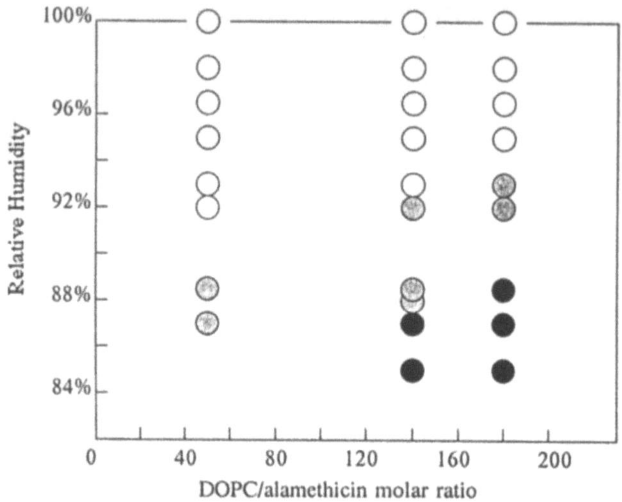

Fig. 4. The phase diagram for the orientation of alamethicin in DOPC bilayer is similar to that in DPhPC.

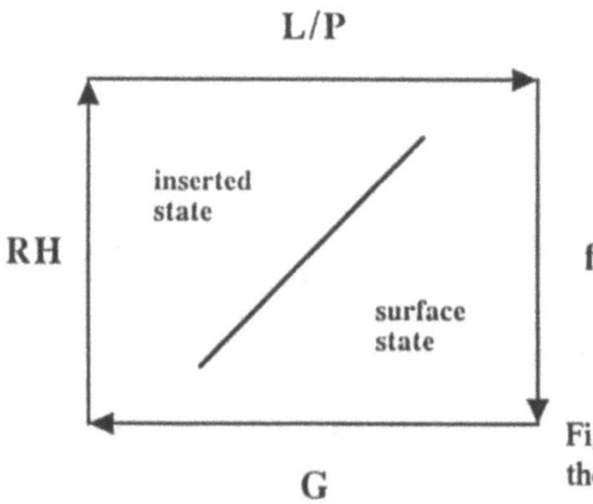

Fig. 5. The schematic phase diagram for the model Eq. (5).

is the abrupt change of state for the majority of alamethicin molecules across the transition region. It is difficult to see how this can happen if the peptide molecules are not all directly or indirectly interacting with each other.

We speculate that the phase-transition-like behavior is a manifestation of membrane-mediated interactions between peptide molecules. Perhaps alamethicin, because of its amphiphilic nature, forms water-filled aggregates in bilayer; in that case, high RH may favor the inserted state of alamethicin. We further speculate that the bilayer is deformed around an inserted alamethicin aggregate; that is, similar to the gramicidin channel, each inserted alamethicin aggregate is the center of a circular area of bilayer deformation. When two deformation areas overlap, the total free energy of deformation decreases; therefore, there is an interaction between the two inserted aggregates. Let us consider a collection of alamethicin aggregates in the inserted state, designated by indices i=1, 2, 3,.... We also designate the state of the ith aggregate by a symbol σ_i. Let $\sigma_i=0$ if the aggregate changes into the surface

state, otherwise $\sigma_i=1$. The energy of the system can be written as

$$H = - \Sigma_{(ij)} G_{ij}\, \sigma_i\, \sigma_j + \Sigma_i f_i\, \sigma_i, \tag{4}$$

where $-G_{ij} < 0$ represents the interaction energy between the ith and the jth aggregates if their areas of bilayer deformation overlap; (ij) represents pairs of interacting aggregates; f_i represents the free energy difference of the ith aggregate between the inserted state and the surface state, provided there are no other peptide molecules present in the system. A qualitative result can be obtained by the approximation

$$H = -G\Sigma_{(ij)}\, \sigma_i\, \sigma_j + f\, \Sigma_i\, \sigma_i. \tag{5}$$

This model can be shown to have a first-order phase transition between the $\langle\sigma_i\rangle\sim 0$ state and the $\langle\sigma_i\rangle\sim 1$ state, where $\langle\ \rangle$ represents the ensemble average [12]. The phase boundary is given by $f = \eta G/2$ (Fig. 5), where η is the average number of neighboring aggregates. Since we expect the average free energy f of an aggregate to be a decreasing function of RH and the average interaction G to increase with the peptide concentration (inverse of L/P), the schematic phase diagram is similar to Figs. 3 and 4.

This research was supported in part by the Office of Naval Research grant N00014-90-J-1020, and the Robert A. Welch Foundation.

References

1. H. W. Huang, Biophys. J. **50**, 1061 (1986)
2. W. Helfrich, Z. Naturforsch. **28c**, 693 (1973)
3. S. H. White, Biophys. J. **23**, 337 (1978)
4. H. P. Duwe, J. Kaes and E. Sackmann, J. Phys. Fr. **51**, 945 (1990)
5. M. Mutz and W. Helfrich, J. Phys. Fr. **51**, 991 (1990)
6. S. B. Hladky amd D. A. Haydon, in *Current Topics in Membranes and Transport.* edited by F. B. Bronner (Academic Press, New York, 1984) **21**, 327.
7. C. Chothia, Nature **248**,338(1974)
8. J. R. Elliott, D. Needham, J. P. Dilger, and D. A.Haydon, Biochim. Biophys. Acta **735**, 95 (1983)
9. R. O. Fox and F. M. Richards, Nature **300**, 325 (1982)
10. Y. Wu, H. W. Huang and G. A. Olah, Biophys. J. **57**, 797 (1990)
11. H. W. Huang and Y. Wu, Biophys. J. Nov. 1991
12. H. W. Huang, Phys. Rev. **B12**,216 (1975)

Interaction of Charged and Uncharged Calcium Channel Antagonists with Phospholipid Membranes. Binding Equilibrium, Binding Enthalpy, and Membrane Location

H.-D. Bäuerle and J. Seelig

Biocenter of the University of Basel, CH-4056 Basel, Switzerland

The membrane location and the binding mechanism of two calcium channel antagonists, amlodipine and nimodipine, in pure lipid membranes were investigated with different physical-chemical methods. Deuterium nuclear magnetic resonance shows that the non-charged, hydrophobic nimodipine is distributed homogenously across the whole hydrocarbon layer. In contrast, the positively charged, amphiphilic amlodipine adopts a well defined position in the lipid bilayer. The charged ethanolamine side group of amlodipine is located at the water-lipid interface, interacting with the dipoles of the headgroup region while the non-polar ring system is located at the level of the hydrocarbon segment interacting specifically with the cis double bond of the lipid, forming a weak association complex. The binding equilibrium of amlodipine to phosphatidylcholine membranes was studied by measuring the electrophoretic mobility of lipid vesicles and with a centrifugation assay. Both experimental methods lead to identical results. The binding of amlodipine to a lipid membrane can be described by a surface partition equilibrium with an intrinsic partition constant $Kp = 15500\ M^{-1}$, yielding a free enthalpy of $\Delta G = -8.1$ kcal/mol. The enthalpy of amlodipine binding to POPC membranes was independently measured with a high sensitivity titration calorimeter yielding $\Delta H = -9.2$ kcal/mol, at 27^0C.The partitioning of the amphiphilic drug into the lipid bilayer is an enthalpy-driven process.

INTRODUCTION AND RESULTS

Amlodipine and nimodipine, two calcium channel antagonists, bind to specific receptors in cardiac and smooth muscle, modulating the transmembrane influx of extracellular calcium involved in the excitation-contraction process [1]. The positively charged amlodipine has been shown to be a potent inhibitor of calcium

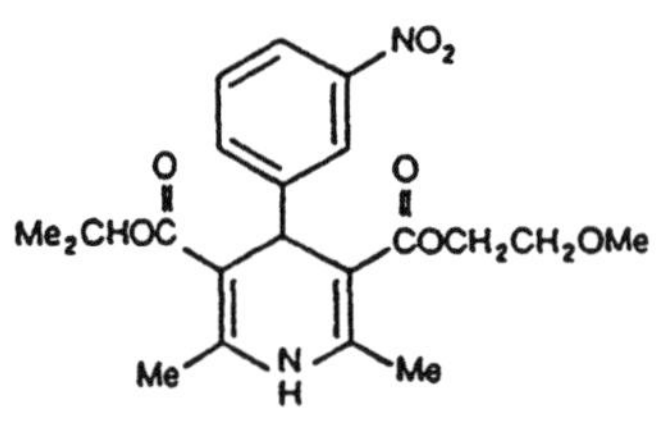

NIMODIPINE

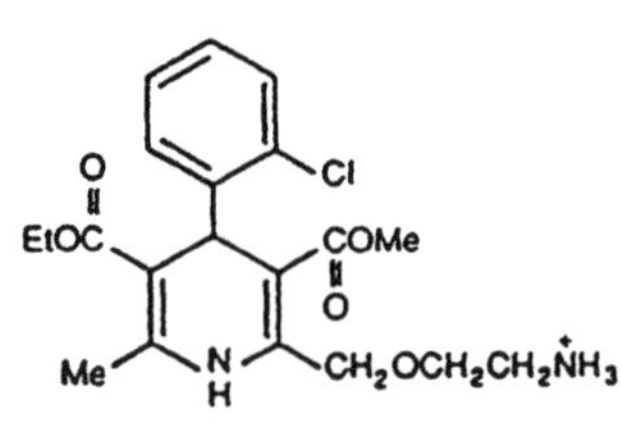

AMLODIPINE

Springer Proceedings in Physics, Vol. 66 The Structure and Conformation of Amphiphilic Membranes
Editors: R. Lipowsky · D. Richter · K. Kremer

induced contraction of K^+ depolarized rat aorta [2]. Due to their structure, the two drugs have different physical-chemical properties. The hydrophobic nimodipine is electrically neutral, almost insoluble in water but highly soluble in lipid. In contrast, the positively charged amlodipine, which is protonated at physicological pH, shows an excellent solubility in both buffer and lipid.

Binding analysis. The analysis of amlodipine binding follows the approach described for the binding of local anesthetics and small peptides to neutral and charged lipid bilayers [3,4]. The binding of amlodipine to neutral POPC bilayers was investigated in the concentration range of 1μM < Ceq < 35μM, with a centrifugation assay and ς-potential measurements [5]. Figure 1 shows the binding isotherm, where the degree of binding X_b (mmol amlodipine/mol lipid) is plotted versus the equilibrium concentration of free drug denoted Ceq. The equilibrium concentration is the drug concentration in the bulk solution. For the centrifugation assay coarse liposome dispersions, which were subjected to several freeze-thaw cycles, and for the ς-potential measurements multilamellar vesicles were used [5]. The ς-potential was calculated by measuring the migration velocity of drug bound lipid vesicles in an applied electric field [5]. An excellent agreement between the results of both methods was obtained. The dashed line in figure 1 corresponds to an average partition coefficient of $Kp = 15500\ M^{-1}$, calculated with $Kp = X_b/Cm$. Cm is the effective drug concentration at the membrane surface. It is reduced compared to the equilibrium concentration, Ceq, because of the electrostatic repulsion between the drug molecule and the positively charged membrane surface. Electrostatic effects were taken into account by means of Gouy-Chapman theory.
With the partition coefficient constant given we have calculated a free energy according to $\Delta G = -RT\ln(55.5\ Kp) = -8.1$ kcal/mol (at 25 ^{0}C). Furthermore, we have measured the binding enthalpy of amlodipine and sonicated small unilamellar POPC vesicles yielding a corrected reaction enthalpy of $\Delta H = -9.2$ kcal/mol per mole of bound amlodipine. This result is in contrast to the so called "hydrophobic effect" (see Tanford 1980 [6]) and supports that the binding of the amphiphilic amlodipine to POPC membranes is a pure enthalpy driven process.

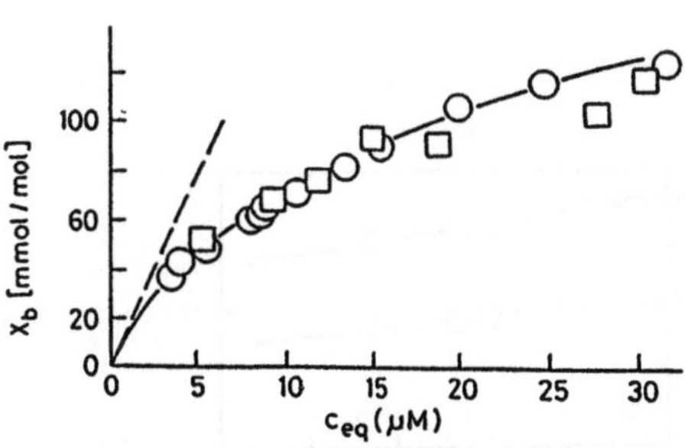

Figure 1: Binding of amlodipine to bilayer membranes composed of 1-palmitoyl-2-oleoyl-sn-glycero-3-phosphocholine (POPC) (○, centrifugation assay; □ microelectrophoretic measurements)

Headgroup interaction and drug location. The effect of amlodipine and nimodipine on the phosphocholine headgroup was studied with deuterium and phosphorus nmr. All spectra were characterized by a single quadrupole splitting and a single phosphorus chemical shielding anisotropy, indicating a liquid-cristalline bilayer structure and time averaged headgroup conformation. Amlodipine and nimodipine have different effects on the phosphocholine head-group of the lipid. Binding of the charged amlodipine induces a counterdirectional response of the two choline segments: the quadrupole splitting of the α-segment decreases, that of the ß-segment increases up on drug in agreement with the "Molecular Electrometer " concept described previously [7]. In contrast, uncharged nimodipine decreases the quadrupole splittings of both choline segments simultaneously. Figure 2 shows the influence of the two drugs to the phosphocholine headgroup with increasing drug concentration expressed by the change of the quadrupole splittings. The counterdirectional change of the quadrupole splitting corresponds to a cooperative change in the tilt angle of the time averaged headgroup conformation with respect to the bilayer normal. The extend of the conformational change of the phosphocholine headgroup is limited to the phosphate region because the chemical shielding anisotropy of all ^{31}P nmr spectra were approximately constant in the concentration range investigated. The influence of amlodipine and nimodipine on the hydrophobic part of the bilayer was investigated with deuterium labelled DOPC,POPC, and DEPC, labelled either at the 9,10 position (cis/trans-double bond) or at the 2 postion of the hydrocarbon chain of the lipid. At all labelled hydrocarbon chain segments investigated, nimodipine induces an increase in $\Delta\nu_Q$ of about 10% at $X_b = 0.15$, which then decreases again at higher nimodipine concentration. Based on the nmr results we suggest a homogenous distribution of nimodipine in the whole volume of the hydrocarbon layer. Both the NO_2 group and the dihydropyridine end of the molecule will probably see the water-/hydrocarbon edge intermittently. In contrast, amlodipine induces little change in the quadrupole splitting in the region of the hydrocarbon chain. Only in the case of 9,10 deuterium labelled POPC (cis-double bond) an addition of amlodipine induces a

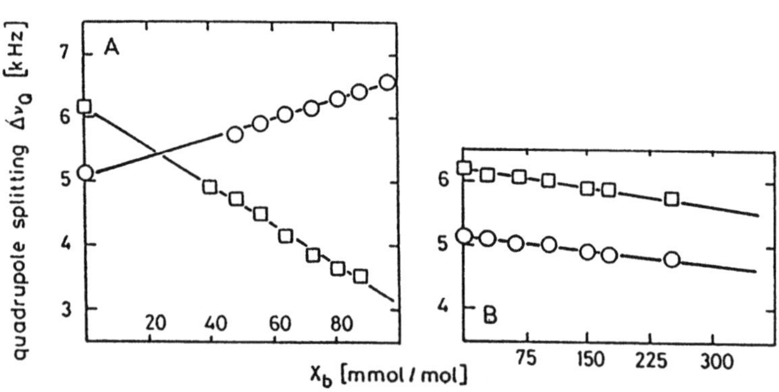

Figure 2: Variation of the phosphocholine headgroup with bound amlodipine (A) and nimodipine (B). (□) [α-CD_2] POPC; (O) [ß-CD_2] POPC.

line braodening in the ^{2}H nmr spectra coupled with a strong decrease in the ^{2}H nmr signal intensity. The higher is the amlodipine concentration the stronger is the decrease in signal intensity. For all amlodipine concentrations ^{31}P nmr spectra show full intensity, characteristic of a liquid cristalline bilayer. The line braodening combined with a decrease of signal intensity can be interpretated as a specific van-der-Vaals interaction between the cis-double bond and one of the non-polar ring system [8]. Taken all nmr results together for amlodipine, a location between the phosphocholine headgroup region, near the lipid/water interface, and the level of the cis-double bond of the hydrocarbon chain of the lipid can be proposed.

References

1. Janis, R.A., Silver, P.J. & Triggle, D.J. (1987) *Adv. Drug Res. 16*, 309-591.
2. Burges, R.A., Gardiner, D.G., Gwilt, M., Higgins, A.J., Blackburn, K.J., Campbell, S.F., Cross, P.E., & Stubbs, J.K. (1987) *J. Cardiovasc. Pharmacol. 9*, 110-119.
3. Seelig, A., Allegrini, P.R., & Seelig, J. (1988) *Biochim. Biophys. Acta 939*, 267-276.
4. Beschiaschvili, G. & Seelig, J. (1990a) *Biochemistry 29*, 52-58.
5. Bäuerle, H.D. & Seelig, J. (1991) *Biochemistry 30*, 7203-7211.
6. Tanford, C.H. (1980) The Hydrophobic Effect : Formation of Micelles and Biological Membranes, Wiley, New York.
7. Scherer, P.G. & Seelig, J. (1989) *Biochemistry 28*, 7720-7728.
8. Bienvenue, A., Bloom, M., Davis, J.H, & Deveaux, P.F. (1982) *J. Biol. Chem. 257*, 3032-3038.

Melittin-Induced Reversible Micelle ↔ Bilayer Transition

B. Sternberg[1] *and C.E. Dempsey*[2]

[1]Electron Microscopy Laboratory, Medical School, Friedrich-Schiller-Universität, O-6900 Jena, Fed.Rep.of Germany

[2]Department of Biochemistry, Bristol University, Bristol, UK

Melittin, the 26 amino acid peptide from bee venom, is known to lyse cell membranes and induce voltage-dependent ion conductance in lipid bilayer/1/. It also induces a reversible micelle <--> bilayer transition at the gel to liquid-crystalline phase transition temperature (Tm) in complexes with saturated phosphatidylcholines /2/. This third membrane activity of melittin and its synthetic analogue (Ala-14) melittin has been studied by freeze-fracture electron microscopy in more detail.

Above Tm bilayer vesicle are visible showing fracture faces decorated with intramembranous particles (IP; Fig.1a). The density of the IP is roughly proportional to the peptide:lipid ratio, but their size is independent of the ratio, indicating that under conditions studied (50 mM Tris-HCl, pH 7.5, 5 mM EDTA) the peptide is organized as discrete aggregates that penetrate deeply into the fluid bilayer/3/.

Below Tm and from an onset concentration of 3 mol% melittin on each sample shows a massive collapse of liposomal structures into micelles. With increasing melittin concentration less and less liposomes with bilayer in the gel state are observed showing ripple structures (Pß') identical to the peptide-free gel phase lipid. These residual bilayer do not exhibit IP on the fracture faces (Fig.2) indicating that the

Springer Proceedings in Physics, Vol. 66 The Structure and Conformation of Amphiphilic Membranes
Editors: R. Lipowsky · D. Richter · K. Kremer

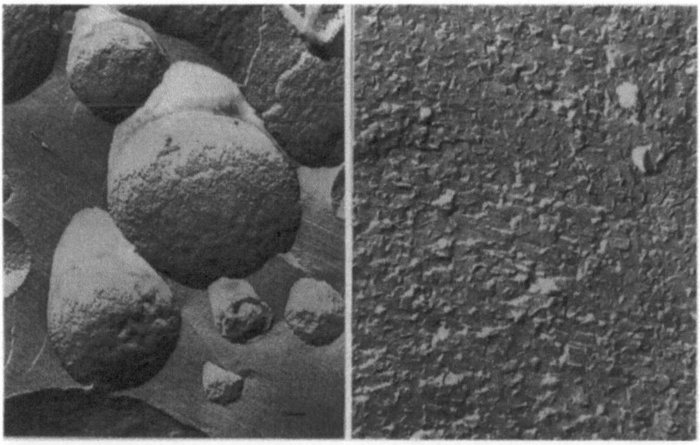

Fig.1:
Freeze-fracture electron micrographs of melittin:DMPC (a) or Ala-14 melittin:DMPC complexes (b) at peptide concentrations of 5 mol% relative to lipid (buffer composition as mentioned above with EDTA (a) but without EDTA (b). The samples were quenched from a temperature well above Tm (55°C).

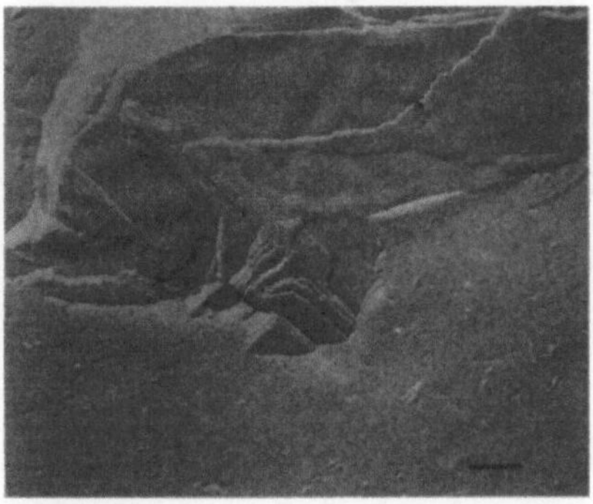

Fig.2:
Freeze-fracture electron micrograph of a melittin:DMPC complex at a peptide concentration of 3 mol% quenched from a temperature below Tm (11°C).

The scale bar represents always 100 nm.
DMPC=1,2-dimyristoyl-<u>sn</u>-glycero-3-phosphocholine.

peptide is disassociated and/or expelled from a bilayer-inserted location in the gel phase lipid/3/.

The synthetic analogue, Ala-14-melittin where the proline at residue 14 near the centre of the sequence is substituted by alanine/4/, shows the same effect on the reversible micelle <--> bilayer transition as melittin (not shown), but the rates of both reversible processes are quite different: It is decreased by 100-fold in Ala-14-Mel:DMPC complexes compared with Mel:DMPC complexes /3/. On the other hand the analogue has enhanced hemolytic activity but is unable to form stable voltage-dependent ion channels in planar lipid bilayer/5/ supporting the conclusion that bilayer disc micellization and ion channel formation do not underlie the hemolytic activity of melittin.

The influence of EDTA on the bilayer stability is not quite clear yet: Without EDTA-inclusion even above Tm a massive collapse of liposomal structures into membrane fragments is noticeable (Fig.1b). Some very few residual liposomes do not show any IP on the fracture faces.

REFERENCES:

1 Dempsey, C.E. (1990) Biochim. Biophys. Acta 1031, 143-161.
2 Dempsey, C.E. and Watts, A. (1987) Biochemistry 26, 5803-5811.
3 Dempsey, C.E. and Sternberg, B. (1991) Biochim. Biophys. Acta 1061, 175-184.
4 Dempsey, C.E., Bazzo, R., Harvey, T.S., Syperek, I., Boheim, G. and Campbell, I.D. (1991) FEBS Letters 281, 240-244.
5 Tosteson, M.T. and Tosteson, D.C. (1981) Biophys.J. 36, 109-116.

Theory of Hydration Forces

A.A. Kornyshev[1;2] *and S. Leikin*[3;2]

[1]Medical Biophysics Research Laboratory,
Universitäts-Kliniken r.d.I., Technische Universität München,
Ismaningerstr. 22, W-8000 München 80, Fed.Rep.of Germany
[2]The A.N. Frumkin Institute of Electrochemistry of the
Academy of Sciences, 117071 Moscow, Russia
[3]Laboratory of Physical Sciences, Division of Computer Research
and Technology, National Institutes of Health,
Bethesda, MS 20892, USA

Abstract. We give a mini-review of main ideas and results of a new theory of interaction between hydrophilic surfaces which describes the observable features of repulsive hydration forces and studies the conditions for a crossover from repulsion to attraction with the variation of system parameters.

1. Introduction

Interaction between two hydrophilic surfaces in water and some other polar solvents at distances smaller than 3 nm is dominated by the so called *hydration forces* [1,2]. These forces do not obey DLVO theory. They decay exponentially with distance between the surfaces with the decay length almost independent of electrolyte concentration but strongly dependent on the nature of the surfaces. These forces are usually repulsive, but there have been found recently few evidences of hydration attraction [2].

Hydration forces are of great importance in the stability of colloids [3], properties of wetting films [4], interaction and fusion of biological membranes and macromolecules [2,5]. Of interest here are the regularities of hydration repulsion (such as the observed dependence of the decay range of the force on the nature of the surfaces) and the conditions of crossover from repulsion to attraction. Interacting surfaces in the case of lipid lamellae or smooth solid surfaces can be considered as plane at the scales of action of these forces. The theory is thus reduced to the problem of interaction of plane parallel surfaces separated by a water film polarized at the boundaries due to hydration of the surfaces.

The first attempt to describe this phenomenon was made by Marcelja and Radic [6]. They used a Landau free-energy gradient expansion, quadratic in fluctuations of the order parameter (later interpreted as local water polarization [5]). The interaction of the oppositely polarized water layers near the surfaces thus led to repulsion. Nonlocal electrostatics with homogeneous boundary conditions at the surfaces was later applied in a number of papers [7]. These works gave an insight on the *nonlocal nature* of hydration forces but failed to explain many of the observable features (e.g. the mode of decay was an intrinsic property of water in these theories, independent of the nature of the surfaces). These theories did not consider the variation of water orientation along the boundaries induced by alternation of positively and negatively charged solvated surface groups. This situation is typical, in particular, for lipid membranes with the zwitterionic heads. It appears that just this feature plays the dramatic role in the picture of hydration interaction.

2. New Theory of Hydration Repulsion

The role of the *random inhomogeneous boundary conditions* has been studied in ref.8. It was shown, in particular, that the preexponential factor and decay length are both determined by the lateral correlations of the values of the order parameter along the boundaries; the decay length is the renormalized quantity which keeps information both on the correlations in the bulk water and the lateral correlation length along the surfaces. Semi-microscopic interpretation of the boundary conditions explained the main observable features of hydration repulsion in lipid lamellae systems [8,9]. The established relationship between the surface structure factors of the solvated surface groups and the value of the force results in a rule: *with the increase of the correlation length of the lateral structure factors the effective decay range of the hydration repulsion and its absolute value decrease.* In other words: *the order is stronger, the force is weaker.*

This rule is qualitatively approved in various lipid membrane systems [8,9]. Quantitative test of the theory would be possible when the direct experimental information would be available for the partial structure factors of the solvated surface groups. Such an information can be obtained by low angle neutron diffraction measurements of the isotopically labelled samples or other diffraction techniques, a subject of a research program in progress at the Laboratory of Neutron Physics (JINR, Dubna). Till that time we have to apply realistic approximations for the structure factors [8,9] or dwell on a qualitative analysis.

3. Dehydration transition

Mechanisms of crossover from a repulsive to an attractive mode of hydration forces are of great interest due to its possible role in the *dehydration transition*, a sudden decrease of the thickness of the water film separating the surfaces. This effect has been observed in some lipid lamellae structures as an abrupt escape of water from the intralamellae space with a temperature decrease or variation of other parameters [2,10].

The crossover from repulsion to attraction may happen when the energetically favorable mutual adjustment of the values of the order parameter at the opposing surfaces take place. A theory of this phenomenon can be developed on the basis of hydration repulsion [8,9]. One may start with an expression for the system free energy obtained for fixed boundary conditions at the surfaces. Then, regard the values of the order parameter at the surfaces as variables, having added to the free energy terms describing their in-plane interaction. The two fields would be coupled by terms of the interaction of their momentary values at the opposing surfaces with the coupling constant determined by the theory of hydration repulsion [8,9]. Water film thickness enters also via the self-energy terms [8,9] which describe the hydration of each of the surfaces by water film of finite thickness. The chemistry and physics which regulates the in-plane correlations of the solvated surface groups (polar heads) is very complicated. In this situation it is justified to use a phenomenological approach: construct typical expressions and investigate their consequences. The problem is then mapped on theory of 2D antiferroelectricity or antiferromagnetizm.

Such a strategy with different approaches has been taken in Ref.10,11. Ref.10 is based on the *spherical model* [10]. It prescribes a crossover from repulsion to attraction under certain conditions, including sufficiently low temperatures and long lateral correlations. The physics of mutual adjustment is hidden into the *spherical constraint* which fixes the mean square fluctuations of the local density of solvated groups averaged along the surfaces. The constraint reflects some quasiperiodical structures in the boundary conditions to be followed in the intersurface adjustment. In the spherical model we do not know what exactly happens at the surfaces, which kind of ordering is born out to provide a favorable adjustment, though we get it it in some integral form.

Attempts to look at the adjustment in a more detailed way has been made in Ref.12. A model with Gaussian fluctuations around the intrinsically periodical substrate induced values was treated there. A strong oversimplification of the real mechanisms, this model is exactly solvable and its analytical results demonstrates the main properties of the crossover.

Neither Ref.10, nor Ref.12 result in any critical phenomena: the crossover proceeds gradually with parameters variation. A study of dehydration as a first order transition from a ferro- to an antiferro-state has been made in Ref.11 by means of a phenomenological Landau theory. The approach has given an interesting phase diagram with a possibility of hysteresis. However, the analysis has yet been limited to the homogeneous fluctuations of the order parameter at the surfaces which are hardly imaginable for the neutral zwitterionic lipid membranes. Thus, this approach is to be extended on the case of inhomogeneous fluctuations which is a subject of our present study.

4. An outlook

Thus far, it was fascinating to obtain that the main features of hydration interaction of lipid membranes can be understood within the linear response description of water. The question whether there are some *bound water* effects, i.e. the influence of the surfaces on the correlation properties of water (as suggested by a number of experimental observations) motivates a special direction of research. A reasonable approach here could be investigations of typical lattice models of water in confined geometry with a study of the influence of different boundary conditions [13]. Field analogs are also relevant.

There is a number of problems however left within the linear response theory which we plan to study in the nearest future. These are the role of the fine structure in the bulk correlation function in water (oscillations), the final size of solvated groups and the tilt of the polar heads together with more accurate approximations for the lateral structure factors.

5. References

1. Chem.Scripta 25 (1985).
2. R.P.Rand and V.A.Parsegian, BBA 988, 351 (1989).
3. B.V.Derjaguin and N.V.Churaev, in *Fluid Interfacial Phenomena*, edited by C.A.Croxton (Wiley, New York,1986), p.633
4. N.V.Churaev and B.V.Derjaguin, J.Colloid Interf.Sci. 103, 542 (1984)
5. D.W.R.Gruen, S.Marcelja, and V.A.Parsegian, in *Cell Surface Dynamics,* edited by A.S.Perelson, C.Delisi and F.W.Wiegel (Marcel Dekker, New York, 1984), p.59.
6. S.Marčelia and N.Radic, Chem.Phys.Lett. 42, 129 (1976).
7. G.Cevc, Ref.1, p.96; M.Belaya, M.Feigelman, and V.Levadny, Chem.Phys.Lett. 126, 361 (1986), A.A.Kornyshev, J.Electroanal. Chem. 204, 79 (1986); P.G.Dzhavakhidze, A.A.Kornyshev,and V.Levadny, Phys.Lett. A 118, 203 (1986).
8. A.A.Kornyshev and S.Leikin, Phys.Rev.A 40 6431 (1989).
9. S.Leikin and A.A.Kornyshev, J.Chem.Phys. 92 6890 (1990).
10. S.Leikin and A.A.Kornyshev, Phys.Rev.A 44 1156 (1991).
11. A.A.Kornyshev, D.A.Kossakowski and S.Leikin, to be published (1991).
12. G.Cevc and A.A.Kornyshev, to be published (1991).
13. V.Kabanovich, A.A.Kornyshev, D.A.Kossakowski and V.B.Priezzhev, in progress.

Force Equilibria Between Charged Surfaces with Confined Polyelectrolyte Chains

R. Podgornik

J. Stefan Institute, Theoretical Physics Department,
P.O.B. 100, 61111 Ljubljana, Slovenia

Abstract. The forces in a system composed of two charged (membrane) surfaces are studied, effected by the addition of an infinitely long, flexible polymeric chain carrying charges of the sign opposite to the sign of the fixed surface charges. The conformation of the polymeric chain as a function of the intermembrane separation clearly displays the onset of the unbinding transition of the polymer analogous to the one observed in a system of deformable, fluid membranes. This transition couples with the usual double layer forces in such a way that the total intermembrane force can become attractive.

Stabilization of colloidal dispersions is an important field not only in chemistry but also in biology where different self - assembly mechanisms, generating aggregates of different geometry such as membranes, micelles or vesicles, are dependent on the nature of the interactions between the aggregates [1]. These interactions are usually subsummed under the heading of the DLVO theory where they are thought of as being composed of an attractive van der Waals component and an often repulsive stabilising double - layer electrostatic component. Only recently has it been realized that the DLVO theory misses an important facet of the (bio)colloidal interactions connected with the flexibility of the interacting surfaces [2]. The thermally excited undulations of the interacting membrane surfaces can lead to drastic modifications in the nature of the intersurface forces introducing an additional stabilizing interaction into the total force equilibrium. This interaction is seen as stemming from the sterical exclusion of all thermal undulation modes of the apposed membranes that would violate the impenetrability of the interacting surfaces and leads to the prediction of the s.c. unbinding transition for a stack of suspended membranes under vanishing external (osmotic) stress [2].

Recently we have further extended the framework of the DLVO theory by considering the modifications in the intersurface forces brought about by exchanging (point) counterions confined between the interacting surfaces with flexible, charged polyelectrolyte chains [3]. Flexible polymers can provide additional stabilization to the (bio)colloidal systems [1] in somewhat the same vein as the steric interaction between the undulating membrane surfaces enhances the repulsion provided by the double - layer electrostatic (or equivalently the hydration) forces [2]. However, with polymers bearing charges of the oposite sign to those residing on the bounding (membrane) surfaces, an attractive component to the total interaction may also be introduced because the polymer segments can adsorb to the surfaces thus enabling the formation of bridges between them. The magnitude of this bridging attraction can be significant if compared to the van der Waals attraction or the double - layer repulsion and can thus alter the force equilibria in a colloidal system [3].

The case of confined flexible polyelectrolyte chains has been treated by various authors on different approximation levels starting from the seminal work of Wiegel [4]. The different approaches mainly belong to the continuous - chain models in external field , lattice self - consistent field (SCF) theories or computer simulations [5]. In order to formulate the problem in a language that would be close to the Poisson - Boltzmann formulation of the confined electrolyte problem we recently introduced the self-consistent field Poisson-Boltzmann approximation for polyelectrolytes [3] that allowed us to compare in a most direct manner the double - layer forces with and without the presence of flexible polyelectrolyte counterions.

Here we consider the statistical properties of a confined polyelectrolyte chain related to the connection between the conformation of the polymeric chain and the nature of forces between the charged bounding surfaces. In the case that a polymeric chain is confined from one side only (semiinfinite confinement) it has been argued recently [6] that it undergoes unbinding transitions as the strength of the wall potential is varied. In the case that the wall potential is electrostatic in nature (screened Debye - Hückel form) the existence of an unbinding transition for semiinfinite confinement has been demonstrated quite some time ago [4]. Most notably the unbinding of the polymer can in this case be triggered by varying the surface charge density.

We consider a continuum (Gaussian) model for a charged polymeric chain confined between two charged surfaces separated by $2a$. The charges of magnitude τ residing on the polymer segments of length l are assumed to be of a sign different from the charges of surface charge density σ fixed on the bounding hard walls. In the limit of the ground - state dominance the transverse polymer Green's function can be written as $G(z,z';N) \approx \psi(z)\psi(z')e^{-E_N N}$ where z is the transverse coordinate limited to $|z| \leq a$, while E_N is the part of the total free energy stemming from the polymer configurations alone. In this case the SCF equations reduce to the one dimensional Hartree equation [3] of a form

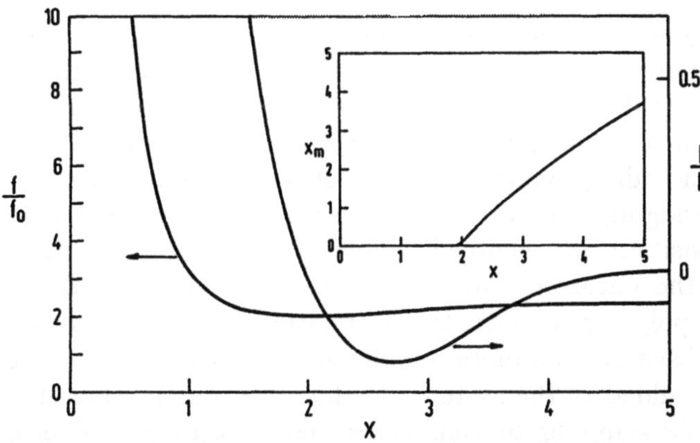

Fig.1 The dimensionless surface free energy density f/f_0, where $f_0 = \sigma^2/(\epsilon\epsilon_0 \lambda_B{}^{\frac{1}{3}})$ and $\lambda_B = 6\beta\tau\sigma/(l^2\epsilon\epsilon_0)$, as a function of dimensionless separation $x = \lambda_B{}^{1/3}a$. The dimensionless pressure is defined as $p/p_0 = -\frac{\partial}{\partial x}(f/f_0)$. The inset shows the position of the maximum of the segment density distribution function (x_m) as a function of x. When $x_m = 0$ the density distribution is monomodal with a single maximum at the midpoint of the intersurface space and depletion layers in the vicinity of the bounding surfaces. On the other hand when $x_m \neq 0$ the polymer is in the bound state with its segment density distribution of a bimodal form with two maxima located close to the bounding surfaces at $x = \pm x_m$ and a broad depletion layer in the middle of the intersurface space. The unbinding transition sets in at $x \approx 1.986$.

$$\frac{l^2}{6}\frac{d^2\psi(z)}{dz^2} + E_N\psi(z) + \frac{\beta\tau\sigma}{\epsilon\epsilon_0}\int_{-a}^{+a} |z - z'|\psi^2(z')dz'\psi(z) = 0 \qquad (1)$$

where β is the inverse thermal energy while the SCF potential has the form $V(z) = \int_{-a}^{+a} |z - z'|\psi^2(z')dz'$. We were able to obtain analytical solutions to this equation in the case where we approximate the SCF potential by a linearized form $V(z) \approx |z|$ which is accurate close to the walls, where the SCF potential is the largest, but fails in the vicinity of $z = 0$, where the expansion starts as $V(z) \approx \langle z \rangle + \psi^2(z = 0)z^2$ [3]. In the linearized case the solution of Eq.(1) can be given in terms of the Airy functions and leads to a form of the free energy and pressure as a function of the dimensionless intersurface separation presented in Fig.1.

The most interesting feature of this dependence can be extracted if we monitor the conformation of the polymeric chain together with the pressure acting between the bounding surfaces. Clearly exactly at the point where the polymer undergoes an unbinding transition, characterised by a transformation of the polymeric segment density distribution from a bimodal to a monomodal form the intersurface pressure changes sign thus going from repulsion to attraction. The unbinding of the polymer thus severs the electrostatically promoted bridges between the bounding surfaces and turns a bridging attraction into an essentially steric repulsion. The unbinding transition of a charged confined polymer chain is thus closely coupled to the nature of forces between the confining surfaces. This feature of the polymer conformation in relation to the intersurface forces appears to subsist even for more complex situations as in e.g. a case where electrolyte is added to the system. In this case, which represents a generalization of the semiinfinite confinement model studied by Wiegel [4], the connection between intersurface pressure and the conformation of the polymeric chain is somewhat more complicated, though. The reason being that the total free energy of the system is composed of a polymeric part as well as of a contribution of the ionic double-layer overlap which has the same form as in the DLVO theory.

References

1. J.N. Israelachvili, *Intermolecular and Surface Forces* (Academic Press New York, 1985) and references therein.
2. S. Leibler, R. Lipowsky, Phys. Rev. **B35**, 7004 (1987).
3. R. Podgornik, J.Phys.Chem. **95**, 5249, (1991); in press (1991).
4. F.W. Wiegel, J.Phys. **A10**, 290, (1977).
5. J.K.J. Opheusden, J.Phys. **A21**, 2739, (1988).; R. Israëls, J.M.H.M. Scheutjens (personal communication); Åkesson, C. Woodward, B. Jönsson, J.Chem.Phys. **91**, 2461, (1989).
6. A.C. Maggs, D.A. Huse, S. Leibler, Europhys. Lett. **8**, 615, (1989).

Part II

Conformation of Membranes

Budding Transition for Bilayer Fluid Vesicles with Area-Difference Elasticity

U. Seifert, L. Miao, H.-G. Döbereiner, and M. Wortis

Department of Physics, Simon Fraser University,
Burnaby, BC, Canada, V5A 1S6

Abstract. We consider a curvature model for bilayer vesicles with an area-difference elasticity or non-local bending-energy term. Such a model interpolates between the bilayer-couple and spontaneous-curvature models. We report preliminary results for the budding transition. The shape transformation between the dumbbell and the pear phases can be continuous or discontinuous depending on the ratio of the non-local to the local bending rigidities.

1. The model

Vesicles exhibit an amazing variety of different shapes. Transformations between these shapes can be induced by changing external parameters such as temperature or osmotic conditions. Our understanding of these phenomena is largely based on the analysis of simple curvature models [1]. In these models the vesicle is described as a two-dimensional surface whose shape is determined by minimizing an appropriate curvature energy under constraints on the enclosed volume V and the total area $A \equiv 4\pi R^2$, where R defines a length scale. Two such models have been studied so far in some detail: (i) In the bilayer-couple or area-difference (ΔA-) model [2,3], the curvature energy is given by

$$G = (\kappa/2) \oint dA (C_1 + C_2)^2, \tag{1}$$

where C_1 and C_2 are the two principal curvatures and κ is the local bending rigidity. In addition, the shape is subject to a constraint on the area difference $\Delta A \approx D \oint dA (C_1 + C_2)$ between the inner and outer monolayers, which have a separation D. (ii) In the spontaneous-curvature (SC-) model [4], the latter constraint is dropped but in the curvature energy (1) $(C_1 + C_2)$ is replaced by $(C_1+C_2-C_0)$, where C_0 is a spontaneous curvature. The two models are related by a Legendre transformation and have, therefore, the same stationary shapes but rather different phase diagrams, which have recently been investigated systematically [5,6]. At present, experiments [7-9] favour the bilayer-couple approach; but, there are still discrepancies between experiment and theory, which are critically reviewed in Refs. 8-10.

Both models can be considered as special cases of a more general curvature model which reads,

$$F = (\kappa/2) \oint dA (C_1 + C_2 - C_0)^2 + (\kappa' \pi / 2AD^2)(\Delta A - \Delta A_0)^2. \tag{2}$$

The second term formally attributes an elastic energy to deviations of the area difference ΔA from an unstretched equilibrium value ΔA_0. This term has been derived long ago from the bending of two unconnected monolayers, which leads to this "non-local bending resistence" [11], compare also Refs. 12 and 13. Using elastic models for the two monolayers, these derivations show that the non-local bending rigidity κ' has the same order of magnitude as κ. Therefore, (2) might be considered more realistic than the ΔA- or SC-models, which are recovered for $\kappa' = \infty$ or $\kappa' = 0$, respectively. Minimization of (2) for fixed A and V leads, again, to the same stationary shapes as the two original models. Indeed, there is a simple mapping between the ΔA-model and the model (2), which is most conveniently expressed in dimensionless variables, $m \equiv \Delta A/(2RD), m_0 \equiv \Delta A_0/(2RD), c_0 \equiv C_0 R$, and the dimensionless ratio of the two bending rigidities $\alpha \equiv \kappa'/\kappa$: Every stationary shape in the ΔA-model with reduced area difference m and energy $G(m)$ is also a stationary shape in model (2) for

$$m_0 = m + (G'/\kappa - 2c_0)/\alpha, \tag{3}$$

where $G' \equiv \partial G/\partial m$ denotes the derivative of the curvature energy (1) in the ΔA-model along the branch of stationary shapes at fixed A and V. This condition follows from the variation of (2) with respect to ΔA. Thus, the knowledge of the energy G as a function of m in the ΔA-model allows, in principle, a derivation of the phase diagram for (2), which depends on the four variables, α, c_0, m_0, and the reduced volume $v \equiv V/(4\pi R^3/3)$.

2. The budding transition

In this contribution, we report preliminary results [14] for the part of the phase diagram of model (2) where budding occurs. This shape transformation involves up/down symmetric prolate shapes or "dumbbells", up/down asymmetric "pears" and fully vesiculated shapes with necks of microscopic diameter. Experimentally, the symmetry-breaking transition has been reported both as a continuous transition [7] and as a discontinuous transition from prolates to vesiculated shapes without pear-shaped intermediates [9]. In addition, even if the symmetry breaking is continuous and leads to stable pears, there is some evidence for a discontinuous shrinkage of the neck and irreversibility (hysteresis) in the final vesiculation process [8,9].

For the model (2), the relevant part of the three-dimensional phase diagram for a typical (fixed) reduced volume ($v = 0.8$) is shown in Fig. 1. This phase diagram is obtained by using the mapping (3) and the energy $G(m)$ of the dumbbells and pears, as displayed in Fig. 4 of Ref. 6 (with $m = 4\pi\Delta a$). For $\alpha = \infty$, which corresponds to the ΔA-model, the transition or bifurcation between the dumbbells and the pears is continuous at $m_0 = m_C$, independent of c_0. With decreasing α, this transition remains continuous and is given by $m_0 = m_C + (G'_C/\kappa - 2c_0)/\alpha$, where the subscript C refers to the bifurcation point in the ΔA-model. The local topology of the bifurcation is preserved as long as the mapping (3) is monotonic. On further decreasing α, it becomes

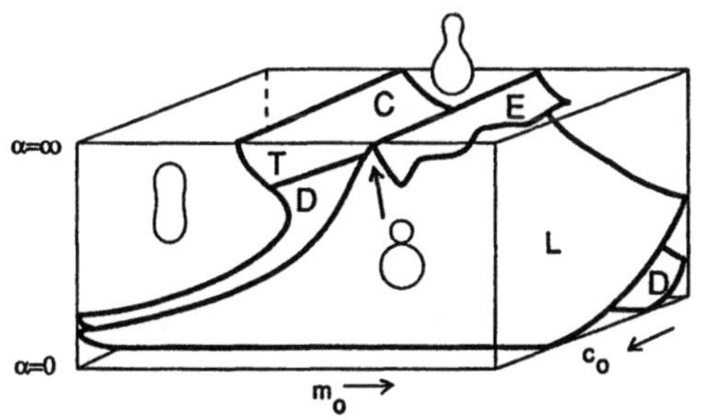

Fig. 1: Schematic phase diagram for model (2) at fixed reduced volume $v = 0.8$. The sheets C and D denote continuous and discontinuous transitions between the dumbbells and the pears. These sheets are separated by the tricritical line T. The sheet L marks the continuous transition to the vesiculated shape consisting of two spheres. For large $\alpha < \infty$ these shapes have a finite region of existence bounded by the sheet E, which denotes the transition to a sphere-plus-ellipse shape.

non-monotonic (first at the bifurcation point) if $\alpha < \alpha_T \equiv -G''_C/\kappa$, where $G''_C < 0$ denotes the second derivative of the energy G along the pear-shaped branch at the bifurcation point. This leads to a wing [6] in the $F(m_0)$-curve and, thus, to a discontinuous transition between the dumbbells and the pears for $\alpha < \alpha_T$. Therefore, $\alpha = \alpha_T (\simeq 2.5$ for $v = 0.8)$ corresponds to a tricritical line independent of c_0 (which becomes a tricritical surface $\alpha_T(v)$ in the full four-dimensional phase diagram). The smaller α the more pronounced becomes the first-order transition. At $\alpha = 0$, one recovers the discontinous transition of the SC-model independent of m_0.

Consider now the transition from the pear to the limiting (L), fully vesiculated shape, which consists of two spheres sitting on top of each other connected by a narrow neck. In the ΔA-model the pears approach this limit continuously as m_0 is increased. Due to its geometrical constraints, such a shape requires a particular $m_0 = m_L$ in the ΔA-model. For $m_0 > m_L$, the fully vesiculated shapes consist of an ellipse and a sphere connected by an infinitesimal neck. For $\alpha < \infty$ the limit shape with the two spheres exists over a finite range of m_0-values, the width of which scales like $\sim 1/\alpha$ (compare Fig. 1). However, with decreasing α additional first-order transitions to vesiculated shapes with more necks or consisting of more complicated segments (such as a dumbbell on top of a sphere) become relevant.

In conclusion, the order of the budding transition and the sequence of vesiculated shapes depends crucially on the ratio of the two bending rigidities. In order to correlate the theoretical results with the various experimental findings beyond mere fitting, an independent measurement of the non-local bending rigidity is required.

Acknowledgments. We thank E. Evans, W. Helfrich, R. Lipowsky and E. Sackmann for informative discussions. This work was funded by the National Science and Engineering Research Council of Canada.

References

[1] For a general review and a comprehensive list of references, see R. Lipowsky, *Nature* **349**, 475, 1991.
[2] E. Evans, *Biophys. J.* **30**, 265, 1980.
[3] S. Svetina, B. Zeks, *Biomed. Biochim. Acta* **42**, 86, 1983.
[4] W. Helfrich, *Z. Naturforsch.* **28c**, 693, 1973.
[5] L. Miao, B. Fourcade, M. Rao, M. Wortis, and R.K.P. Zia, *Phys. Rev. A* **43**, 6843, 1991.
[6] U. Seifert, K. Berndl, and R. Lipowsky, *Phys. Rev. A* **44**, 1182, 1991.
[7] K. Berndl, J. Käs, R. Lipowsky, E. Sackmann, and U. Seifert, *Europhys. Lett.* **13**, 659, 1990.
[8] H.G. Döbereiner, W. Rawicz, M. Wortis, and E. Evans, preprint.
[9] J. Käs and E. Sackmann, *Biophys. J.*, to be published.
[10] M. Wortis, U. Seifert, K. Berndl, B. Fourcade, L. Miao, M. Rao, and R.K.P. Zia, in *Proceedings of the workshop on "Dynamical phenomena at interfaces, surfaces and bilayers"*, edited by D. Beysens, N. Boccara and G. Forgacs.
[11] E. Evans, *Biophys. J.* **14**, 923, 1974.
[12] W. Helfrich, *Z. Naturforsch.* **29c**, 510, 1974.
[13] S. Svetina, M. Brumen, and B. Zeks, *stud. biophysica* **110**, 177, 1985.
[14] H.G. Döbereiner, L. Miao, U. Seifert, and M. Wortis, to be published.

Some Remarks on the Shape of Toroidal Vesicles

R. Mosseri[1], *J.F. Sadoc*[2], *and J. Charvolin*[3]

[1]L.P.S.B., CNRS, F-92195 Meudon Cedex, France
[2]Laboratoire de Physique des Solides, Bât. 510, Université Paris-Sud, F-91405 Orsay, France
[3]Institut Laue–Langevin, 156 X, F-38042 Grenoble Cedex, France

The shape of toroidal vesicles is considered in the light of the conformal invariance of the bending energy in R^3. Viewed from the spherical space S^3, this symmetry amounts to a rotation. Upon a stereographic map onto R^3, the toroidal shapes conformally equivalent to the Clifford torus are parametrized by a single angle. Explicite coordinates are given. Furthermore, it is shown that these toroidal vesicles should exhibit pulsations near the Clifford torus, in analogy with a 2 dimensional oscillator.

1 - Introduction

Vesicles with toroidal topology are predicted to be stable for standard free energy models [1][2], and have been observed in partially polymerized phospholipid membranes [3]. Furthermore, calculations and experiments agree on the specific toroidal shape, the so-called "Clifford torus", an axial-symmetric torus such that the ratio of its generating circles is $\sqrt{2}$. From the theoretical side, this results from the consideration of the bending energy F_b which is proportional to the so-called "Willmore functional" T [4][5]:

$$T = \int H^2 dA \tag{1}$$

where H is the mean curvature. A second term in the energy comes from an integration over the gaussian curvature which only depends upon the surface genus, and is therefore constant here. Willmore has conjectured that $T \geq 2\pi^2$ for any toroidal surface in R^3, the equality holding for the Clifford torus. A second important point is that the Willmore functional is invariant under a conformal transformation (CT) in R^3. This property was recalled by Duplantier [6] in the context of vesicle geometry and extensively used by Seifert [2] in deriving stability phase diagrams. Both authors note that this CT invariance imply fluctuations in the ground state configuration.

In the present paper we show the following points:

- By lifting the vesicles into the spherical space S^3, embedded in R^4, the study of CT invariance is largely simplified. The set of toroidal shapes conformally equivalent, in R^3, to the Clifford torus is parametrized by a single angle α in the range $[0, \pi/2]$.

- If an area(or volume) dependant term is added in the free energy, one expects configurational oscillations around the Clifford torus. For small enough α, it is analogous to a 2 dimensional oscillator.

In the course of this work, we shall see that the Clifford torus is a particular stereographic projection of the "spherical torus" in S^3. This is very interesting in the context of the curved space approach to frustration in amphiphilic bilayers. Two of us (J.F.S. and J.C. [7]) have stressed the central role played by the spherical torus as a frustration free ideal membrane. From this point of view, the Willmore functional gives indications for choosing the best mapping from the ideal surfaces in curved space to the real, euclidean one. For instance, as one is interested in the way the bilayer surfaces can increase their genus (even up to infinite genus configurations), the following theorem might prove useful [4]: there exist surfaces of arbitrary genus in R^3 which are extrema of the Willmore functional; these surfaces are the stereographic projection of minimal surfaces in S^3.

2 - The torus foliation of S^3 and the Hopf map

Points in the unit radius S^3 have coordinates x_i in a toroidal system [8]:

$$\begin{array}{lll} x_1 = \cos\theta \sin\phi & x_2 = \sin\theta \sin\phi & \omega, \theta \in [0, 2\pi[\\ x_3 = \cos\omega \cos\phi & x_4 = \sin\omega \cos\phi & \phi \in [0, \frac{\pi}{2}] \end{array} \tag{2}$$

The angle ϕ parametrizes a continuous foliation of torus with two exceptional circles ($\phi = 0, \frac{\pi}{2}$) as axes, three torus being displayed in figure 1 under a stereographic projection in R^3. Here the pole of projection is on one of the two exceptional circles. Consider the following Hopf map:

$$\begin{array}{l} \psi : S_3 \longrightarrow S_2 \\ (\theta, \phi, \omega) \longrightarrow (\alpha, \eta) \\ \qquad \alpha = 2\phi, \quad \eta = \theta - \omega \quad \lfloor 2\pi \rfloor, \qquad \alpha \in [0, \pi], \quad \eta \in [0, 2\pi[\end{array} \tag{3}$$

where α and η are standard polar coordinates on S^2 (figure 1-b). This map is such that the inverse image of a point on S^2 is a great circle on S^3. The full pre-image of S^2 is a fibration of S^3 by great circles, called the Hopf fibration; S^2 is the "base" of the fibration. Each torus of the above foliation maps onto a "parallel" circle (constant α), the spherical torus ($\phi = \frac{\pi}{4}$) being mapped onto the equator.

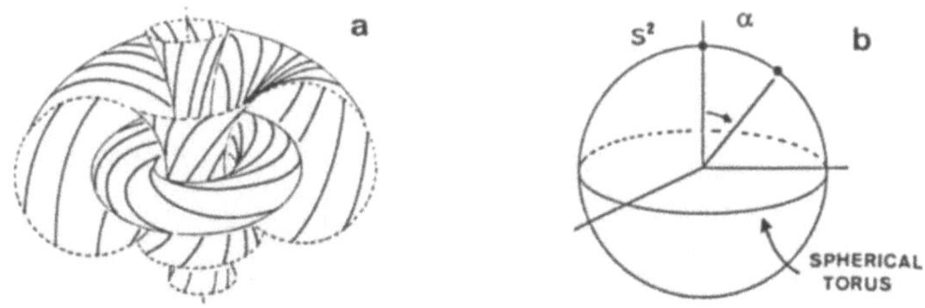

Figure 1: a) A limited view of the torus foliation under stereographic map. Circles of the Hopf fibration are drawn. b) The base S^2 of the Hopf fibration. The equator is the image of the spherical torus.

3 - Stereographic map and conformal transformations

Now, consider the stereographic map of the spherical torus. By symmetry, all points of the same Hopf fibre, when used as a pole for this projection, gives the same torus in R^3. Furthermore, the different shapes only depends on the distance, in the base space S^2, between the image of the spherical torus (the equator) and the image of the pole. Hence a given torus of the foliation is the locus of points in S^3 which, as projection pole for the spherical torus, gives the same toroidal shape in R^3. The stereographic map, which is also a CT, is such that othogonal transformations in S^3 corresponds to inversions in R^3. Therefore, as far as the shape is concerned, changing the pole in S^3 will produce all the conformally equivalent objects in R^3. When the projection pole is on one the two exceptional circles (north or south pole in the base), the spherical torus maps onto the Clifford torus in R^3. As far as the shapes of the torus in R^3 are concerned (modulo rotations and dilatations), they are parametrized by the angle α in the range $[0, \frac{\pi}{2}]$. The calculation gives the following coordinates (X, Y, Z) for the torus in R^3 (recall $\alpha = 2\phi$):

$$X = \frac{\sin\phi\cos\omega - \cos\phi\cos\theta}{\sqrt{2} - W} \quad Y = \frac{\sin\phi\sin\omega + \cos\phi\sin\theta}{\sqrt{2} - W}$$

$$Z = \frac{\sin\phi\sin\theta - \cos\phi\sin\omega}{\sqrt{2} - W} \qquad \text{with } W = \sin\phi\cos\theta - \cos\phi\cos\omega \tag{4}$$

For $\alpha = 0, \pi$, the Clifford torus (as given in [1]) is recovered. The figure 2 shows three CT equivalent torus in R^3.

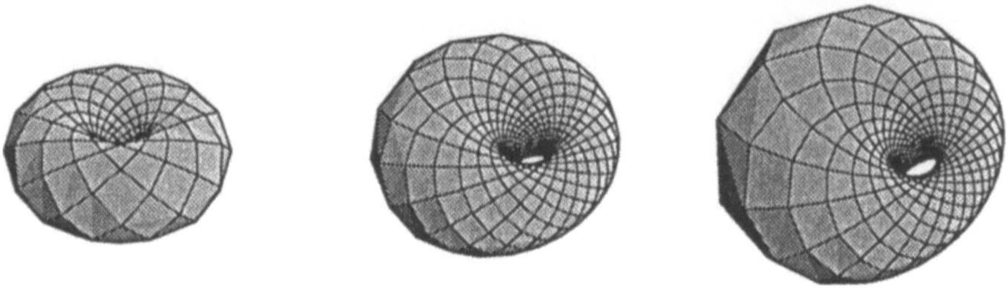

Figure 2: Some stereographic maps in R^3 of the spherical torus when the projection pole is suitably moved in S^3.

4 - Shape oscillations

Suppose that a term function of the area, or even of the volume, is added to the energy. Since the area is minimum for the mapped Clifford torus ($\alpha = 0$), it is the equilibrium configuration (for some range of the volume to surface ratio). If a small conformal deformation is imposed, a force will oppose this deformation. At a first order (α small enough), the force is linear with the deformation, like in the Hooke law, and the physics is analogous to that of the harmonic oscillator. One expects a mode of shape pulsation, whose frequency is directly linked to the physical parameters. The symmetry of the problem is best seen by locating the projection pole onto the base space S^2. Once α is given, the angle η remains free, which shows that it is indeed like a 2-dimensional oscillator. Upon increasing α, anharmonic effects are expected. Whether these modes can be experimentally measured is an open question. Note finally that we have only considered oscillations in a continuous set of CT equivalent torus; we expect that they are the first to be excited for small deformations.

Acknowledgements : it is a pleasure to thank E. Dubois-Violette, D.Bensimon, M.Mutz and B.Fourcade for valuable discussions.

References

[1] Ou-Yang Zhong-can,Phys. Rev A **41** (1990) p.4517
[2] U. Seifert, Phys. Rev. Lett. **66** (1991) p.2404
[3] M. Mutz, D. Bensimon, Phys. Rev A **43** (1991) p.4525
[4] T.J. Willmore, "Total Curvature in Riemannian Geometry" (Ellis Horwood Ltd, Chicester 1982)
[5] U. Pinkall, I. Sterling,Math.Intelligencer **9**(1987)p.38
[6] B. Duplantier, Physica **168A** (1990) p.179
[7] J.F.Sadoc, J.Charvolin , J. de Physique **47** (1986) p.683
[8] "Geometry in Condensed Matter Physics", ed. by J.F.Sadoc, World Scientific 1990

The Effect of Membrane Elasticity on Shapes of Nearly Spherical Phospholipid Vesicles

F. Sevšek, S. Svetina, and B. Žekš

Institute of Biophysics, Medical Faculty, Lipičeva 2,
61105 Ljubljana, Slovenia and
J. Stefan Institute, Jamova 39, 61000 Ljubljana, Slovenia

Abstract. The equilibrium shapes of nearly spherical phospholipid vesicles were determined taking into account finite elasticity of the membrane and of the enclosed liquid. Using spherical harmonics expansion and the second order approximation, the equilibrium shape of a vesicle was found to be either spherical or quadrupolarly distorted, depending on the volume of the vesicle and the areas of the two leaflets of the membrane in the unstretched state.

1. Introduction

The analysis of shapes of phospholipid vesicles and their thermal fluctuations may serve as a means to determine the elastic properties of the membrane. Usually, the mean square amplitudes of normal-mode displacements are determined as a function of the wave number from the Fourier analysis of the shapes of a fluctuating vesicle [1]. Taking into account only the bending energy of a nearly spherical membrane simplifies the theoretical treatment, as the spherical harmonic modes are then not coupled, but the area and the volume of the vesicle are not preserved. Since this approximation was often found not to be sufficient to describe the experimental data a tension parameter is introduced in the data analysis [2]. Considering the shape fluctuations at constant volume and area of a vesicle is quite complicated as the modes are coupled. Anyhow, it was found that for nearly spherical vesicles in this case the equilibrium shapes were the quadrupolarly distorted ones with $L = 2$ [3]. Our aim was thus to determine equilibrium shapes of the vesicle, taking into account the finite elasticity of the vesicle interior and its enclosing membrane, and to investigate the energy surface near the minima as a function of the parameters of the vesicle.

2. Elastic Energy of a Vesicle

The energy of a giant phospholipid vesicle, as described by the continuum elasticity theory for the membrane, consists of several contributions: the energy due to isotropic lateral extension of the internal and the external monolayer constituting the membrane, the energy related to the compressibility of the vesicle content and pure bending energy of the bilayer. For a closed symmetric bilayer composed of two layers it was shown by Svetina et al. [4] that the energy of the isotropic lateral extension could be expressed as a function of the area of the neutral layer of the bilayer (A) and of the area difference between the areas of the external and the internal leaflet constituting the membrane (ΔA).

It is convenient to write the variable part of the elastic energy of a nearly spherical vesicle in terms of dimensionless variables for the area (a), volume (v) and the area difference (Δa) relative to a reference sphere with the radius R_s. The energy normalized to the bending energy of a sphere ($8\pi k_c$) is:

$$w = w_b + \kappa_a \frac{1}{2a_0}(a - a_0)^2 + \kappa_v \frac{1}{2v_0}(v - v_0)^2 + \kappa_\Delta \frac{1}{a_0}(\Delta a - \Delta a_0)^2, \tag{1}$$

where w_B is the relative bending energy of the bilayer and a_0, v_0 and Δa_0 the parameters of the unstressed vesicle. The normalized elastic constants can be estimated for a vesicle with the radius $R_s = 20\mu m$ by taking for the bending rigidity of the phospholipid bilayer the value $k_c = 10^{-19}J$, and assuming $K = 0.15 Jm^{-2}$ [5], $K_V = 2 \times 10^{-19} Jm^{-3}$ and $k_r = 4 \times 10^{-19}J$ [6] for the membrane expansivity, the volume compressibility and the non-local bending moduli, respectively. This yields: $\kappa_a \approx 3 \times 10^8$, $\kappa_v \approx 3 \times 10^{13}$ and $\kappa_\Delta \approx 4$.

As this treatment is limited to nearly spherical vesicles their shapes relative to a chosen sphere will be expressed in terms of spherical harmonics. Considering only the terms up to the second order in all the expansion coefficients (u_l), the variables are:

$$w_b - 1 = \frac{1}{16\pi} \sum_{l=0}^{l_{max}} l(l+1)[l(l+1) - 2]u_l^2, \qquad \Delta a - 1 = \frac{u_0}{\sqrt{4\pi}} + \frac{1}{8\pi} \sum_{l=0}^{l_{max}} l(l+1)u_l^2,$$

$$a - 1 = \frac{2u_0}{\sqrt{4\pi}} + \frac{1}{8\pi} \sum_{l=0}^{l_{max}} [l(l+1) + 2]u_l^2, \qquad v - 1 = \frac{3u_0}{\sqrt{4\pi}} + \frac{6}{8\pi} \sum_{l=0}^{l_{max}} u_l^2. \tag{2}$$

We see that in this approximation a, v and Δa are not independent so that for nearly spherical vesicles the two leaflets of the membrane are in general both stretched. Namely, from eq.(2) the following relation is obtained: $(\Delta a - 1) = (a - 1) - (v - 1)/3$, as also noted by Seifert et al. [7].

3. The Equilibrium Shapes

In thermal equilibrium, the vesicle should attain the shape that corresponds to the minimum of the energy in eq.(1). Using the eq.(2), the conditions for the extremal energy: $(dw/du_0) = 0$ and $(dw/du_{l\neq 0}) = 0$ are calculated. Introducing the effective tension parameter (Λ):

$$\Lambda \equiv -\frac{2}{a_0}\left[\kappa_a(a - a_0) + \kappa_\Delta(\Delta a - \Delta a_0)\right], \tag{3}$$

the condition for the extremal energy can be obtained as:

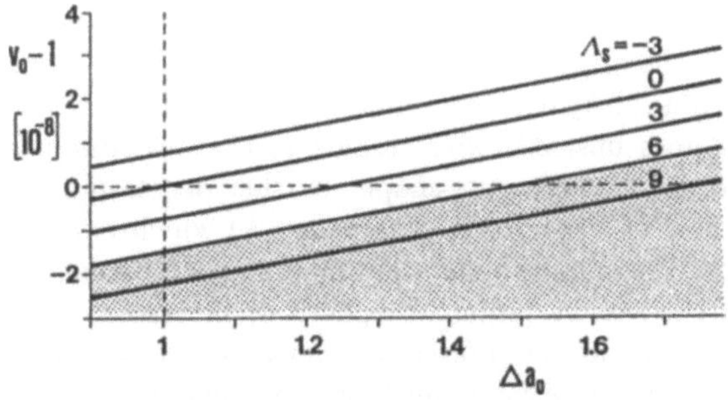

Figure 1: The line $\Lambda_s = 6$ represents the limit of the stability of spherical shapes depending on v_0 and Δa_0 which are normalized so that $a_0 = 1$. The vesicles are spherical above that line and quadrupolarly distorted below it. The tension parameters Λ_s are calculated as if vesicles were spherical.

$$\frac{1}{8\pi}[l(l+1)-2][l(l+1)-\Lambda]u_l = 0. \tag{4}$$

It can be fulfilled for $u_l = 0$ or for $l(l+1) = \Lambda$. The latter may be, of course, the case for only one value of $l = L$. Thus, besides u_0, only one nonzero coefficient u_L may appear in the expression for the equilibrium shape.

To determine which of the solutions are stable the second derivatives of the energy of the vesicle in the equilibrium state are calculated. As $(d^2w/du_l^2) = [l(l+1)-2][l(l+1)-\Lambda]/8\pi$ for $l \neq 0, L$, we see that if there exists one $u_L \neq 0$ it must be the one with the lowest value of L i.e. $L = 2$. The equilibrium shapes of nearly spherical vesicles are in this approximation either spherical, where all the coefficients $u_{l>0}$ are zero, or quadrupolarly distorted ones with $\Lambda = 6$ and only $u_0, u_2 \neq 0$.

It is interesting to investigate the case when the spherical shape becomes unstable with respect to the deformations with $L = 2$. This happens when for a spherical vesicle the value of Λ becomes equal to 6. On the other hand, this instability can be also approached from nonspherical shapes requiring that the previously non-zero coefficient u_2 vanishes.

For convenience the reference sphere is chosen so that it has the same unstretched membrane area (a_0) as the vesicle. Then $a_0 = 1$ and the points of the instability of spherical shapes may be represented by a line in the $v_0/\Delta a_0$ diagram. For the spherical vesicles the value of the tension parameter Λ_s can be calculated from the condition $(dw/du_0) = 0$ and eq.(3). The bifurcation line is then obtained by requiring that $\Lambda_s = 6$. It is shown in fig.(1) together with lines for some other values of the tension parameter Λ_s. For the vesicles where the elastic constants and the parameters of the unstressed state are such that they are situated above the instability line ($\Lambda_s < 6$) in fig.(1) the equilibrium shape is spherical, whereas the ones below that line (shaded part of fig.(1)) are quadrupolarly distorted.

As an example we can consider, to start with, spherical vesicle without any membrane stress. If the volume of such a vesicle is decreased, the value of the spherical tension parameter Λ_s, which was at the beginning zero, increases. In fig.(1) this means moving from the origin of the coordinate system down along the ordinate axis. When the value of Λ_s becomes equal to 6, the instability line is reached and the shape becomes unstable with respect to the deformations with $L = 2$ resulting in the vesicle changing its shape. Further decreasing the volume does not affect the value of the tension parameter (Λ) which remains equal to 6, but only increases the quadrupolar deformation (u_2). Anyhow, Λ_s which is calculated as if the vesicle remained spherical, increases. The value of the volume parameter (v_0) at the bifurcation can be estimated from eq.(3). Using the above values for the elastic constants one gets that: $\Lambda_s \approx -\frac{4}{3}(v_0 - 1)\kappa_a$, when: $a_0 = \Delta a_0 = 1$.

The bending rigidity for a phospholipid bilayer is thus so small that for a spherical vesicle the decrease of its relative volume for about 1.5×10^{-9} already changes the equilibrium shape from the spheric to the quadrupolarly deformed one. On the other side, quite large increase of Δa_0 is possible for an unstressed spherical vesicle before the bifurcation line is reached.

4. References

[1] I.Bivas, P.Hanusse, P.Bothorel, J.Lalanne, O.Aguerre-Chariol: *J.Phys. France* **48**, 855 (1987)

[2] H.P.Duwe, E.Sackmann: *Physica A* **163**, 410 (1990)

[3] M.A.Peterson: *Phys.Rev.A* **39**, 2643 (1989)

[4] S.Svetina, M.Brumen, B. Žekš: *Studia biophys.* **110**, 177 (1985)

[5] E.Evans, D.Needham: *J.Phys.Chem.* **91**, 4219 (1987)

[6] S.Svetina, B.Božič, J.Song, R.E.Waugh, B.Žekš: this issue

[7] U.Seifert, K.Berndl, R.Lipowsky: *Phys.Rev.A* **44**, 1182 (1991)

Electron Microscopy of Biological Model Membranes

B. Klösgen and W. Helfrich

Freie Universität Berlin, Fachbereich Physik,
Arnimallee 14, W-1000 Berlin 33, Fed.Rep.of Germany

Abstract. Cryo-electron microscopy was used to study the appearance of vesicles of phosphatidylcholine. From the micrographs we concluded that lateral tension was produced during the freezing process. Membrane undulations were flattened and also the tension forced the vesicles to change their shapes. However, proper preparation techniques led to the finding of vesicles with grainy membrane surfaces or with pronounced edges or bulges. This supported the idea that unstressed membranes may tend to develop an optically unresolvable superstructure capable of consuming large amounts of membrane area.

1. Introduction

Fluid model membranes of pure lipids are not as smooth as they are often assumed to be. Because of their bending stiffness k_c of about 10^{-19}J - 10^{-20} J [1], [2], [3], [4] they exhibit thermally excited membrane fluctuations [5], [6] the amplitudes of which reach up to 10% of the vesicle radius. They result in a steric repulsion between adjacent membranes. In membranes of electrically neutral lipids like phosphatidylcholines (at pH ≈ 7.0) this undulation interaction is the only repulsive component of the total interaction potential. It may thus result, if it overcomes the van der Waals attraction in strength, in the separation of two otherwise adhering membranes. Lateral tension decreases the amplitudes of the fluctuations. Thus it may lower the undulation repulsion and induce the adhesion of formerly separated membranes [6], [7]. Both the membrane tension σ and the contact angle φ could be evaluated from experiments. Thus the adhesion energy g_a could be calculated from a Young-equation. Another value for g_a could be appreciated by a crude application of the undulation theory and came out to be too low by about a factor of 100 [6]. No experimental errors could account for this discrepancy. Also, the consideration of the area stored in the undulations could not resolve the problem. This and other experimental results [7], [8], [9] resulted in to the postulation of an extra membrane roughness of unstressed membranes that is not resolvable with light microscopy [10]. A model to depict such a superstructure has already been formulated [11]. Now we present results from a series of experiments where we combined a careful way of vesicle preparation to avoid destructive tensions with the application of transmission electron microscopy in order to find further hints that could substantiate the existence of such a superstructure.

2. Methods

The lipid vesicles to be investigated were prepared of electrically neutral egg-phosphatidylcholine starting from a ca. 1% (wt/wt) suspension of lipid in pure water. Different ways of handling resulted in populations of different shapes and sizes (see below) [12].

The vesicle suspension was then applied onto a copper grid to form a thin lamella of aqueous medium between the bars of the grid [13], [14]. By use of a plunger device it was rapidly frozen in propane at the temperature of liquid nitrogen (76 K) within less than 5 ms [14], [15]. This cryo-fixation resulted in lipid vesicles that are embedded in vitrified water. Then the sample was transferred into a precooled electron microscope and directly investigated at 100 kV in the transmission mode. Phase contrast was achieved by defocussing the electron beam to about - 1μm [16].

3. Results

All samples studied were made from similar basic suspensions and differed only in the way how the vesicles were formed and the subsequent incubation conditions. Some of the dispersions were prepared in 10 mM NaCl solution; they did not differ in appearance from the other samples.

3.1 Spherical vesicles

The simplest method to produce vesicles consists in the sonication (5 min, 308 K, interrupt mode) of the basic suspension. It results in vesicle populations with spherical groundshape like the one shown in fig. 1. On forming the lamella on the grid before freezing the vesicles have been sorted by size, the larger ones being expelled from the thinner regions near the center of the lamella. All vesicle contours look rather smooth, no membrane fluctuations are visible. However, even at the expense of a deformation the vesicles do not touch each other but seem to avoid contact. In all micrographs of vesicles frozen from the fluid state we found this behaviour. We interpret it as a trace of the undulation repulsion that was effective when the membranes were fluid. During the cooling process the membranes contract and the water expands. Both results in a relative deficiency of membrane area as compared to the enclosed volume. Thus, a lateral membrane tension is produced by freezing. The fluctuations of the spherical vesicles are completely flattened. Only sufficiently small non-spherical vesicles can avoid to build up high lateral tensions by changing their shape. The small tubular vesicle of fig. 2 has presumably become shorter and thicker during freezing, but the overall shape was conserved and the lateral tension was

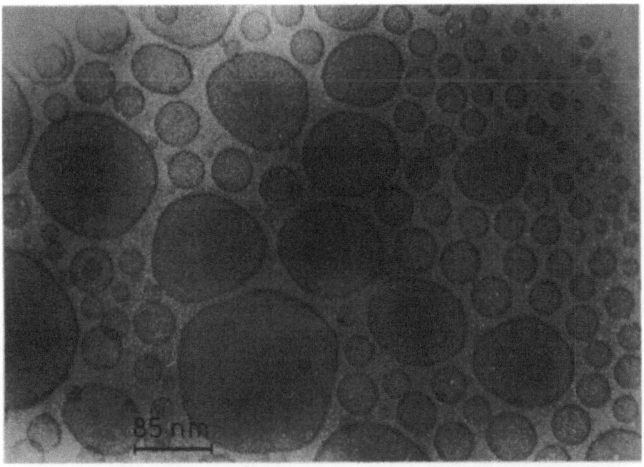

Fig. 1: Spherical vesicles of egg-phosphatidylcholine (EYPC) in pure water. Defocus: Δ= - 1.0 µm. Lamella thickness: d ≈ 78 nm.

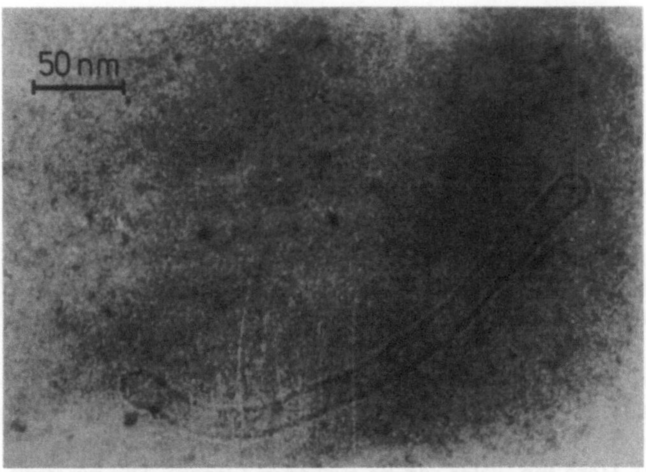

Fig. 2: Small tubular EYPC vesicle 10 mM NaCl solution, still exhibiting fluctuation characteristics in the frozen state. Δf = 1.0 µm; d ≈ 63 nm.

kept small. This can be concluded from the distinct bending fluctuation that has been captured in the frozen state and that is a direct proof of the fluctuations that were effective in the fluid state.

3.2 Non-spherical vesicles

Vesicles like the tubular one are produced by very gentle rotation of the basic suspension by hand or in a rotatory shaker (200 cycles/min, 8 h). The

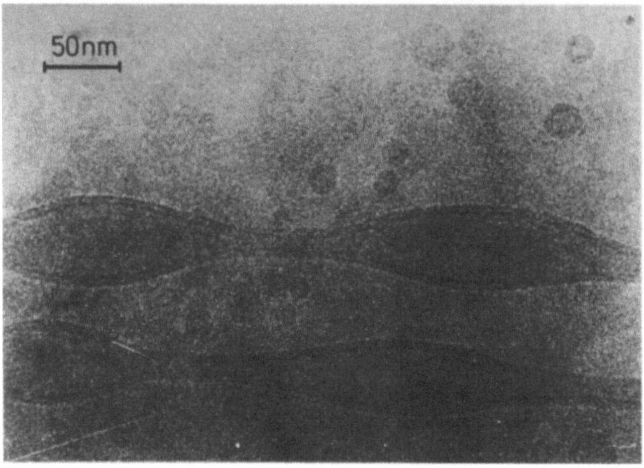

Fig. 3: Pair of tubular EYPC vesicles in pure water with a tension induced spindle deformation. $\Delta f = 1.0\ \mu m$; $d \approx 46$ nm.

obtained vesicle suspensions contain as well spherical as tubular vesicles of varying size. Small tubes as the one in fig. 2 are rare; most tubes are very extended as those in fig. 3. This almost parallel pair of tubes extends over the whole micrograph. The lipid double membrane is very well resolved. Also of interest is the spindle-like shape which we found very often in extended tubes. These long tubes cannot conserve their cylindrical shape on cooling as did the little one of fig. 3, but they can respond to the membrane tension by undergoing a local shape transformation.

Among other vesicles of different sizes and shapes there are two pairs of stacked vesicles in fig. 4. Again, the contours look very smooth; the membranes of each pair seem to run almost parallel to each other. This is again attributed to the undulation repulsion. One of the tubular pairs exhibits the spindle modulation of the cylindrical form. But the other has essentially conserved its former shape. The inner tube shows several sites of membrane rupture (see arrows). Here the water can be exchanged freely. The pressure produced inside by freezing can thus be reduced very fast and no shape transformation is necessary in order to avoid a high membrane tension.

3.3 Irregularly formed vesicles

In rare cases we obtained slightly elliptical vesicles with a somehow grainy surface like the one of fig. 5 amongst other vesicles looking quite normal. Maybe this is an example for a loose arrangement of the saddles that were proposed as the basic element of the membrane superstructure [11].

We thus were encouraged to try another modification of our preparation in order to produce more vesicles with high surface and low enclosed volume, a

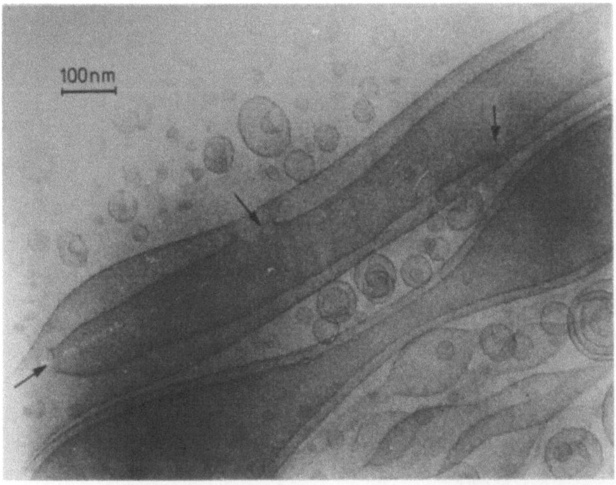

Fig. 4: Mixed population of EYPC vesicles in 10 mM NaCl. Tubular vesicles show either rupture sites see (arrows) or spindle deformations because of lateral tension. $\Delta f = -2.8\ \mu m$; $d \approx 260$ nm.

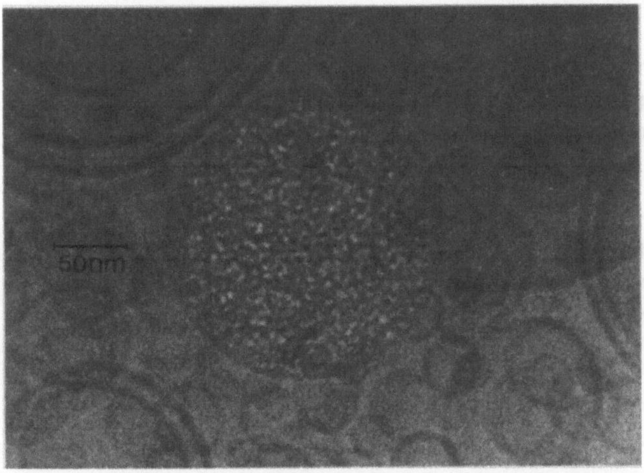

Fig. 5: EYPC vesicle with a grainy membrane surface in a population of others with smooth membranes. $\Delta f = 0.8\ \mu m$; $d \approx 195$ nm.

condition which should lead to very low tension and thus favour the buildup of the presumed membrane roughness. These vesicles were prepared like the ones described in the previous chapter but then they were incubated for about two weeks at 281 K in a refrigerator. After 2 h of reheating at room temperature they were frozen and observed. These vesicles (like the one in fig. 6) differ very much from all the others because of their very irregular shapes. Again, the small undulation modes are flattened by freezing, but the relatively constant distance of adjacent membranes strongly hints at the

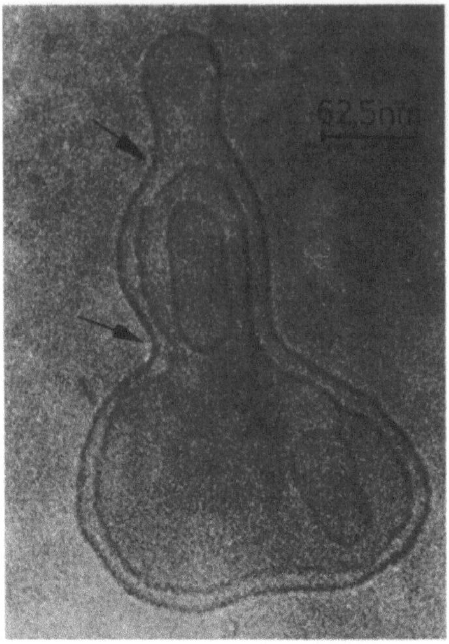

Fig. 6: Strange EYPC vesicle in 10 mM NaCl, incubated at 281 K for 12 d. The extra ordinary bumpiness (see arrows, e.g.) may indicate a novel membrane roughness. $\Delta f = -1.4\ \mu m$; $d \approx 165$ nm.

action of repulsive forces. The contours themselves cannot be explained by undulations or be attributed to spontaneous curvature. We take them as a product of the cooperative interaction [11] of the saddle structure thus leading to a pronounced warping of the membrane which consumes a considerable amount of area.

4. Discussion

The membrane tension produced by freezing may be crudely estimated from the relative area change by use of [5]

$$\frac{\Delta A}{A} = \frac{\sigma}{\lambda} + \frac{k_B T}{8\pi k_c}\left[\ln\frac{A}{a^2} - \ln\frac{\pi^2/a^2+\sigma/k_c}{\pi^2/A+\sigma/k_c}\right]$$

Here σ is the membrane tension, λ the stretching modulus, k_B the Boltzmann constant, T the temperature, k_c the bending elasticity, a a molecular length and A the membrane area. The density of vitrified water is 0.94 g cm^{-3} and the thermal expansivity modul for egg-phosphatidylcholine is $2.4 \cdot 10^{-3}\ K^{-1}$ [17]. Thus, the relative area decrease of the vesicle membranes produced by freezing is about 17%. The 6% of area maximally stored in the undulations are not sufficient to compensate for this. With a stretching modulus of $\lambda = 140$ mN/m [18] the tension induced by Hookean expansion is estimated to be ca. 16mN/m. This would be higher than the limiting tension of ca. 3 mN/m [18] of membrane rupture. In any case, small wavelength fluctuation modes are completely suppressed, which leads to the formation of smooth vesicle contours. Long wavelength deformations can in principle persist provided the system is free to undergo a form conserving size adjustment without building up high tensions (see fig. 2 and fig. 6). With the very short freezing times extended systems can only rupture to allow for free water exchange or transform into another shape. The time to form a spindle can be roughly estimated from

$$\tau_{sp} \approx \eta \frac{r}{\sigma}$$

to be about 10^{-7} s. Here η is the viscosity of water, r the radius of the cylinder and σ (= 1 mN/m) a lateral tension below the rupturing limit. Obviously, this time is short enough to allow for the spindle formation.

Presumably, a tension sensitive superstructure can only be observed in vesicles of medium size and with sufficient membrane area. Maybe they can preserve their initial shape after freezing by proportionate adjustment of their shape. An example is the vesicle of fig. 6 the contour of which has been flattened concerning the small fluctuation modes. However, the pronounced bulges, which could also be thought of as long wavelength deformations, remain observable.

5. Conclusion

The method of very rapid cryo-fixation is not as free from artefacts as was formerly presumed. The area contraction of the membranes and the volume expansion of the water may induce very high membrane tensions. These in turn alter the appearance of the vesicles under investigation. In spite of undulations being effective before freezing all vesicles look rather smooth. Also, tension may enforce very distinct shape transformations. Proper preparation techniques can in part balance these artefacts and careful consideration of the effects produced renders the evaluation of the micrographs possible. A lot of work remains to be done but we think that the grainy membranes and also the bulged vesicles are further evidence of the membrane microroughness that was postulated.

Acknowledgement. The electron microscopy was done at the Fritz Haber Institut of the Max-Planck-Gesellschaft. We are grateful to Prof. E. Zeitler for his support. B.K. thanks J. Jäger for excellent technical assistance. This work was supported in part by the Deutsche Forschungsgemeinschaft through Sfb 312.

[1] G. Beblik, R.M. Servuss, W. Helfrich: J. Physique. **46**, 1773 (1985)
[2] J.F. Faucon, M.D. Mitov, P. Méléard, I. Bivas, P. Bothorel.: J. Phys. France **50**, 2389 (1989)
[3] M.B. Mutz, W. Helfrich: J. Phys. France **51**, 991 (1990)
[4] M. Kummrow, W. Helfrich: Phys. Rev. Letters **A**, (1991) (in press)
[5] W. Helfrich, R.M. Servuss: Il Nuovo Cimento **3**, 137 (1984)
[6] R.M. Servuss, W. Helfrich: J. Phys. France **50**, (1989)
[7] W. Harbich, W. Helfrich: J. Phys. France **51**, (1990 a)
[8] W. Harbich, W. Helfrich.: Chem. Phys. Lipids **55**, (1990 b)
[9] W. Helfrich, B. Klösgen: In *Proceedings of the workshop on Dynamical phenomena at interfaces, surfaces and membranes*, ed. by D. Beysens, N. Boccara, G. Forgas (1991) (in press)
[10] W. Helfrich. B. Klösgen: Fourth Nishinomiya-Yukawa Symposium, ed. by A. Onuki, Springer Proc. Phys. N.Y. (1990).
[11] W. Helfrich: Liquid Crystals **5**, 1647 (1989 b)
[12] B. Klösgen, W. Helfrich: (1991) (submitted)
[13] J. Dubochet, M. Adrian, J.-J. Chang, J.-C. Hano, J. Lepault, A.W. McDowall, P. Schultz: Quart. Rev. Biophys. **21**, 129 (1988)
[14] Y. Talmon: Colloids Surf. **19**, 237 (1986)
[15] R. Groll: PHD-Thesis, Berlin (1986)
[16] M. Adrian, J. Dubochet, J. Lepault, A.W. McDowall: Nature **308**, 32 (1984)
[17] D. Needham, E. Evans: Biochemistry **27**, 8261 (1988)
[18] E. Evans, D. Needham: J. Phys. Chem. **91**, 4219 (1987)

Erythrocytes Membranes: Tethered Shells with Fluid-Like Deformation Regime

A. Zilker, H. Strey, and E. Sackmann

Physik Department, Biophysics Group E22, Technische Universität München, James Franck-Str., W-8046 Garching, Fed. Rep. of Germany

Abstract. The erythrocyte flickering was studied by phase contrast microscopy at wave lengths, λ , comparable to the cell diameter L and by RIC-microscopy in combination with fast image processing at $L \gg \lambda$ (to 220nm). The mean square amplitudes $< u_q{}^2 >$ scale with the 4th power of the wavevector, q, showing that the erythrocyte compound membrane behaves as a fluid membrane at small deformations. The bending stiffness is an order of magnitude smaller than expected. Accompanying model membrane studies suggest a solute effect as a possible explanation. A quasihexagonal array of the cytoskeleton-coupled band III is visualized in freeze fracture electronmicrographs and the average length of the spectrin tetramer is determined to be $\approx 80nm$.

1 Introduction

Erythrocytes are the simplest mammalian cells. Owing to remarkable similarities in the composition and architecture of mammalian cell plasma membranes the erythrocyte shell is a rather universal model system to study the physical basis of self organization and function of biomembranes. Since erythrocytes lack genetic control mechnisms they are better suited than other cells to gain insight into the biochemically mediated adaption of physical properties of biological systems to their biological function.

Owing to their compound-like composition of a fluid-like lipid bilayer and a twodimensional spectrin/actin network erythrocyte membranes are natural models of tethered membranes (that is 2D-analogues of macromolecules). In recent years substantial progress has been made on the statistical mechanics of elastic properties of soft interfaces, in particular hexatic films, fluid and tethered membranes [1, 2, 3]. Various scaling laws concerning the wavevector dependence or frequency dependence of the dynamic surface roughness of soft membranes have been postulated [1] ; transitions between flat and crumpled states of tethered membranes have been predicted [2, 3] and the essential role of undulation induced repulsion forces for the interaction of soft interfaces has been established [4, 5, 6].

This theoretical progress has not been matched by the experimentalists; with the exception of extensive studies of shape fluctuations and shape transitions of fluid bilayer vesicles showing that these obey well predicted laws of fluid membranes [7]. Models of fully polymerized membranes could not be prepared yet [11]. Erythrocytes are therefore still the best examples of tethered membranes [8].

Springer Proceedings in Physics, Vol. 66 The Structure and Conformation of Amphiphilic Membranes
Editors: R. Lipowsky · D. Richter · K. Kremer

In the present contribution we summarize recent progress of our efforts to elucitate the static and dynamic scaling laws of erythrocyte flickering. By application of phase contrast microscopy and the newly developed reflection interference contrast microscopy in combination with fast image processing we analysed the cellular excitations in the whole optical wavelength region.

Three types of methods were applied to determine the wavevector dependence of the flicker amplitudes and the characteristic frequencies (or decay times) ω_q : (1) Measurement of the power spectrum $P(\omega)$ (Eq. 4 below) of the thickness-fluctuations of the cell (called flicker spectroscopy) [8], (2) comparison of the (long wavelength) mean square thickness fluctuation $< \delta d^2 >$ across the cell diameter measured by phase contrast microscopy with calculated $< \delta d^2 >$-profiles [14], (3) the reflection interference contrast (RIC) microscopy [12].

The first method enables the most precise measurements of the frequency power laws: $P(\omega) \propto \omega^{\alpha}$. The second technique is best suited to evaluate flickering in the $qL < 1$ regime. The RIC-microscopy allows very precise measurements of the wave vector dependence of the mean square amplitudes $< u_q{}^2 >$ in the short wavelength regimes ($qL > 1$); up to wavevectors of $q \approx 4\mu m^{-1}$.

In the third part we report our attempts to visulize the lateral organization of the spectrin/actin network by evaluation of the two dimensional pair correlation function of the band III-proteins (serving as anchors for the spectrin-spectrin joints) in freeze fracture electron micrographs.

2 Structure of the compound membrane

Our present view of the erythrocyte plasma membrane is presented in Fig. 1. Judged from Electron-Microscopy (EM) studies of bilayer depleted cytoskeleta the spectrin/actin network exhibits basically a hexagonal (triangular) topology. The sides consist of spectrin tetramers (formed by head-head association of dimers) and the vestiges of actin oligomers (made up of about 13 monomers). Judged from the numbers of spectrin tetramers and actin oligomers the average side length of the triangles is $\approx 700nm$ [8, 15].

One decisive factor determining the elasticity of the membrane is the defect structure of the cytoskeleton depicted in Fig. 1a.

The second important factor is the coupling of the cytoskeleton to the lipid/protein bilayer. It is well established that (at least part of) the spectrin-spectrin joints are coupled to the integral membrane protein, band III via ankyrin (a dalton protein). There is further strong evidence [17] that part of the (multivalent) actin cross-links are coupled to the membrane proteins glycophorin A to C (which function simultaneously as blood group antigen). The coupling is medialid by the binding protein 4.1.

Model membrane studies ([8] for references) suggest as a further possible coupling mechanism the direct interaction of spectrin with the inner lipid monolayer, mediated by the negatively charged lipid components (phosphatidylserines and -inositols). This binding requires, however, the penetration of hydrophobic domains of spectrin into the bilayer and is thus expected to occur in

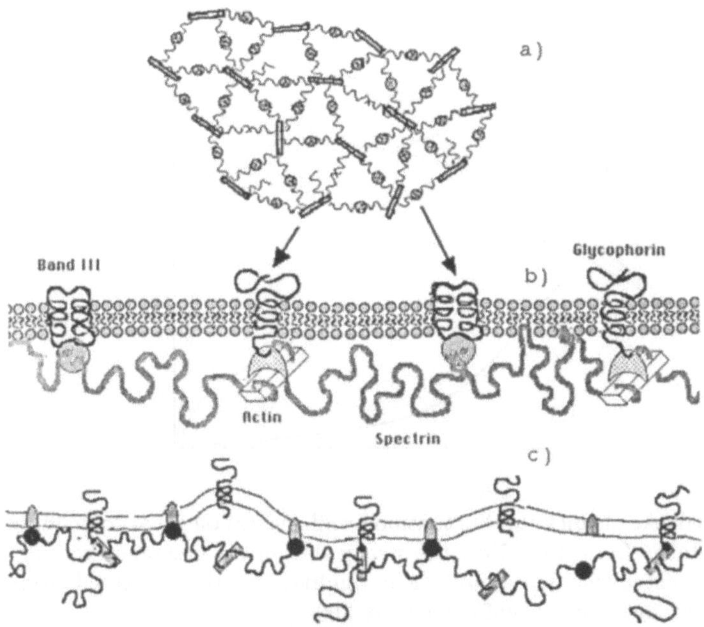

Figure 1: Schematic representation of present view of erythrocyte plasma membrane organization. a) Lateral organization of spectrin actin network. The spectrin tetramers (head-head associated dimers) form the sides and actin oligomers the vestiges of a triangular network. Besides to a large number of +60° and −60° disclinations the network exhibits furthermore a substantial number of dangling spectrin bonds suggesting an uncomplete connectivity. b) Coupling of cytoskeleton to lipid/protein bilayer. Well established is the coupling of the bivalent spectrin joints to band III via ankyrin and of the multivalent actin cross-links to glycophorin via band 4.1. A further possibility suggested by model membran studies consists in the electrostatic interaction of spectrin with negative lipids and partial penetration of flexible side chains into the semipolar lipid bilayer surface. c) Partial decoupling of domains of cytoskeleton from bilayer associated with pre-undulated surface of the latter.

the neigbourhood of defects. No evidence was found yet for such a coupling in erythrocytes.

3 Theoretical Background of Flickering

The quantitative evaluation of the flickering is hampered by the small size of erythrocytes and the constraint of finite volume-to-area ratio. Both introduce geometry effects and we have to consider three undulation wavelength regimes (with respect to two characteristic lengths: cell diameter L and cell thickness d): (1) wavevectors, q, comparable than the inverse cell diameter L ($qL \leq 1$); (2) q comparabel to cell thickness ($dq \leq 1$) and (3) $dq \geq 1$.

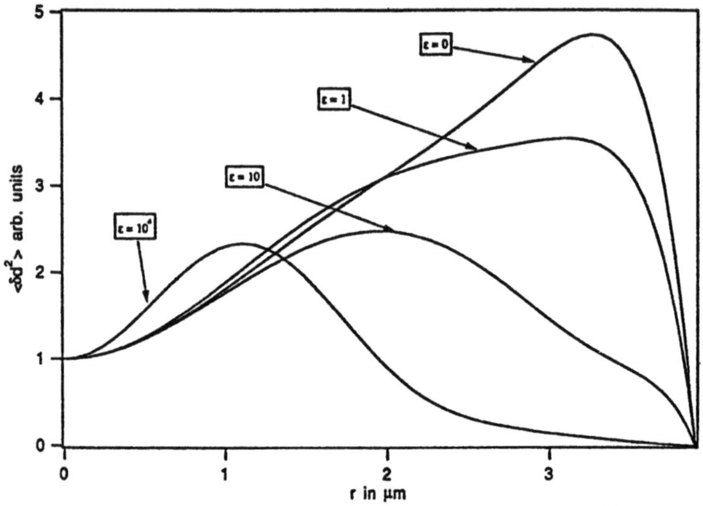

Figure 2: Dependence of thickness fluctuation profile $\delta d^2(r)$ on shear to bending moduli ratio ϵ (taken from [13]). Note the strong shift of the maximum coming-up towards the center with increasing shear rigidity. For the measured shear modulus of erythrocytes one would expect $\epsilon \approx 2000$ [21].

For $qL \leq 1$ the undulations have to be described in terms of functions belonging to symmetry group $C_i \times D_{\infty h}$ and therefore spherical harmonics Y_{lm} provide the right basis [13, 18]. The mean square amplitudes of the cell thickness (which can be evaluated by phase contrast microscopy ;cf. Fig. 2) can be expressed in terms of a truncated series of squares of spherical harmonics since the amplitudes do not necessary obey the equipartition theorem but are restricted by two factors: 1) By the condition that the two monolayers of the membrane exhibit a fixed area ratio which is determined by the resting shape. This "bilayer coupling constraint" [9] is essentially equivalent [10] to the Helfrich model of spontaneous curvature (since spontaneous curvature, c_0, can be regarded as Lagrange multiplier enforcing fixed average curvature). 2) Secondly, the condition of fixed average surface area restricts the long wavelength modes owing to the fact that these would imply an area expansion (or lateral tension) [7, 18]. The theory of the quasispherical model including both bending and shear elasticity has been worked out by Peterson [13]. He showed that the mean square thickness correlation function of thickness fluctuations $D(\tau) =< \delta d(\tau)\delta d(0) >$ can be expressed in terms of only one dimensionless parameter (cf. section 4 below):

$$\epsilon = \mu R^2/k_c \qquad (1)$$

(μ: shear modulus, k_c, bending modulus, R equivalent radius = radius of sphere with some volume as cell).

For the other two regimes ($qL > 1$), the flickering can be described in terms of the classical Brochard-Lennon model of two independent bilayers of dimension L^2 with free boundaries [19].

For the case of pure bending (= fluid membranes), the time correlation function of the undulations expressed in terms of a plane wave expansion is given by:

$$< u(\vec{r},t)u(\vec{r},0) >= \sum_q \frac{k_B T}{k_c q^4 L^2} e^{-\omega_q t} \tag{2}$$

The associated frequency spectrum is [18]

$$P(q,\omega) = k_B T L^2 \frac{L_{eff}(q)}{\omega^2 + (k_c q^4 L_{eff}^2(q))^2} \tag{3}$$

with $L_{eff}(q) = 1/\eta q$ for $qd > 1$ and $L_{eff}(q) = q^2 d^3/24\eta$ for $qd < 1$.

With respect to membranes as 2D-analogues of polymers it is interesting to realize that the above spectrum corresponds to the case of Zimm dynamics of linear polymers: that is a situation where the membrane segments are hydrodynamically coupled.

The cross over from the decoupled ($qd > 1$) to the coupled ($qd < 1$) case is most conveniently studied by considering the integrated flicker spectrum

$$P(\omega) = \int_{q_{min}}^{q_{max}} P(q,\omega) q \, dq \propto \omega^{-\delta} \tag{4}$$

with $\delta = 5/3$ for $qd > 1$ and $\delta = 4/3$ for $qd < 1$.

An analytical expression for $< u(\vec{r},t)u(\vec{r},0) >$ or $P(q,\omega)$ has not been derived yet for the case of solid membranes. Monte Carlo simulations predict the following mean square amplitudes ??

$$< u_q^2 >= \frac{k_B T}{k_c q^4 + 2.6\sqrt{\frac{\mu k_B T}{k_c^2}} q^3} \tag{5}$$

(Here and in the following q is defined as $q = 1/\lambda$ not $q = 2\pi/\lambda$). The simulations predict thus a cross-over from fluid-like to solid-like behaviour at a wave vector $q = 2.6\sqrt{\mu k_B T/k_c^2}$. For membranes under lateral tension an additional term γq^2 (γ = lateral compression modulus) would have to be added to the denominator.

4 Flickering in the regime $qL < 1$: Contribution of Shear and Monolayer Coupling

As noted above the flickering is best described in terms of spherical harmonics. One important consequence is that the excitations become strongly quantized. This regime has been studied by measurement of the mean square thickness fluctuation

$$\overline{\delta d^2(\vec{r})} =< \delta d(\vec{r},t)\delta d(\vec{r},0) >_t \tag{6}$$

across a cell cross-section and comparison with calculated thickness fluctuation profiles. By comparison of experimental data with different models one can get detailed insight concerning the contributions of shear and/or the monolayer-coupling to the flickering.

A first rigorous quasispherical model of thermal excitations of (incompressible) solid shells was established by Peterson [13]. The total elastic energy is

$$F(V, A, c_0) = \frac{k_c}{2} \oint (c_1 + c_2 - c_0)^2 \, dA + \frac{\mu}{2} \oint u_{ik}^2 \, dA \tag{7}$$

where c_0 is the spontaneous curvature and u_{ik} is the 2-dimensional shear strain tensor of the deformation. A first basic assumption is that the equilibrium shape is determined by the bending elastictity alone. This implies that the cytoskeleton can easily relax to the shape determined by the minimum of the bending energy by rearrangement of the elements. The experimental results justify this assumption. The equilibrium shape can be determined by minimizing the total bending energy, using the spontaneous curvature model or the bilayer coupling model [21, 9]. In the former model c_0 is introduced as an adaption parameter while in the second it arises as a Langrange multiplier enforcing fixed average curvature $< c_1 + c_2 > =$ const. In the coordinate free Peterson model of flickering one starts from the equilibrium surface M and describes flickering in terms of an expansion of the total energy to second order in deformation. The variation of the bending energy is determined by displacements, δu_n, along the local normal, $\vec{n}$, to M, which must satisfy the conditions of fixed volume and area. The shear is determined by local tangential deformations β fulfilling the conditions of zero local extensions. The total elastic deformation energy, δF, can be expressed as a purely quadratic form in δu_n and β. Expansion of these displacements into spherical harmonics $Y_{lm}(\vartheta, \varphi)$ yields for the mean square displacement in the normal direction δu_n [14]

$$\overline{\delta u_n^2(r, \vartheta, \varphi)} = \sum_{l,m} \frac{k_B T}{E_{lm}} Y_{lm}^2 \tag{8}$$

where E_{lm} are the eigenvalues of the deformation energy δF. Only modes for which l-m is even contribute to the mean square thickness fluctuation. It is given by the projection onto the rotation axis

$$\overline{\delta d^2(r, \varphi)} = 4\overline{\delta u_n^2(r, \vartheta, \varphi)} \; sin^2 \, \vartheta \tag{9}$$

The major predictions of the model are: (1) The volume constraint eliminates Y_{00} and the area constraint Y_{20}. Since the $m = 0$ modes contribute to thickness fluctuations in the center these are expected to be largely suppressed. (2) The contribution of the lowest energy even mode, Y_{40}, depends on the type of model. It is largely suppressed in the bilayer coupling model leading to a further supression of the central excitations. (3) The average thickness fluctuation profile is determined by the normalized ratio of the shear to bending modulus: $\epsilon = \mu R^2 / k_c$. As shown in Fig. 2 increasing ϵ leads to a shift of the radial position of maximum thickness fluctuaion towards the cell center. This strong effect of m is a benefit since it allows us to elucidate the contribution of shear elasticity

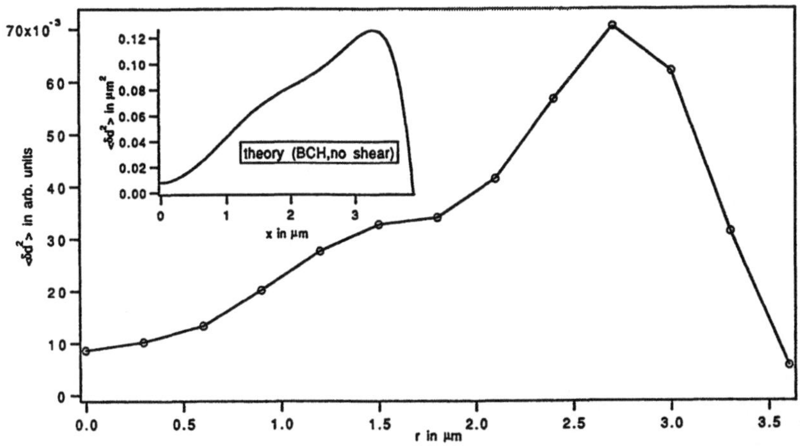

Figure 3: Comparison of mean square thickness fluctuation profile of discocyte at 200 mOsm with theoretical curve. The latter was calculated by the bilayer coupling model and for a fluid membrane ($\epsilon = 0$).

to long wavelenght undulations. (4) Since the mean square amplitudes decay with the 4th power of l, the deformation is mainly determined by the Y_{22}-modes and the Y_{31}-modes.

In Fig. 3 we present a comparison of the experimental and calculated thickness fluctuation profiles. Good agreement is obtained for $\epsilon \leq 1$ corresponding to a shear elastic modulus which is three orders of magnitude smaller than the value measured by static experiments ($\epsilon \approx 2000$). The ratio of the $\delta d^2(r)$ value at the maximum to the value at the center of the cell is $r = 8$. The theoretical prediction is $r = 4.6$ for the spontaneous curvature and $r = 15$ for the bilayer coupling model. This suggests, that the bilayer coupling contraint is partially relaxed. The absolute value of the thickness fluctuation in the center is $\sqrt{\delta d^2(r)} = 60nm$. Comparison with the model yields a bending modulus of $k_c = 1 \cdot 10^{-12} erg$. This is by factor of three larger than the value found by the RIC-technique. Since the measurements were perfomed at 200 mOsm tension effects cannot be excluded.

5 RIC-microscopic evaluation of short wavelength regime ($qL > 1$)

Fig. 4 shows an example of RIC-micrographs of a discocyte and a stomatocyte. In principle, the instantaneous surface profiles $S(\vec{r},t)$ and its time average $< S(\vec{r},t) >$ (the resting shape) can be reconstructed from the Newtonian-like diffraction pattern by fast image processing. However, a more elegant deconvolution procedure was recently developed [23] by which the mean square (plane wave) Fourier components of excitations, $< u_q^2 >$, can be directly obtained from the diffraction pattern by fast image processing.

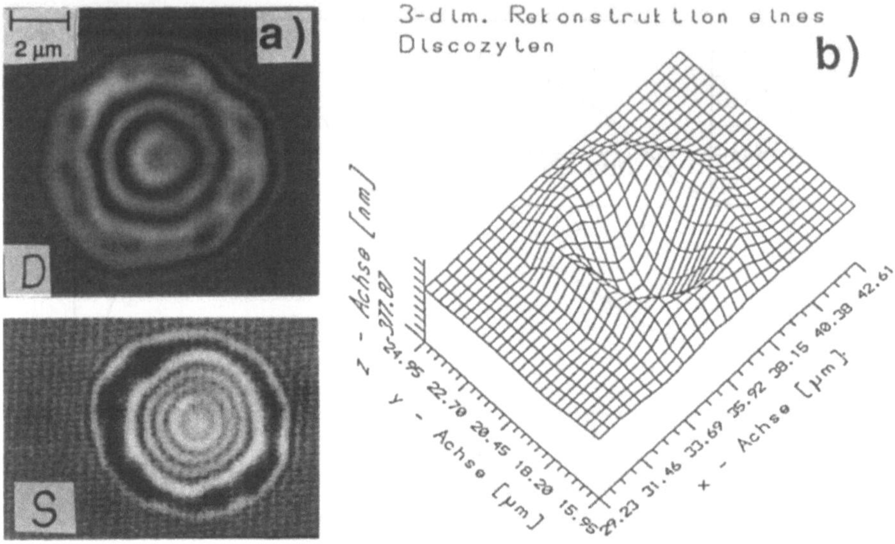

Figure 4: a) Reflection interference contrast (RIC)-micrograph of discocyte (D) and stomatocyte (S). b) Surface profile reconstructed from reflection interference pattern.

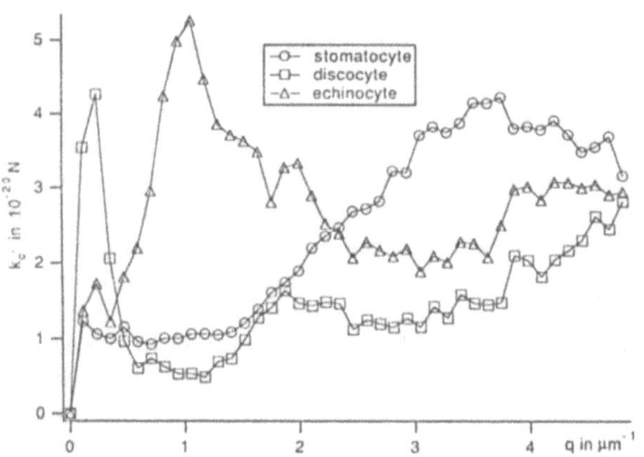

Figure 5: Wavevector dependence of apparent bending stiffness k_c^* (defined by Eq. 10) of three basic erythrocyte shapes.

Fig. 4 to Fig. 6 show some pertinent results. Starting from the assumption that erythrocyte flickering is primarily determined by bending an apparent bending stiffness defined by

$$k_c^*(q) = \frac{k_B T}{< u_q^2 > q^4 L^2} \tag{10}$$

is plotted as a function of q in Fig. 5 for cells of different shapes. One finds two

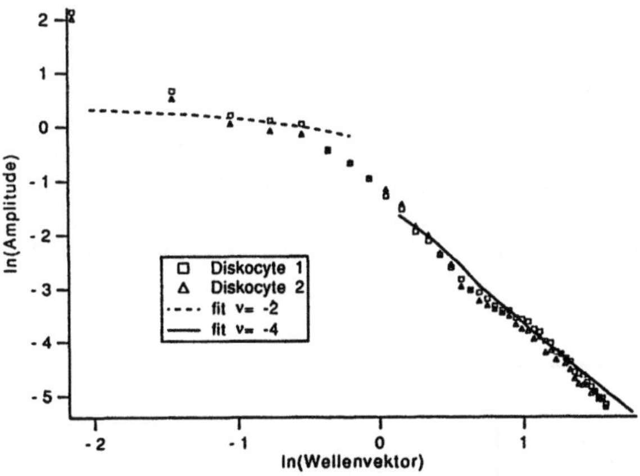

Figure 6: Double logarithmic plot of mean square amplitude $< u_q^2 >$ as function of q. The straight lines () and () correspond to powers q^4 and q^2, respectively. The transition regions marked by () are an artifact of the filtering technique.

remarkable differences: 1) The apparent bending stiffness of the stomatocyte is about equal to that of the discocytes in the long wavelength domain ($q^{-1} > 0.5mm$) but is by a factor of two to three larger at short wavelengths. 2) The k_c^*-value for echinocytes agrees with that of the discocyte at high q-values but is by a factor of four to five larger at $1.5 < q < 2\mu m^{-1}$. After a further ascent at $q < 1.5\mu m^{-1}$, k_c^* drops rapidly at $q < 1\mu m^{-1}$ to the value characteristic for discocytes. The remarkable agreement of k_c^* at high wave vectors shows that the structure of the membrane is hardly changed by the spicule formation. This is further verified by the agreement at $q \leq 0.3\mu m^{-1}$. This is attributed to the fact that the small membrane area stretching between the inner spicules (and facing the substrate) swims on a larger area excited by large wavelenght undulations. ($\lambda \approx L$). The latter are expected to exhibit the same amplitudes as discocytes provided the membrane softness is not remakably decreased by spicule formation. This result agrees with that obtained by the classical flicker spectroscopy [8].

In order to determine the law $< u_q^2 > \propto q^\nu$ we present in Fig. 6 a double logarithmic plot of ln $< u_q^2 >$ versus ln q for several discocytes at 300 mOsm. Ignoring the dashed regions we find three straightline regions. In the region $0.7 \leq \ln q \leq 1.5$ and $-0.4 \leq \ln q \leq 0.2$ the slope is $\nu = -4$ (that is $< u_q^2 > \propto q^{-4}$) and at $-1.5 \leq \ln q \leq -0.4$ the slope is $\nu = -2$. The slope in the dashed region is uncertain, since it is altered by the filtering procedure (to be published). The transition from the slope $\nu = -4$ to $\nu = -2$ is rather abrupt and we do not find an extended region of slope $\nu = -3$. This leads to the conclusion that the erythrocyte behaves as a fluid membrane over the whole wavelenght region (or exhibits a regime of very small shear rigidity).

6 Dynamic crossover between $qd < 1$ and $qd > 1$

Provided the cross-over from the uncoupled to the coupled membranes occurs in the regime $qL >> 1$ it should be revealed by the change in power law for $P(\omega)$ (cf Eq. 4)[19, 20]. Fig. 7. shows the power spectrum $P(\omega)$ of the thickness fluctuation taken by the flickerspectroscopy on a double logarithmic plot clearly two regimes are observed. At $3 \leq \omega \leq 15Hz$ $P(\omega)$ scales as $P(\omega) \propto \omega^{-1.43}$ and at $\omega > 15Hz$ as $P(\omega) \propto \omega^{-1.68}$. This is in very good agreement with the theoretical predictions.

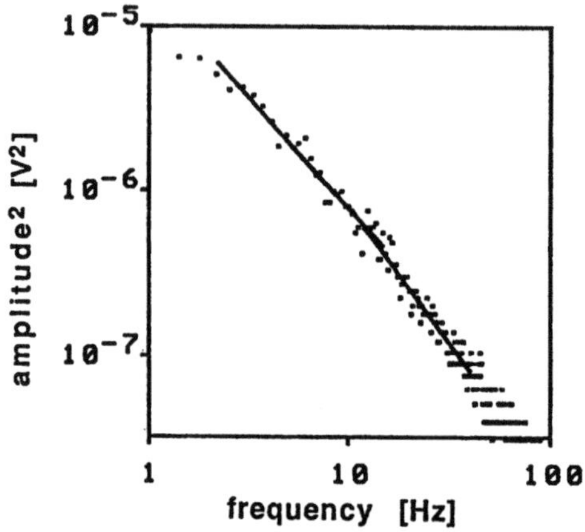

Figure 7: Double logarithmic plot of power spectrum $P(\omega)$ of erythrocyte thickness fluctuation. The slopes of the straight lines are $\delta = -1.43$ and $\delta = -1.68$.

7 Visualization of Cytoskeleton by Evaluation of angular distribution function of band III in Freeze Fracture Electron Micrography.

Band III is immobilized by its coupling to the cytoskeleton (cf. Bennet [17]). Provided the network exhibits a quasi-hexagonal structure (short range spatial order but some long range bond orientational order) the fixed band III should exhibit a lateral organization as illustrated in Fig. 1a. The band III proteins can be visualized in by freeze fracture electronmicroscopy (cf. Fig. 9). Evaluation of a large number of cells yield a particle density of $n_p \approx (5 \pm 1) \cdot 10^3 \mu m^{-2}$. the average density at the center of the cell (negative curvature) is remarkably higher (by a factor of 1.3) than at the buldge suggesting a slight heterogeneity of the membrane organization. The measured average density agrees well with the density of band III-molecules, obtained by biochemical analysis [16] of $7 \cdot 10^3 \mu m^{-2}$ showing that most of the particles in EM images are indeed band III.

We attempted to visualize the underlying network in the freeze fracture electron micrographs by determination of the particle pair correlation functions, $\rho(\varphi, r)$, in the radial and angular direction. In order to distinguish between free and spectrin-attached band III, ρ, was evaluated at different temperatures. As is well known phase separation sets in within the lipid/protein bilayer at $T < 20°C$

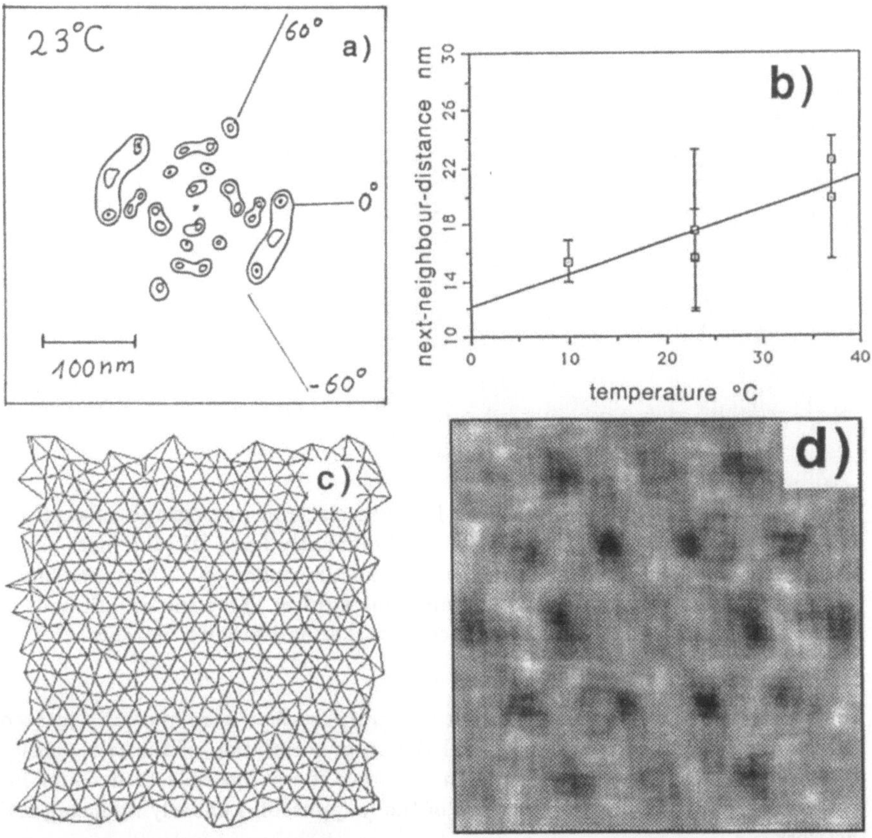

Figure 8: a) density-autocorrelation $\rho(x,y)$ function of particels within area of $4X4\mu m^2$ in image plane containing about 250 particles. The rings define lines of equal density (lying by 1% (outer rings) and 2% (inner rings) above average)(after P.Karg,Diploma Thesis, Technical University Munich 1987). b) Plot of radial distance of inner ring as function of temperature. Besides the decrease of the particle distance the intensity of the maximum increases. The outer maxima are not affected by T-changes. c) Distorted hexagonal lattice generated by computer simulations. d) Density autocorrelation function of lattice in (d).

which is expected to cause clustering of the free protein fraction but not of the fixed one (cf [17]).

The EM-images were scanned with a Joyce-Loebel MDMG densitometer with 10mm step width. The location of the particles were stored. Pair correlations functions $\rho(x,y)$ were determined in the image plane (x, y) over areas of $4X4\mu m^2$ (containing 250 particles) using periodic boundary conditions. Areas in the center of the cells were selected in order to avoid systematic errors caused by distortions owing to obligue orientations of the cell surface with respect to the image plane. The correlation functions obtained by starting from different

Figure 9: Rotationshadowing Freeze fracture electron micrograph of erythrocyte. The particles are expected to be predominantly band III (the most abundant membrane protein) since the glycophorins are pulled out of the membrane at least as demonstrated by model membrane studies. Note the undulations of the bilayer indicated by the appearence of darker and brighter patches.

centers were finally superimposed. For that purpose they were mutually rotated (about the normal to the plane) until the cross-correlation of the two density distributions reached a maximum. In this way it was possible to account for the variations in the bond orientation of the (supposedly) underlying hexagonal network.

Fig. 8a shows the 2-dimensional pair-distribution function $\rho(x, y)$ of a cell frozen from 23 C. We find five distinct maxima in the radial direction at distances $R_1 = 25nm$; $R_2 = 40$; $R_3 = 70nm$; $R_4 \approx 100nm$ and $R_5 \approx 150nm$. The radius of the innermost ring increases from $R_1 = 15nm$ at $10°C$ to $R_1 = 25nm$ at $37°C$ (cf. Fig. 8) whereas the radius of the outer rings remains constant. We therefore assume that the former corresponds to the nearest neighbour distance of the mobile and thus randomly distributed band III fraction whereas the outer maxima are due to cytoskeleton-coupled band III. For the hexagonal network of Fig. 1a the three smallest distances should be related as $a : \sqrt{3}a : 2a : 2.7a$ where the lattice constant a is equal to half the bond length. About the same ratios hold indeed for the distances R_2 to R_5. Analogous distributions of particles where found for some 10 cells evaluated. The radius of the maximum R_1 varies from 33 to 43nm. According to Fig. 7b the distribution of the maxima, of the inner ring, exhibits roughly hexagonal symmetry. We conclude that we can indeed locate the band III fixed to the cytoskeleton in freeze fracture images and find a bond length of the spectrin tetramers of $2a = 75nm$. The bond length observed in the electron micrographs of triton-isolated cytoskeleta was $2a = 190nm$. During preparation the area of the network is expanded by a factor of three corresponding to an extension of the bonds by a factor of 1.7.

In order to get a better feeling for hidden artefacts in such an analysis of the pair distribution function of superimposed random and symmetric populations of particles we perfomed some computer experiments. A distorted hexagonal lattice was generated by Monc Carlo simulation (cf. Fig.8c). A random distribution of particles with twice the density as the lattice points was superimposed. The pair distribution function of this system is presented in Fig.8d. We can clearly distinguish between the population of particles fixed to the cytoskeleton and the mobile fraction.

8 General Discussion: Softness of Erythrocyte Membranes Requires Metabolic Energy

All measurements show conclusively, that, with respect to flickering, erythrocytes behaves as fluid membranes. This could be explained in terms of a shear free deformation regime. The RIC measurement in the regime $qL > 1$ confirm previous findings of a surprisingly low value of the bending stiffness ($k_c \approx 2.5 \cdot 10^{-20} Nm$). The value obtained by the thickness fluctuation analysis at $qL < 1$ is considerably larger ($k_c \approx 10^{-19} Nm$). We do not have an explanation for this discrepancy yet. Considering the high colesterol content of erythrocytes membranes (50 mole %) however, the bending stiffness is nevertheless much smaller than expected. The k_c-value of DMPC bilayers containing 30 mole % cholesterol is $4 \cdot 10^{-19} J$.

Thus, there are two likely explanations for the low bending rigidity: (1) a preundulation of the lipid bilayer with wavelength of some 1000Å[8] and (2) a reduction of the bending stiffness of the lipid/protein bilayers by membrane proteins or other solutes. Recent studies provide evidence for the latter explanation. In a parallel RIC-study of vesicles we found that 15 mol % of the small cyclic peptide valinomycin reduces k_c of DMPC vesicles by a factor of 5. A still larger effect is exerted by a bipolar lipid, 5 mole % of which reduce k_c to some 0.5 k_BT [7]. Another possible explanation for the k_c-anomaly is that the undulations are driven by random chemical forces within the cytoskeleton [8]. Thus local changes in the connectivity of the cytoskeleton caused by phosphorylation-dephosphorylation reactions (cf. below) could lead to substantical local fluctuations in the lateral density of the network. As estimated previously the observed flicker amplitudes would only require fluctuations in the relative density of the order of 10^{-4}.

There are two likely explanations for a shear free deformation regime: (1) an only point-like and dynamic coupling of the cytoskeleton to the bilayer associated with a strong folding of the cytoskeleton and (2) a dynamic partial connectivity of the cytoskeleton. As is well known, the shear rigidity of a fractal network goes to zero at the percolation limit (cf [24]). Several observations favour the first possibility. (1) ATP is required for the membrane softness. (2) Small stretching-deformations of the cells in HF electric fields [22] are associated with shear. (3) Freeze fracture electron microscopic studies suggest that the lipid/protein bilayer exhibits a corrugated surface [8]. This could be the consequence of a negative tension exerted fy the folded cytoskeleton coupled to band III in a point-like manner.

The requirement of ATP to maintain the membrane softness suggests that the connectivity of the cytoskeleton and its degree of coupling to the bilayer are dynamic processes. It appears indeed, that a large fraction of the metabolic energy consumed by erythrocytes is required to maintain this softness. Spectrin and practically all of the regulating coupling proteins (ankyrin, band 4.1, band 4.9) have one or several groups which are continuously phosphorylated by various kinases and dephosphorylated by phosphatases. Following Bennet [17] phosphorylation (corresponding to negative charging) reduces (1) the binding of band 4.1 to spectrin (five fold); (2) of band 4.9 to actin; (3) of spectrin to actin and (4) of ankyrin to both band III and spectrin (also by about a factor of five). Spectrin phosporylation does, however, not affect the tetramer formation.

This suggests that the phosphorylation-dephosphorylation reactions do not affect the bonds of the network but its connectivity and coupling to the bilayer (both via band 4.1 to glycophorin and of the spectrin-spectrin joint to band III via ankyrin). Now let us assume that the cytoskeleton can be treated as a 2D network of entropy springs. The shear modulus would then be of the order $\mu \approx N_c k_B T$ where N_c is the number of spectrin tetramers per cm^2. The latter is $N_c = 105$ per cell and since the area per cell is $A = 140 \mu m^2$: one obtains $\mu = 2 \cdot 10^{-3} dyn/cm$. This is only by a factor of three smaller than the experimental shear modulus value. It appears thus unlikely that the cytoskeleton is near the percolation threshold. This favours the model of a folded cytoskeleton. Further support for this comes from the finding that cross-linking of spectrin by Diamid leads to an increase of the shear modulus by a factor of four whereas k_c increases only by some 50%.

References

[1] NELSON D.R., PELITI L., J. Physique 48(1987)1085.

[2] LEIBLER S.,SINGH R., FISHER M., Phys. Rev. Lett. 59(1987)1989.

[3] LIPOWSKY R., GIRADET S., Phys. Rev. Lett. 23(1990)2893.

[4] LIPOWSKY R., Nature 349(1990)475.

[5] HELFRICH W., SERVUSS R.M., Il Nuovo Cimento 3D(1984)137.

[6] PARSEGIAN E., EVANS E., in Surface Phenomena in Homorheology : Copley,A.L. and Seaman,G.V.F.,Eds.,N.Y. Acad. Sci (1983)13.

[7] DUWE H.P., KAES J., SACKMANN E., J. Phys. France 51(1990)945.

[8] ZEEMAN K., ENGELHARDT H., SACKMANN E., Eur. Biophys. J., 18(1990)203.

[9] SVETINA and ZĚKS, Eur. Biophys. J. 17(1989)101.

[10] BERNDL K.,KAES J.,LIPOWSKY R.,SACKMANN E.,SEIFERT U.,Europhys. Lett. 13(1990)659.

[11] EGGL P., RINGSDORF, SACKMANN E., Ber. Bunsenges. Phys. Chem.

[12] ZILKER A., ENGELHARDT H., SACKMANN E., J. Physique 48(1987)2139.

[13] PETERSON M., Mol. Cryst. Liq. Cryst. 127(1982)257.

[14] PETERSON M., STREY H., SACKMANN E., to be submitted.

[15] SHIK-CHUNG, DERICK L., PALEK L.H., J. Cell Biol. 104(1987)527.

[16] STECK T.L., Cell Shape, Academic Press (1989)205.
[17] BENNET V., Physiological Rev. Vol.70 4(1990)1029.
[18] MILLNER S.T., SAFRAN S.A., Phys. Rev. A 36(1987)4371.
[19] BROCHARD F. and LENNON J.F., J. Physique 36(1975)1035.
[20] NELSON D.R., FREY E., to be published.
[21] EVANS E., Biophys. J. 14(1974)923.
[22] ENGELHARDT H., SACKMANN E., Biophys. J. 54(1988)495.
[23] ZILKER A., Doctoral Thesis, Munich 1991.
[24] PLISCHKE M., FOURCADE B., Phys. Rev. A 43(1991)2056.

The Isolated Human Red Blood Cell Skeleton: An Example of a Flexible Tethered Membrane

C.F. Schmidt[1,2], *K. Svoboda*[1,2], *N. Lei*[3], *C.R. Safinya*[3], *S.M. Block*[2,4], and *D. Branton*[2]

[1]Lyman Laboratory of Physics, Harvard University, Cambridge, MA 02138, USA
[2]Department of Cellular and Developmental Biology, Harvard University, Cambridge, MA 02138, USA
[3]Exxon Research and Engineering Company, Annandale, NJ 08801, USA
[4]Rowland Institute for Science, 100 Cambridge Parkway, Cambridge, MA 02142, USA

Abstract. Red blood cell (RBC) membrane skeletons were observed during and after detergent extraction from RBC ghosts, using video microscopy, light scattering, and synchrotron-based small angle X-ray scattering. Optical tweezers were employed to trap and manipulate single cells under the light microscope. Immediately after extraction, the skeletons, which are two-dimensional protein networks, exhibited large thermal undulations and deformed strongly in response to small shear flow. Shrinkage by a factor of about two could be induced by screening the electrostatic repulsion in the network. Skeletons ceased to shrink at a monovalent salt concentration larger than 100 mM, which leads to an estimate of ≈120Å for the intrinsic persistence length of spectrin tetramers. Intriguingly, inspite of their large microscopic flexibility, the skeletons maintained roughly spherical shapes, and the static structure factor, measured over four orders of magnitude in length scale in the scattering experiments, displayed a power law exponent close to 2. This provides experimental evidence for the theoretically predicted flat phase of elastic tethered membranes, although a crumpled phase with a fractal dimension of D_f=2.2±0.2 is also within the error limits of the data.

1. Introduction

The multilayer membrane structure of red blood cells is believed to be optimized to preserve the cell's integrity under the extreme mechanical stresses encountered in blood circulation, while at the same time providing flexibility to ensure efficient blood flow in small capillaries. The lipid bilayer of the cell membrane exhibits a bending rigidity 10 to 50 times larger than k_bT [1], large enough to guarantee smooth cell shapes. A shear elastic triangulated protein network, consisting mainly of spectrin tethers and actin nodes (fig. 1), is anchored to the cytoplasmic side of the fluid lipid membrane [2]. Understanding the thermodynamics of this polyelectrolytic polymer network is a crucial step in explaining the properties of the composite membrane.

Two-dimensionally linked polymers, as a generalization of linear polymers, have been studied intensively in recent years, in theory [3], computer simulations [4] and experiments [5]. Theory predicts that the more pronounced influence of excluded volume interactions and the coupling of out-of-plane undulations to in-plane shear

Springer Proceedings in Physics, Vol. 66 The Structure and Conformation of Amphiphilic Membranes
Editors: R. Lipowsky · D. Richter · K. Kremer

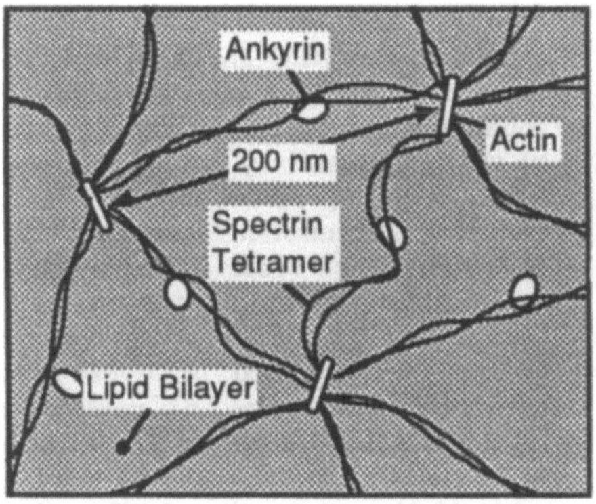

Figure 1: Cartoon of the spectrin/actin membrane skeleton of red blood cells, showing the main components and the average triangular structure. For clarity, the skeleton is drawn in an expanded state. *In vivo,* the average node distance is about 1/3 of the contour length of spectrin tetramers (200 nm).

deformations leads to a flat average configuration at low temperatures. This "flat phase" has a microscopic roughness, but, in contrast to linear polymers, an infinite large-scale persistence length [6]. At high temperatures, the polymer sheets are expected to be in a "crumpled" phase with a fractal dimension of 2.5 [7]. Simulations suggest that excluded volume effects in real 2D-polymers might prevent crumpling at all temperatures [4]. Lattice defects, however, may restore a crumpled phase by destroying the long range order of the flat phase [8].

We present an experimental study of RBC membrane skeletons during and after isolation by detergent treatment of RBC ghosts, using high resolution video enhanced microscopy [9], static light scattering and high resolution small angle X-ray scattering. We observed the process of skeleton extraction and the response to varying buffer conditions under the microscope directly on single cells, which were suspended in a flow chamber with the help of a focused infrared laser beam ("optical tweezers") [10]. Combination of light and small angle X-ray scattering, using synchrotron radiation, enabled us to measure the static structure factor of membrane skeletons over four orders of magnitude in length scale, from 8 μm down to ≤10Å.

2. Results and Discussion

Immediately after extraction in 5 mM Na-Phosphate buffer, skeletons expanded from 6.3 μm to 8.5 μm average diameter and showed a roughly spherical average shape (fig. 2a). The skeletons exhibited large thermal fluctuations and deformed in response to weak flow fields (fig. 2b). Skeletons shrank in response to increasing salt concentration, but maintained their spherical shell-like shape and flexibility. The decrease in diameter leveled off at ≈100 mM NaCl. At 1.5 M NaCl, the diameter was 4.3±0.3 μm. Under continued exposure to high ionic strength buffer, a slow process of structural rearrangement (beginning within minutes after detergent extraction) led to further shrinkage and increasing rigidity of the skeletons (for details see [11]), which finally resembled those prepared in bulk (fig.2c), using established recipes [12]. The relative salt-induced shrinkage of skeletons is comparable to that of isolated spectrin dimers [13]. An intrinsic persistence length of ≈120 Å for spectrin can be estimated from the leveling-off in the salt response at 100 mM NaCl by applying Odijk's model of polyelectrolytic worm-like chains [14]. This intrinsic persistence length of spectrin is much smaller than a tether contour length of 2000 Å.

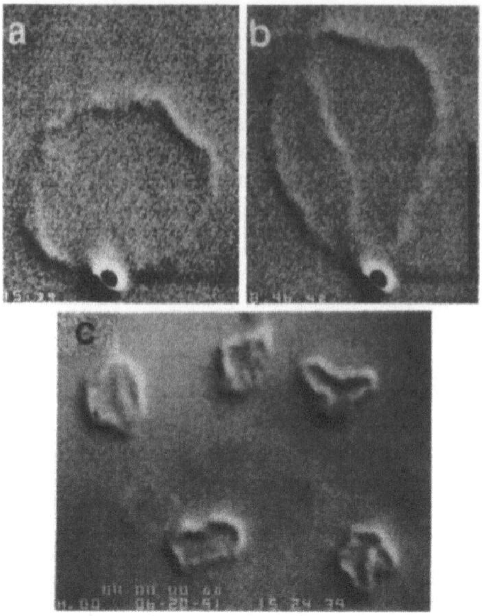

Figure 2: RBC skeletons after detergent extraction from RBC ghosts under video enhanced differential interference contrast microscopy. The cells in a) and b) were suspended by optical tweezers (Nd:YAG-laser, 1W, CW [10]) in a temperature controlled flow cell [15] (T=30°C). Silica beads (0.4 μm), non-specifically bound to the skeletons, were used as "handles" for the optical tweezers. Ghosts were prepared in 5 mM Na-phosphate buffer (pH 7.6), containing 0.1 mM $MgSO_4$, 0.1 mM EGTA, 0.1 mM ATP, 1 mM DTT, 1 mg/ml BSA and 0.2 mM NaN_3. Extraction buffer also contained 1% Triton X-100. Bar, 10 μm.
a) Freshly extracted skeleton in 5 mM buffer without buffer flow.
b) The same skeleton with buffer flow of ≈40 μm/s. Note distortion.
c) Bulk prepared skeleton [11,12] in 1 M NaCl, T=20°C.

Using static light scattering and small angle X-ray scattering, we measured the skeleton structure factor, S(q), which shows three distinct regimes (fig.3). At small angles, S(q) displays diffraction minima and maxima. The first minimum is clearly delineated and corresponds to an average skeleton diameter of 2.9 μm, which is consistent with the value measured under the microscope, 2.6±0.3 μm. Polydispersity obscures higher order extrema. In the q-range from 7×10^{-4}Å^{-1} ($2\pi/q$=9000Å) to 0.1Å^{-1} ($2\pi/q$=60 Å), S(q) displays approximate power law behavior, $S(q)\propto q^{-\nu}$, with ν=2.2±0.2. The X-ray experiments consistently led to an exponent closer to 2 (ν=2.0±0.1) than the light scattering experiments (ν=2.3±0.1). At scattering vectors $q\geq0.1$ Å^{-1}, where the molecular structures of actin and spectrin are expected to determine the structure factor, the curve levels off again, but the decreasing signal/noise ratio makes the data inaccurate in this region. The proximity of the value of the scaling exponent to 2 (the predicted value for a flat sheet) complements the microscopic observation of rather expanded, shell-like configurations. An exponent larger than 2 in the light scattering data might be due to large scale static defects in the skeletons, which we observed in the microscope (fig. 2c). The high-q X-ray data are consistent with ν=2. On the other hand, in an anisotropic flat phase, the small scale roughness should appear as a regime of steeper slope in S(q) [4]. A possible power-law exponent larger than 2 and the absence of structure in the curve down to small length scales are also consistent with a fractally crumpled phase. Its fractal dimension of $D_f\approx$ 2.2±0.2 would be smaller than the one predicted from a Flory estimate, D_f= 2.5 [7], but comparable to what Hwa et al. found for thin graphite oxide membranes in solution [5]. Alternatively, the absence of a region in S(q) reflecting the microscopic roughness of the flat phase, i.e. with a slope steeper than -2, could be due to the finite size of our objects, where we begin to see linear chain behavior when $2\pi/q$ becomes smaller than the mesh size of the

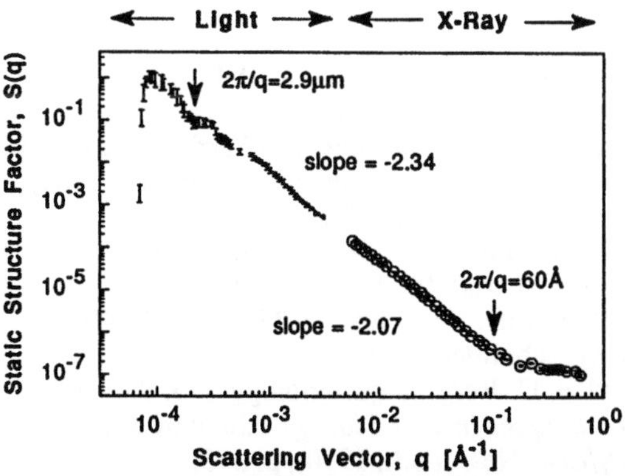

Figure 3: Static structure factor of detergent-extracted RBC skeletons (bulk method [11,12]) measured by light scattering and small angle X-ray scattering, using synchrotron radiation. The power law exponents were determined by linear regression through part of the data points as indicated by the drawn lines.

Light scattering was performed at λ=488 nm and room temperature with scattering angles between 2.5° and 160°. The X-ray experiments were carried out at the National Synchrotron Light Source on the Exxon beam line X-10A. The monochromator consists of a double-bounce Ge (1,1,1) crystal, set at 8 keV. The analyzer is a triple-bounce Ge (1,1,1) channel cut crystal. The configuration yields a very sharp in-plane Gaussian resolution function with very weak tail scattering with Half Width at Half Maximum (HWHM) 1.6×10^{-4} Å^{-1}, and allowed us to take data at a minimum scattering vector of ≈0.002 Å^{-1}. The out-of-plane resolution was set by narrow slits which yielded an out-of-plane Gaussian resolution function with HWHM = 0.001 Å^{-1}. The samples were contained in sealed quartz capillaries with a diameter of 1.5 mm (further details in [16]).

network. From the average mesh size, calculated from the measured surface areas of skeletons and the number of meshes, we can estimate an upper limit for the amplitude, d, of the roughness of a flat phase: d≤400Å.

The results illustrate the experimental difficulty of differentiating between a flat and a crumpled phase when the difference in dimensions is only small. Further experiments, probing the structure and also the dynamics of similar 2D-polymers, will help to elucidate the properties of these macromolecular networks in more detail.

Acknowledgements

We would like to acknowledge helpful discussions with David Nelson. We thank Paul Janmey, Toyoichi Tanaka and Michal Orkisz for the use of their apparatus. This work was supported by grants from the Deutsche Forschungsgemeinschaft (C.F.S.), the NIH (K.S., Biophysics Training Grant; D.B., HL17411), the Rowland Institute for Science (S.M.B.), and Exxon Research and Engineering (N.L and C.R.S.). The Synchrotron small angle X-ray scattering experiments were carried out on the Exxon beam line X-10A at the National Synchrotron Light Source of Brookhaven National Laboratory, which is supported by the U.S. Department of Energy.

References

[1] A. Zilker, H. Engelhardt, and E. Sackmann, J. Phys. (Paris) **48,** 2139 (1987).
[2] V. T. Marchesi, Ann. Rev. Cell Biol. **1,** 531 (1985).
[3] A recent review is: *Statistical Mechanics of Membranes and Surfaces,* edited by D. R. Nelson, T. Piran, and S. Weinberg (World Scientific, Singapore, 1989).
[4] F. F. Abraham and D. R. Nelson, Science **249,** 393 (1990).
[5] T. Hwa, E. Kokufuta, and T. Tanaka, Phys. Rev. A **44,** 2235 (1991).
[6] D. R. Nelson and L. Peliti, J. Phys. (Paris) **48,** 1085 (1987).
[7] Y. Kantor, M. Kardar, and D. R. Nelson, Phys. Rev. A **35,** 3056 (1987).
[8] D. R. Nelson and L. Radzihovsky, submitted (1991).
[9] S. M. Block, K. A. Fahrner, and H. C. Berg, J. Bacteriology **173,** 933 (1991).
[10] S. M. Block, in *Noninvasive Techniques in Cell Biology,* edited by J. K. Foskett and S. Grinstein (Wiley/Liss, New York, 1990), p. 375.
[11] K. Svoboda, C. F. Schmidt, S. M. Block, D. Branton, in preparation (1991).
[12] J. Yu, D. Fischman, and T. L. Steck, J. Supramol. Struct. **1,** 220 (1973).
[13] A. Elgsaeter, Biochim. Biophys. Acta **536,** 235 (1978).
[14] T. Odijk and A. C. Houwaart, J. Polym. Sci. Polym. Phys. Ed. **16,** 627 (1978).
[15] H. C. Berg and S. M. Block, J. General Microbiol. **130,** 2915 (1984).
[16] C. R. Safinya, in *Phase Transitions in Soft Condensed Matter,* edited by T. Riste and D. Sherrington (Plenum, New York, 1989), p. 249; and references therein.

Dynamics of Flat Membranes and Flickering in Red Blood Cells

E. Frey and D.R. Nelson

Physics Department, Harvard University, Cambridge, MA 02138, USA

Abstract. A theory of the dynamics of polymerized membranes in the flat phase is presented. We discuss the renormalization of the kinetic coefficients for the undulation and phonon modes due to hydrodynamic "backflow". The long time behavior of the dynamic structure factor is given by stretched exponentials. We also study the dynamics of the thickness fluctuations in red blood cells (flicker phenomenon) taking into account the underlying polymerized spectrin skeleton. Qualitatively different dynamical behavior is predicted for spectrin skeletons isolated from their natural lipid environment.

Whereas the statics of polymerized membranes has been studied quite extensively [1], the dynamics of polymerized membranes have been studied only in the crumpled phase [2,3]. These investigations follow closely the concepts known from polymer dynamics [4]. The simplest approach to polymer dynamics is to neglect the hydrodynamic interaction between different segments (Rouse dynamics (RD)). It is known, however, that the long-ranged hydrodynamic interaction between different monomers, mediated by the intervening solvent, strongly influences the dynamics (Zimm dynamics (ZD)). The dynamics of thickness fluctuations of red blood cells has been studied by Brochard and Lennon [5] using concepts known from the dynamics of surface waves.

Here we consider the dynamics of polymerized flat membranes. What distinguishes the static universality class of tethered membranes from liquid membranes is their fixed connectivity. The statics of the flat phase of polymerized membranes is characterized by non-local bending rigidity $\kappa(k) \sim k^{2+2\zeta}$ and in-plane elastic constants $\lambda(k)$, $\mu(k) \sim k^{\omega}$ [1].

In dynamics, additionally, the permeability of the membrane to solvent molecules plays a crucial role. For very large mesh size the velocity of a particle is determined only by the local hydrodynamic forces acting on it. The bare viscous drag coefficients $\zeta_{u,h}$ relating the force on the particle to its in-plane (u) and out-of-plane (h) velocity is then simply $\zeta_{u,h} = 6\pi\eta a$, where η is the dynamic viscosity of the solvent. In general, however, the velocity of a particle depends in a complicated way on the forces acting on all other particles, because of the long range solvent velocity field generated by a localized force. The importance of this fluid backflow increases as the mesh size of the polymerized network decreases. If the membrane is completely impermeable to the fluid, the out-of-plane modes become slaved to the fluid motion perpendicular to the plane of the membrane. One can think of highly permeable and impermeable membranes as constituting two different dynamic universality classes of polymerized membranes. As an example of highly permeable polymerized membranes one may think of isolated spectrin networks, whereas red blood cells themselves represent impermeable membranes.

The bare friction of the membrane with the solvent is determined by the structure (permeability) of the membrane. The hydrodynamic interaction between different monomers causes the renormalized mobility to become inversely proportional to the wave vector k. One explanation of this effect is that in the limit of long wavelength the motion of the membrane becomes slaved to the dynamics of the solvent. In this limit the dynamics becomes equivalent to a classical hydrodynamic problem with the boundary condition that the viscous stress of the solvent equals the elastic forces of the membrane.

The crossover from a wave vector independent friction ζ (RD, highly permeable membranes) to $\zeta \sim k$ (ZD, impermeable membranes) is determined by the ratio of the friction coefficients ζ_u and ζ_h for the in-plane and out-of-plane motion and the dynamic viscosity η of the solvent. The crossover vector for the out-of-plane undulation mode is $k_h = \zeta_h/4\eta$, and for the transverse and longitudinal phonon modes one obtains $k_\perp = \zeta_u/2\eta$ and $k_\parallel = \zeta_u/4\eta$. For large viscosity η and/or for low friction ζ (high permeability) the crossover wave vectors become very small or even less than the smallest accessible wave vector $k_{min} = \pi/L$. In this case the dynamics is Rouse-like in the entire wave vector regime. In the opposite regime of very small viscosity and/or small permeability the crossover to ZD sets in already for very large wave vectors. The location of the crossover point can thus be tuned by the viscosity of the solvent and/or the permeability of the membrane.

The wave vector dependences of the characteristic frequencies for the (overdamped) undulation and phonon modes are $\Gamma_h \sim k^{z_h}$ and $\Gamma_u \sim k^{z_u}$ with the critical dynamic exponents $z_h = 2 + 2\zeta$, $z_u = 2 + \omega$ for RD, and $z_h = 1 + 2\zeta$, $z_u = 1 + \omega$ for ZD. A dynamic renormalization theory shows indeed that there are no other than static renormalizations [6]. Estimates of ζ from simulations are in the range $\zeta = 0.5 - 0.67$ [2,7,8], while ω is now believed to be rather small [8].

A quantity which can be directly measured in experiments is the dynamic structure factor $S(\vec{k}, t)$. In the regime where the wave length is much larger than the linear dimension L of the membrane one finds an exponential decay $S(k,t) \sim \exp\left[-\mathcal{D}k^2 t\right]$, where the diffusion constant is $\mathcal{D} \sim 1/L^2$ for RD and $\mathcal{D} \sim 1/L$ for ZD. This result should be compared with $\mathcal{D} \sim L^{-4/5}$, obtained for ZD in the crumpled phase [3]. In the opposite regime ($kL >> 1$) one is probing the internal motion of the membrane. The long time behavior is given by *stretched exponentials* $S(k,t) \sim \exp[-Ck^2t^\alpha]$, if the scattering vector lies in or orthogonal to the membrane plane. For out-of-plane scattering vector the stretching exponent is given by $\alpha_{Rouse} = 2\zeta/(2+2\zeta)$ in the regime of RD and $\alpha_{Zimm} = 2\zeta/(1+2\zeta)$ in the Zimm regime, where ζ is the roughness exponent. For in-plane scattering vector we let $\alpha \rightarrow \beta$, where $\beta_{Rouse} = \omega/(2+\omega)$ and $\beta_{Zimm} = \omega/(1+\omega)$. Since the exponent ω is supposed to be small number, the latter decay represents an enormous stretching which approximates an algebraic decay.

An important application of the dynamics of flat membranes are the thermal thickness fluctuations of red blood cells. The membrane of an erythrocyte essentially consists of a lipid bilayer (believed to be in a liquid phase) with a spectrin polymer network attached. The presence of the spectrin implies that unlike the phospholipid component of a biological membrane, the composite red blood cell membrane exhibits a shear modulus [9]. Under normal physiological conditions the red blood cells show a remarkable flicker phenomenon [5] due to thermal fluctuations of the cell thickness. Isolated spectrin networks differ from composite red blood cells not only in the magnitude of their elastic constants but also -and more importantly- in their permeability for solvent molecules. Hence we expect qualitatively different behavior for isolated spektrin networks as opposed to red blood cells as a consequence of their different permeability and elastic properties.

Upon considering a simplified model of two membranes, separated by an average distance d and neglecting edge effects we find the following crossover scenario for the *thickness fluctuations*. For *highly permeable membranes*, like the isolated spectrin network there is a crossover from RD to ZD at a wave vector k_h followed by a reentrant crossover to RD when $kd \leq 1$. The ZD is restricted to a wave vector regime $\zeta_h/4\eta > k > 1/d$. For highly permeable membranes this becomes a very narrow regime or even vanishes if $k_h < 1/d$.

A completely different crossover scenario is obtained for *impermeable* membranes, like the composite red blood cell. There one has to distinguish between large and small ratio of stretching to bending energy y. For $y >> 1$, which is the case for red blood cells, one obtains a crossover from ZD with a kinetic coefficient proportional to $1/k$ to a kinetic coefficient proportional to k^2 when $kd \leq 1$. In the regime $y << 1$ the linewidth shows a crossover from ZD to a kinetic coefficient proportional to k. The results for the case of a very large ratio y of stretching to bending energy reduce to the results of Brochard and Lennon [5] for fluid like membranes provided we assume a liquid-like roughness exponent $\zeta = 1$. But, in addition to the lipid bilayer there is also a spectrin network attached to the bilayer leading to a solid-like structure of the composite object. This implies that there is a crossover from fluid- to solid-like behavior which has observable static [12] as well as dynamical consequencies.

The crossover scenario of the fluctuations of the relative *in-plane* modes is similar to that of the thickness fluctuations in the case of highly permeable membranes, whereas for impermeable membranes the crossover is from ZD to RD [6]. The in-plane modes have up to now not been studied experimentally. It would be interesting to design experiments which measure the in-plane dynamics of membranes.

We thus conclude that the crossover scenario, upon passing from $kd << 1$ to $kd >> 1$, depends sensitively on two factors, the permeability of the membrane and the ratio of stretching to bending energy. This becomes evident if one considers the two extreme cases (i) an impermeable fluid ($\zeta = 1$) lipid bilayer (with $y >> 1$) [5] and (ii) polymerized ($\zeta \approx 0.5$) isolated spectrin networks with high permeability. The line width for the flicker modes are $\Gamma_h^{(i)} \sim k^6$ and $\Gamma_h^{(ii)} \sim k^3$, respectively, i.e., they differ by three powers in the wave vector!

References.

[1] Statistical Mechanics of Membranes and Surfaces, Eds. D.R. Nelson, T. Piran and S. Weinberg , World Scientific (1989); and refs. therein.
[2] Y. Kantor, M. Kardar and D.R. Nelson, Phys. Rev. **A35**, 3056 (1987).
[3] M. Muthukumar, J. Chem. Phys. **88**, 2854 (1988); J. Niel, Europhys. Lett. **9**, 415 (1989).
[4] M. Doi and S.F. Edwards, The Theory of Polymer Dynamics, (Clarendon Press, Oxford, 1986).
[5] F. Brochard and J.-F. Lennon, J. Phys. (France) **36**, 1690 (1975).
[6] E. Frey and D.R. Nelson, to be published.
[7] F.F. Abraham and D.R. Nelson, J. Phys. (France) **51**, 2653 (1990).
[8] R. Lipowsky and M. Giradet, Phys. Rev. Lett. **65**, 2893 (1990).
[9] R. Waugh and E.A. Evans, Biophys. J. **26**, 115 (1979).

A New Cell Model – Actin Networks Encaged by Giant Vesicles

M. Bärmann, J. Käs, H. Kurzmeier, and E. Sackmann

Physik Department, Biophysics Group E22, Technische Universität München, James Franck-Str., W-8046 Garching, Fed. Rep. of Germany

Abstract. DMPC vesicles were swollen in a buffer containing actin monomers. Protein outside the vesicles was eliminated by enzymatic digestion and ultrafiltration. Polymerization of actin within vesicles was attained by introducing K^+ or Mg^{2+} ions via suitable ionophores. Investigation by DSC showed, that there was no interaction when actin was distributed symmetrically on both sides of the membrane, but that actin interacted well with the lipid when located on one side of the membrane only, with a clear difference between the effects of monomeric and polymeric actin. Microscopic observation and image analysis of thin-walled vesicles demonstrated a correlation between actin polymerization and characteristic shape transitions.

1. Introduction

Giant vesicles with defined lipid composition have been studied as a model system for the physical principles governing membrane curvature and bending elasticity.[1] A step forward in understanding the complex mechanical behavior of biological membranes may result from the reconstitution of cytoskeletal gels within these vesicles. At the same time, the rheological behavior of the included gels will be accessible for studies on a microscopic scale in the dimension of natural mammalian cells. Therefore, actin-containing giant vesicles represent an interesting experimental model for the future.

2. Materials and Methods

Preparation of actin-filled giant vesicles: DMPC, together with an appropriate ionophore, was dissolved in chloroform/methanol (75/25, v/v) in a glass beaker containing a PTFE disk at the bottom, and dried thoroughly first in a nitrogen stream and then under vacuum. After addition of (monomeric) G-actin in G-buffer (2 mM Tris/HCl pH 7.5, 1 mM ATP, 0.05 mM $MgCl_2$, 3 mM NaN_3, 1 mM DTT), vesicles formed spontaneously within at least two hours at a temperature above the phase transition. Protein outside the vesicles was digested by chymotrypsin (50 U/ml) within a few minutes. The peptide containing buffer was exchanged against fresh G-buffer by ultrafiltration under low pressure (0.5 bar) over an Amicon XM-300 membrane. Concentrations were 7.2 µM for actin, and 1 mg/ml for the lipid. In fluorescence experiments, 5 % of the actin was labeled with NBD.

Other Methods : Actin was prepared according to Pardee and Spudich [2], followed by gel filtration on Sephacryl S-200 [3]. For the determination af G-actin concentration, an extinction coefficient of 0.63 mg^{-1} ml was used.

Actin polymerization was monitored by the increase of NBD fluorescence intensity (λ_{ex} = 480 nm, λ_{em} = 530 nm) [4] with a SPEX fluorescence spectrometer Fluorolog 2T1.

For DSC (differential scanning calorimetry), a Microcal MC-2 Scanning calorimeter was used.

Microscopic examination of vesicles was performed on a Zeiss Axiovert 10 with epifluorescence and phase contrast equipment, recorded on videotape with a Hamamatsu C2400 SIT camera, and analysed on a Macintosh IIfx computer, supplemented with a Pixel Pipeline frame grabber (Perceptics), with the aid of a self-modified version of the IMAGE public domain software (Wayne Rasband, NIH).

3. Results and Discussion

As can be seen from fig.1, after chymotrypsin treatment and ultrafiltration , NBD-fluorescence was solely located within vesicles. SDS polyacrylamide gel electrophoresis of the vesicle suspension showed that actin concentration was very

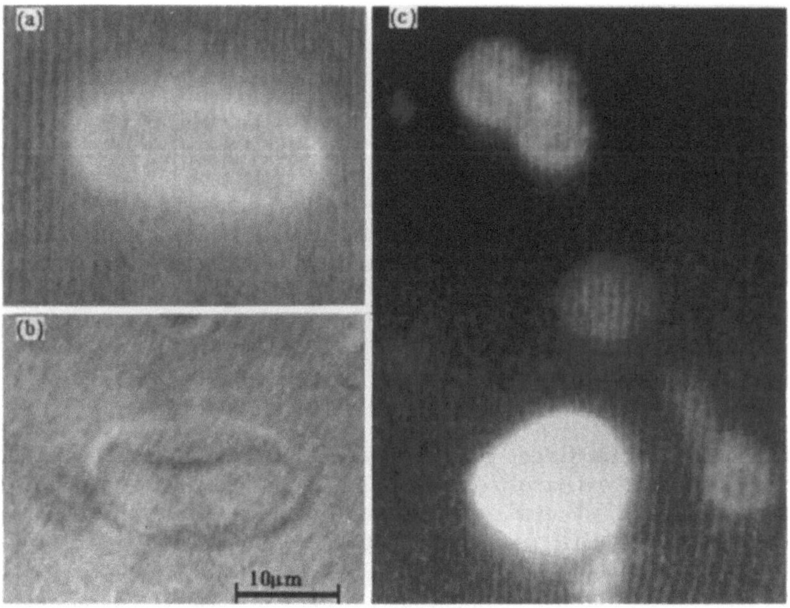

Fig 1. Actin-containing vesicles: (a) NBD-fluorescence of a G-actin containing giant vesicle (95 % DMPC, 5% A23187, T = 32.1 C), note that fluorescence seems brightest beneath the membrane; (b) phase contrast image of the same vesicle showing the typical doughnut-like shape of a dicocyte; (c) F-actin containing vesicles, one of them with very bright fluorescence characteristic for multilamellar vesicles. The reason for this phenomenon is unclear.

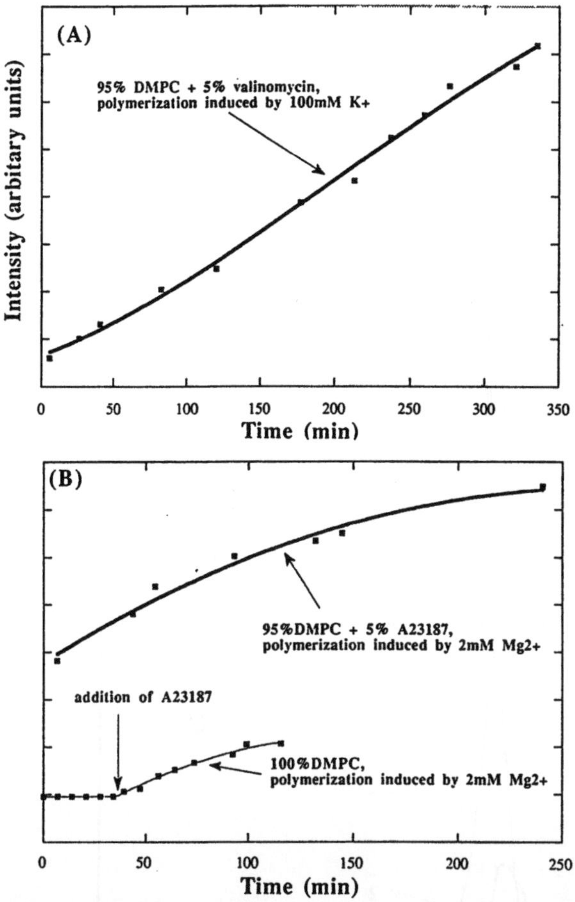

Fig. 2. Actin polymerization (increase of NBD fluorescence): (A) steady increase after addition of 100 mM K^+, the lipid membrane contained 5% valinomycin (molar ratio); (B) upper trace, increase after addition of 2 mM Mg^{2+}, 5 % A23187; lower trace, ionophore added only after 30 min.

low, so that no actin polymerization outside the vesicles seems possible (not shown). Thus, the polymerization observed by fluorescence increase must be due to actin included within vesicles and kept at the original concentration (7.2 µM, the critical concentration for polymerization being 0.2 µM). Polymerization could be induced either by 100 mM K^+ or by 2 mM Mg^{2+}, transported via the membrane by the ionophore, valinomycin or A23187, respectively. Without ionophore, no polymerization was observed, when salts were added to the medium (Fig.2).

In many actin-containing vesicles inspected in the micoscope, fluorescence was strongest near the membrane. Since this was suggestive of preferential membrane binding of the protein (an optical artifact is possible, too), we investigated the actin/lipid interaction by DSC. Fig. 3 clearly demonstrates, that both G- and F-actin influence the shape and location of the phase transition peak. It appears that, given a

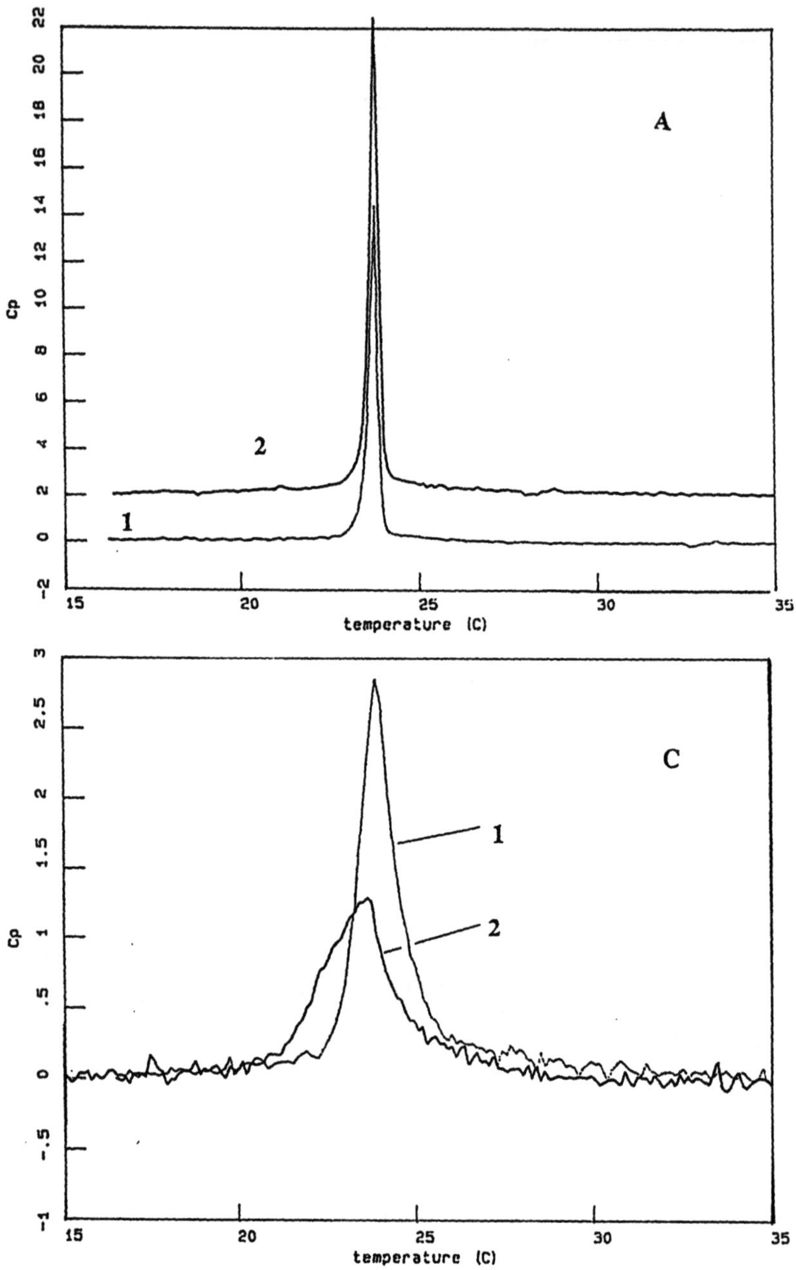

Fig. 3.

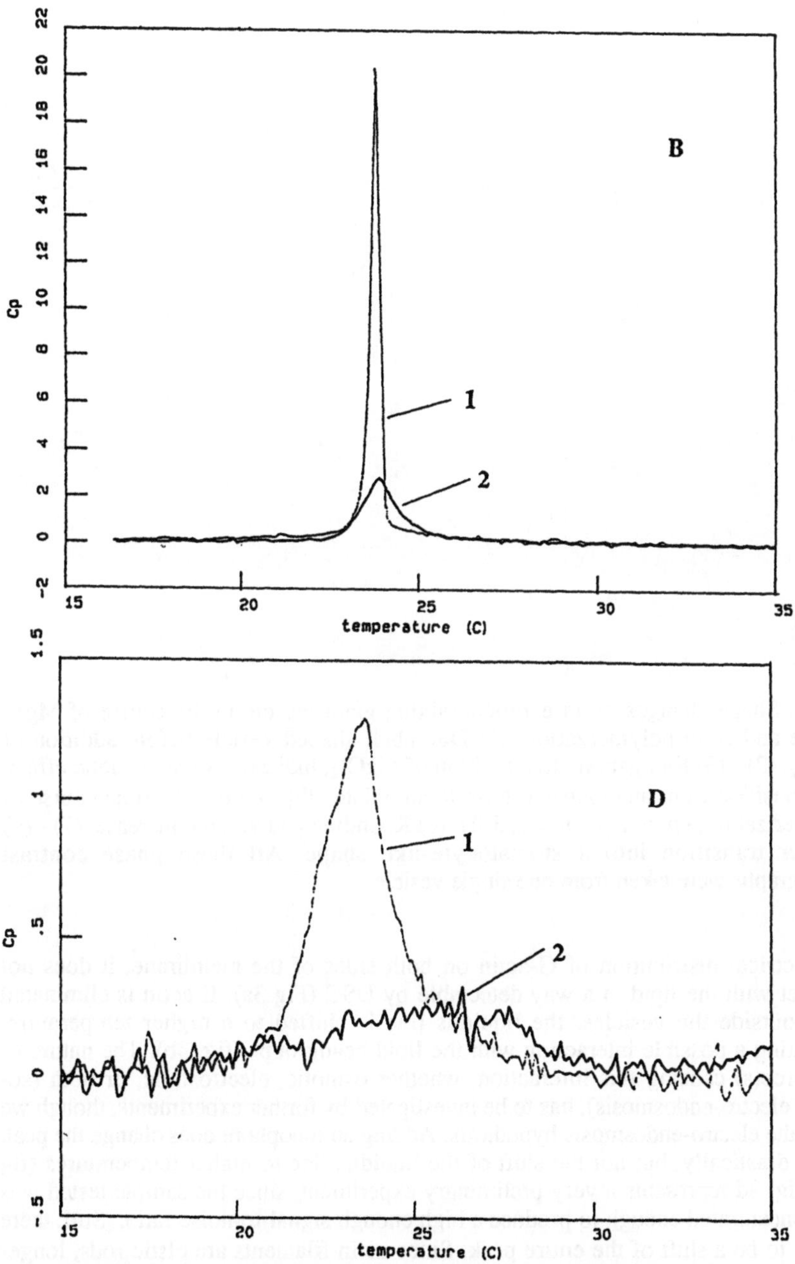

Fig. 3. Differential scanning calorimetry of the actin/lipid interaction in actin-containing vesicles. (A) 1, pure DMPC vesicles in G-buffer (1 mg/ml); 2, DMPC vesicles, swollen in the presence of G-actin. (B) 1,: same as A, 2; 2, the same preparation after chymotrypsin treatment and ultrafiltration. (C) 1, same as B, 2; 2, 100 mM A23187 added. (D) 1, same as C, 2; 2, F-actin-containing vesicles.

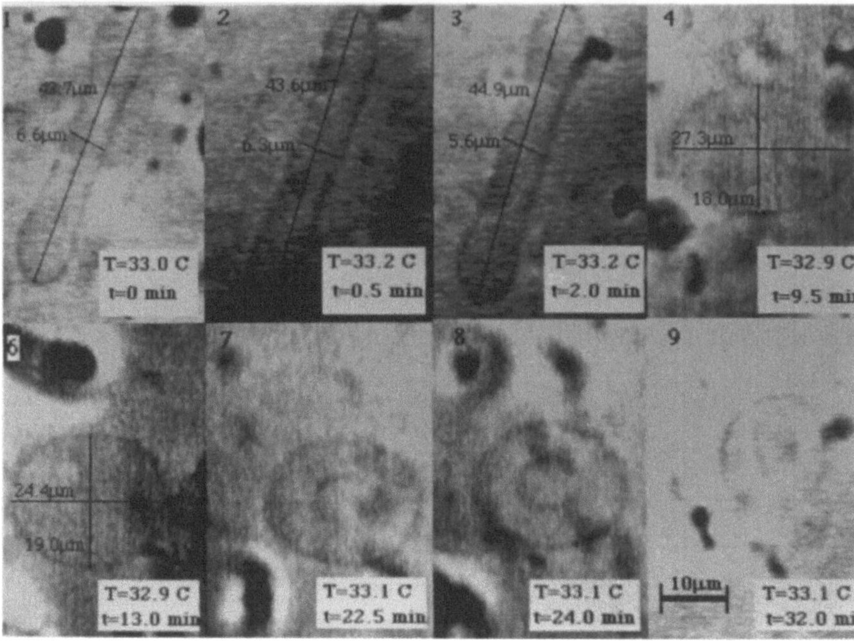

Fig. 4. Shape changes of an actin-containing giant vesicle in the course of Mg^{2+} uptake and actin polymerzation. (1) Dumbbell shaped vesicle before addition of $MgCl_2$. (2) - (3) Elongation after addition of $MgCl_2$, indicative of an osmotic effect. (4) Transition from an elongated form to an oblate ellipsoid (cf. the time course of polymerization, upper trace in fig. 3 B). (6) Rounding and volume increase. (7) - (9) Further transition into a stomatocyte-like shape. All these phase contrast micrographs were taken from one single vesicle.

symmetrical distribution of G-actin on both sides of the membrane, it does not interact with the lipid in a way detectable by DSC (Fig 3a). If actin is eliminated from outside the vesicles, the liquidus line is shifted to a higher temperature, indicating a possible interaction with the lipid headgroups (fig. 3b). The nature of the force(s) causing this interaction, whether osmotic, electrostatic, or both (so-called electro-endosmosis), has to be investigated by further experiments, though we favor the electro-endosmosis hypothesis. Adding an ionophore does change the peak shape drastically, but not the shift of the liquidus line to higher temperatures (fig 3c). Fig. 3d represents a very preliminary experiment, since the sample tested was not concentrated enough to produce a high enough signal-to-noise ratio. Still, there seems to be a shift of the entire peak. Since actin filaments are elstic rods, longer than the vesicle diameter, we suppose that - in addition to electro-endosmosis - an equilibrium of forces between filament bending elasticity and membrane curvature can explane this phenomenon.

The sequence of shape changes shown in fig. 4 demonstrates that not all effects observed can be related to actin alone. The vesicle elongation seen in the first time period after $MgCl_2$ addition is due to osmotic effects because of an increase in small ion concentration. This is in contradiction to an earlier interpretation by Cortese et al. [5]. But then, instead of relaxation of this initial osmotic effect, the vesicle rounded up in parallel to the polymerization time course, thus gaining volume, and finally passed through a transition into a stomatocyte shape.

References

1. Käs, J., and E. Sackmann, Biophys. J. **60**, 1-21 (1991)
2. Pardee, J.D., and J.A. Spudich, Meth. Enzymol. **85**, 164-181 (1982).
3. MacLean-Fletcher, S., and T.D. Pollard, Biochem.Biophys.Res.Comm.**96**, 18-27 (1980).
4. Detmers, P., et al., J.Biol.Chem.**256**, 99-105 (1981).
5. Cortese, J., et al.,Proc.Nat.Acad.Sci.USA **86**, 5773-5777 (1989).

Numerical Simulations of Vesicular and Red Blood Cell Shapes in Three Dimensions

J. Hektor[1], *W. Form*[1], *R. Grebe*[1], *and M.J. Zuckermann*[2,*]

[1]Institute of Physiology, Klinikum, RWTH-Aachen, W-5100 Aachen, Fed. Fed. of Germany

[2]Department of Physical Chemistry, Technical University of Denmark, DK-2800 Lyngby, Denmark

*Permanent address: Centre for the Physics of Materials and Department of Physics, McGill University, Rutherford Building, 3600 University Street, Montréal, Québec, Canada, H3A 2T8

Abstract We present the results of 3d-computer simulations of red blood cell and vesicular morphologies using optimization methods based on simulated annealing. Two membrane models were used in the simulations, the bilayer couple model and our own closed fluid lamina model. The free energy functionals corresponding to these models include contributions from compressive elasticities, both volume and surface, and curvature elasticity. No pronounced differences have so far been found in the simulation results for these models. In the stomatocytic and discocytic domain 3d-simulations show similar axially symmetric shapes as those found in our earlier 2d-simulations.

1. Introduction

Red blood cells (RBC) exhibit a great variety of shapes which can be produced by changing the physical and chemical nature of their environment. Typical RBC shapes include discocytes and stomatocytes with axial symmetry, sphero-echinocytes and spherostomatocytes showing local symmetry and echinocytes which exhibit a more complex symmetry. Shapes resembling stomatocytes have also been observed for large unilamellar lipid vesicles. In our earlier work we showed that typical erythrocyte shapes can be characterized by their mean mean curvature (MMC) [1]. In addition we developed a new model, the *closed fluid lamina model* [2], which is based on curvature homogeneity.

We used the *closed fluid lamina model* in conjunction with Monte Carlo methods plus simulated annealing to simulate axially symmetric erythrocyte shapes [3]. However the constraint of axial symmetry does not allow the simulation of more complex shapes such as echinocytes. It also does not constitute a proof that shapes corresponding to global minima in two dimensions also correspond to global minima in three dimensions.

These considerations prompted us to initiate simulations in three dimensions. For the simulations presented here we have used both our closed fluid lamina model and the bilayer couple model [4, 5].

 Springer Proceedings in Physics, Vol. 66 The Structure and Conformation of Amphiphilic Membranes
Editors: R. Lipowsky · D. Richter · K. Kremer

2. Models and Methods

Our studies involve curvature elasticity models for a two dimensional surface embedded in a three dimensional space. The free energies for the various elasticities are given as follows for a surface of this type

$$E_C = \frac{1}{2}\kappa \oint dA(c_1 + c_2 - c_0)^2, \qquad E_H = \kappa_H \frac{(H_0 - \overline{H})^2}{H_0} A$$

$$E_A = \kappa_A \frac{(A_0 - A)^2}{A_0}, \qquad E_V = \kappa_V \frac{(V_0 - V)^2}{V_0}$$

$$E_{Hom} = \kappa_{Hom} \oint dA \ (\nabla_A H)^2$$

where $c_{1,2}$ are the principal curvatures, c_0 is the spontaneous curvature, H is the local mean curvature, A is the surface area and V is the volume. κ is the bending rigidity, κ_{Hom} is related to curvature homogeneity and κ_V, κ_A and κ_H correspond to volume, surface area and MMC constraints respectively. Using these elastic free energies we can write the total free energy for the various membrane curvature models as follows:

Bilayer Couple Model [4, 5]:	$E = E_C + E_A + E_V \ [+E_H \ ]$,	$c_0 = 0$
Spontaneous Curvature [6]:	$E = E_C + E_A + E_V \ [+E_H \ ]$,	$c_0 \neq 0$
Closed Fluid Lamina [2]:	$E = E_A + E_V + E_H + E_{Hom}$	

[$\cdots$]: optional energy term

We use a simulated annealing method which was adapted for parallel computing (transputers) to minimize the free energies presented above in order to find equilibrium shapes for the membrane. Simulated annealing is a Monte Carlo optimization procedure in which the temperature is successively decreased according to an annealing schedule. The parallel versions of simulated annealing used in this work are the systolic algorithm and the cluster algorithm described in [7]. We found that the systolic algorithm is usually faster than the cluster algorithm whereas the latter gives better optimization results.

3. Results

The inital configuration for the numerical simulations was chosen to be an icosahedron which was then further divided by means of a triangulation of the faces. The icosahedron plus triangulation was next mapped onto a sphere and the triangulation was preserved as much as possible. However this results in the appearance of a small but well defined number of defects i.e. the icosahedron originally had 12 vertices with 5 neighbours and each additional vertex generated by triangulation has 6 neighbours.

The shapes corresponding to global minima were found using the parallelized systolic simulated annealing and the cluster simulated annealing procedure.

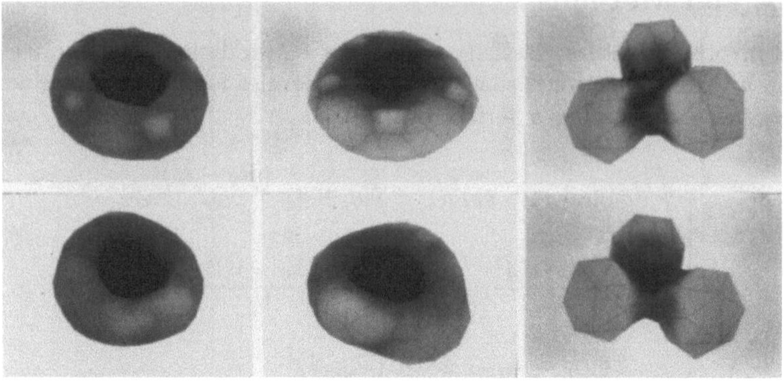

Figure 1: Shapes exhibiting global free energy minima obtained by optimization with constrained values of $v = 0.6$, $c_0 = 0$ and MMC values $\overline{h} = 0.88, 1.02, 1.40$ from left to right. The upper and lower figures show shapes obtained from the bilayer couple model and the closed fluid lamina model respectively.

In order to restrict ourselves to the region of phase space corresponding to normal erythrocyte shapes, we chose $v = 0.6$ and $c_0 = 0$. The value of the MMC $\overline{h}$ was different for every simulation.

The figure gives selected examples of shapes representing global minima for both the bilayer couple and the fluid lamina model for the region of the phase space corresponding to RBC volume and surface area [5]. If we increase the value of $\overline{h}$ starting with that corresponding to discocytes $\overline{h} = 1.02$ up to $\overline{h} = 1.40$ we find a continuous transition from the discocyte to three interconnected spheres.

4. Discussion

- We have so far found no fundamental differences between the bilayer couple model and the closed fluid lamina model in the 3d-simulations.
- In the stomatocytic and discocytic domain our 3d-simulations give axially symmetric geometries similar to those generated in two dimensions.
- Neither of the models gives results corresponding to echinocytes or spheroechinocytes in the appropriate region of phase space. Instead we find shapes composed of interconnected spheres.

We propose that the difference between erythrocyte shape and the geometries found in the simulation is due to the influence of the spectrin layer at the cytoplasmatic face of the phospholipid bilayer. Evidence for this proposal is found in our experimental results. Denaturation of spectrin by Urea causes the RBCs to change their shape to less stable geometries. The configurations found in these experiments are in fact close to those found in our 3d-simulations for a MMC of $\overline{h} \approx 1.4$.

In conclusion we emphasize that the properties of the spectrin layer have to be taken into account in the modelling of RBC shapes. They have not been included to a sufficient extent in the models used in our simulations.

References

[1] R. Grebe, Doctoral thesis, RWTH Aachen (1986).

[2] R. Grebe and M.J. Zuckermann, Biorheology **27**(5), 735-746, (1990).

[3] G. Viehöver and R. Grebe, in *Parallele Datenverarbeitung mit dem Transputer*, edited by R. Grebe and C. Ziemann (Springer, Berlin, Heidelberg, 1991), IFB **272**.

[4] E.A. Evans and R. Skalak, CRC Crit. Rev. Bioeng. **3**, 181-418 (1979).

[5] S. Svetina and B. Žekš, Eur. Biophys. J. **17**, 101-111, (1989).

[6] W. Helfrich, Z. Naturforsch. **28**, 639-703 (1973).

[7] E.H.L. Aarts, F.M.J. de Bont, E.H.A. Habers and P.J.M. van Laarhooven, INTEGRATION, VLSI J. **4**, 209-238 (1986).

Dynamic Coupling and Nonlocal Curvature Elasticity in Bilayer Membranes

E. Evans[1], *A. Yeung*[1], *R. Waugh*[2], *and J. Song*[2]

[1]Pathology and Physics Departments, University of British Columbia, Vancouver, Canada

[2]Department of Biophysics, University of Rochester, NY 14642, USA

1. Introduction

Surfactant membranes and interfaces represent important classes of self-assembled structures with a rich set of phases and phase transitions. Because of the exceptional flexibility, thermal excitations lead to surface "roughening" and dynamic shape fluctuations [1,2]. These special features have stimulated many theoretical developments over the past twenty years beginning with simple elastic models [3-4] and shape predictions [5], followed by thermodynamic and dynamic analyses of fluctuations [2,6], studies of adhesion [7-8] and unbinding transitions [9-11], and - most recently - phase transition models for the unusual shape changes in vesicles [12-13]. Almost all of this theory has been conceptualized on the basis that the membrane can be treated as a "unit" surface structure. However, the two (outer/inner) molecular layers of bilayers are held together only by weak van der Waals attraction. Hence, it is expected that the layers can be slowly displaced relative to one another without change in energy provided that the surface geometry remains constant; but, rapid displacements should create forces as the result of interlayer-viscous drag (dynamic coupling). The significance of coupling in bilayer membranes is to introduce a new level of physics into the consideration of molecularly-stratified liquid interfaces. In the discussion to follow, we outline the special features of weakly-bonded <u>bi</u>layer membranes in relation to recent experiments that clearly demonstrate "nonlocal" elastic coupling and provide direct estimates of the viscous dissipation at the molecular interface between monolayers.

2. Experiment: Microtube Extrusion from Macroscopic-Bilayer Vesicles

Nonlocal curvature elasticity and viscous drag between monolayers are major effects in the mechanical extrusion of a microscopic bilayer tube from the surface of a macroscopic vesicle as shown in Fig. 1. Here, the microtube is pulled-out by the buoyant weight of a macroscopic-size bead of glass

Springer Proceedings in Physics, Vol. 66 The Structure and Conformation of Amphiphilic Membranes
Editors: R. Lipowsky · D. Richter · K. Kremer

attached to the membrane. During extrusion, membrane tension is controlled by micropipet suction applied to the vesicle. The vesicle is slightly dehydrated before aspiration to provide a reservoir of membrane material which meters formation of the microtube; the internal volume of the vesicle is held fixed by osmotic activity of solutes and low membrane permeability. Assuming the mid-surface is "impermeable" to exchange of molecules, the differential area $\Delta A_{\pm}$ between outer and inner monolayers is driven to increase with the length L_t of the microtube, $\Delta A_{\pm} \sim 2\pi h L_t$ (to first order in thickness h). Membrane elasticity implies that a bilayer moment (stress "couple") $M_{\pm}$ also increases with L_t. Because the radius r_t of the microtube in the experiment is so much smaller than the radius R_o of the spherical segment of the vesicle outside the pipet ($r_t/R_o \sim 10^{-3}$), the net bilayer tension $\bar{\tau}_o$ is essentially set by the pipet suction P, i.e. $\bar{\tau}_o \sim R_p P$. However, for the cylindrical microtube, curvature elasticity (k_c) and monolayer tensions dominate the balance of axial force f_z,

$$f_z \simeq 2\pi r_t \bar{\tau}_o + 2\pi M_{\pm} + 2\pi k_c / r_t \tag{1}$$

as well as normal stress across the tube wall,

$$\bar{\tau}_o / r_t \simeq k_c / r_t^3 \tag{2}$$

Thus, the net bilayer tension - scaled by the monolayer bending stiffness - sets the size of the microtube. The excess microtube force (above that given by the net bilayer tension) produces differential "stretch" between the two monolayers (i.e. the bilayer moment $M_{\pm}$ is proportional to the differential dilation $\alpha_{\pm}$ field).

Growth of the microtube in Fig. 1 is controlled by a special procedure: First, the microtube is held close to the vesicle body by large tension (i.e. large pipet suction). Then, the tension is dropped to a low value which creates an excess-axial force; this initiates extrusion of the microtube as shown by the measurements plotted in Fig. 2. Because the total vesicle area A remains constant (to better than 1:1000), the transfer of membrane material from the pipet reservoir - over the vesicle body - to the microtube provides an "amplifier" for measuring the radius of the unobservable microtube, i.e. $r_t \approx -R_p(1-R_p/R_o)\Delta L_p/\Delta L_t$. Growth of the microtube at constant radius is clearly demonstrated by the results plotted in Fig. 3. The results also show that the product $r_t^2 \bar{\tau}_o$ is a constant ($\sim 0.5 \times 10^{-12}$ erg), i.e. the intrinsic bending modulus of a monolayer.

As shown in Fig. 2, approach to equilibrium following a reduction in bilayer tension is exceedingly slow and not a single exponential relaxation. Ideally, if the monolayers could freely "slip" past one another, the response would be limited by hydrodynamic drag on the glass

Fig.1

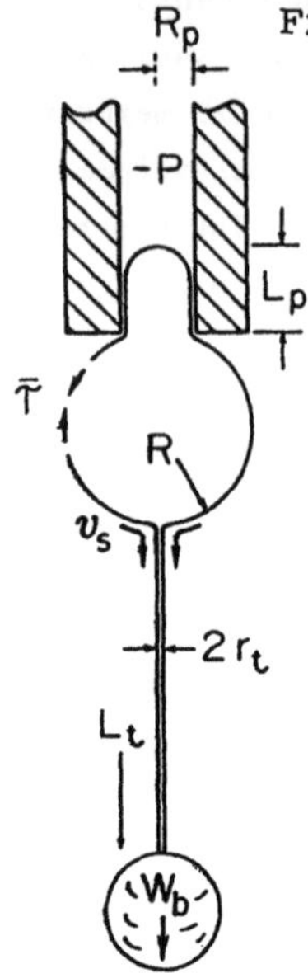

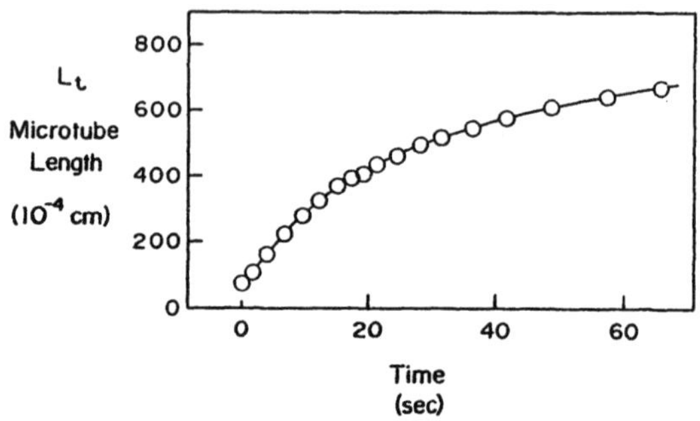

Fig.2

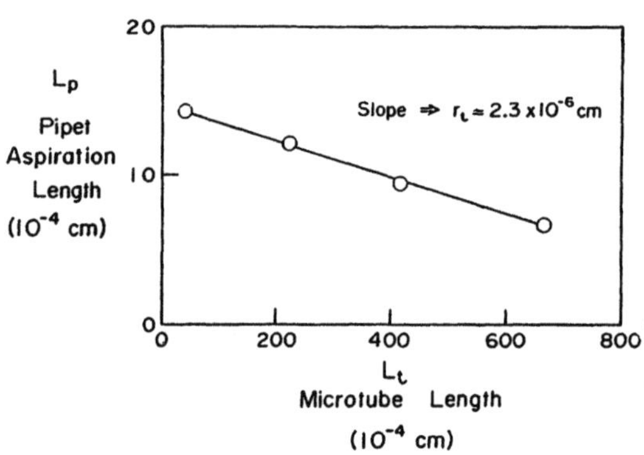

Fig.3

bead (and microtube) plus the fluid shear forces induced in the membrane by convergent flow to the microtube. The major dissipative contribution is the Stokes drag on the bead ($6\pi R_b \eta_w \dot{L}_t$) where R_b is the radius of the glass bead and η_w is the coefficient of viscosity for water. Based on the observed rate of microtube growth, the Stokes drag on the glass bead opposes only about 25-30% of the extrusion force ($\approx W_b - 4\pi\sqrt{k_c \cdot \bar{\tau}_o}$) at the most rapid phase of growth and quickly falls to a few percent. Hence, the origin of the relaxation process must be dissipation at the monolayer-monolayer interface.

3. Force Relaxation and Dissipation in Bilayers

If there is no "slip" between monolayers of a bilayer, lateral diffusion and interlayer exchange ("flip-flop") will eventually relax surface density gradients in the monolayers. Conservation of the constituents in each monolayer leads to the following evolution equation for the differential dilation field $\alpha_{\pm}$:

$$\frac{\partial \alpha_{\pm}}{\partial t} + v_s \cdot \nabla_s \alpha_{\pm} = D \cdot {\nabla_s}^2 \alpha_{\pm} + h \left\{ \frac{\partial \bar{c}}{\partial t} + v_s \cdot \nabla_s \bar{c} \right\} - c_p \alpha_{\pm} \tag{3}$$

where v_s is the local velocity of the "mean" surface flow and ∇_s is the surface gradient along the meridional direction. Thus, dynamic changes in curvature - or flow between regions of high and low curvature - produce local transients in the differential dilation field which are dissipated by surface diffusion and interlayer permeation (modeled by the phenomenological coefficient c_p). When the layers are forced to "slip", the relative motion will be opposed by viscous drag at the layer-layer interface. Assuming a first order model for viscous drag, mechanically-driven slip between the layers leads to a collective diffusion-like process where the coefficient of diffusion is given by the ratio of the monolayer elastic expansion modulus K_a to the interfacial drag coefficient b, $D_r = K_a/2b$.

In microtube extrusion, the "mean" surface flow is given simply by $|v_s| = L_t r_t/r$. Therefore, except for a small microscopic region close to the microtube, relaxation of the differential dilation field reduces to an ordinary diffusion process over the vesicle body, driven by a "source" at one pole. The value of the "flux" is determined by the kinematics of flow over the microscopic region proximal to the microtube. Here, the dominant term is the curvature "flux" $|v_s \nabla_s \bar{c}|$ which is the differential-dilation produced by flow from the low-curvature ($\bar{c} \sim 10^3 cm^{-1}$) vesicle onto the high-curvature ($\bar{c} \sim 10^6 cm^{-1}$) tube. Hence, the diffusion flux $D|\nabla_s \alpha_{\pm}|$ toward the source feeds the curvature flow $h\dot{L}_t/r_t$,

$$\begin{aligned} \dot{L}_t &\simeq 2D \cdot \hat{\alpha}_{\pm} e^{-t/t_D} / h \, \ell n \left(A/\pi {r_t}^2 \right) \\ t_D &\equiv A \cdot \ell n \left(A/\pi {r_t}^2 \right) / 4\pi D \end{aligned} \tag{4}$$

Neglecting hydrodynamic flow (e.g. bead drag) and recognizing that the gradient of the differential dilation field will be small over the region of high curvature close to the microtube, the "temperature" $\hat{\alpha}_{\pm}$ (which drives the diffusion process) is determined by the value of the bilayer-couple stress $M_{\pm}$ in the microtube,

$$\hat{\alpha}_{\pm} = \left(W_b - 4\pi\sqrt{k_c\bar{\tau}_o}\right)/\pi K_a h \tag{5}$$

Hence, a single exponential relaxation to the equilibrium length is predicted where the time constant t_D scales as the vesicle area divided by the coefficient for diffusion, essentially independent of the size of the microtube. Further, the growth rate early in the relaxation process ($t < t_D$) depends only on the excess force scaled by the ratio of diffusivity to the nonlocal curvature elasticity.

4. Observations

Except for rapid growth of the microtube at short times (<20 sec), most of the relaxation is a slow exponential decay to the final equilibrium length. The time constant for the long-term growth is ~ 10^2sec which implies a coefficient of diffusivity on the order of 5-10 x 10^{-8}cm^2/sec. Similar diffusivity values have also been obtained from kinetic measurements of lipid surface diffusion in bilayers which supports the conclusion that the slow approach to equilibrium is limited by molecular diffusion. More interesting, however, is the initial-rapid growth of the microtube over the first 20 seconds. Here, the rates of growth are still much too slow to create enough hydrodynamic force to off-set the excess-axial force. As such, viscous drag at the monolayer-monolayer interface remains the dominant resistance to microtube growth. It is likely that the initial-rapid phase of growth involves slip between the monolayers. Based on the observed growth rates, the mechanically-driven diffusivity is estimated to be an order of magnitude larger than molecular diffusion, i.e. D_τ ~ 10^{-6}cm^2/sec. Since the value of the elastic expansion modulus is known (~ 10^2dyn/cm), the interfacial drag coefficient can be calculated from the diffusivity, i.e. ~ 5 x 10^7 dyn-sec/cm^3. Simple "top-bottom" surface shear of a thin liquid layer implies a drag coefficient equal to the viscosity divided by thickness, i.e. η_ℓ/h. Based on the microviscosity deduced for the acyl chain core of the bilayer (η_ℓ ~ 1 dyn-sec/cm^2), the interfacial drag coefficient should be only ~ 3 x 10^6 dyn-sec/cm^3 for an equivalent film. Thus, momentum exchange appears to be concentrated at the bilayer midplane as expected for interaction between two dense "brush-polymer" surfaces. The penetration depth of the shear is only about 3 Å, comparable to the size of the terminal methyl group.

5. References

[1] R. Lipowsky, Phys. Scr. T29, 259 (1989).

[2] M.D. Schneider, J.T. Jenkins, and W.W. Webb, J. Phys. (Paris) 45, 1457, (1984).

[3] W. Helfrich, Z. Naturforsch. 28c, 693 (1973).

[4] E. Evans, Biophys. J. 14, 923 (1974).

[5] H.J. Deuling and W. Helfrich, Biophys. J. 16, 861 (1976).

[6] W. Helfrich and R.-M. Servuss, Nuovo Cimento D3, 137 (1984);

[7] E. Evans, Colloids and Surfaces 43, 227 (1990).

[8] U. Seifert and R. Lipowsky, Phys. Rev. A 42, 4768 (1990).

[9] W. Helfrich, Z. Naturforsch. 33a, 305 (1978).

[10] R. Lipowsky and S. Leibler, Phys. Rev. Lett. 56, 2541 (1986).

[11] E. Evans, Langmuir (in press, 1991).

[12] K. Berndl, J. Käs, R. Lipowsky, E. Sackmann, and U. Seifert, Europhys. Lett. 13, 659 (1990).

[13] L. Miao, B. Fourcade, M. Rao, and M. Wortis, Phys. Rev. A. (in press, 1991).

Phospholipid Membrane Local and Non-Local Bending Moduli Determined by Tether Formation from Aspirated Vesicles

S. Svetina[1], *B. Božič*[1], *J. Song*[2], *R.E. Waugh*[2], and *B. Žekš*[1]

[1]Institute of Biophysics, Medical Faculty, Lipičeva 2, 61105 Ljubljana, Slovenia and J. Stefan Institute, Jamova 39, 61000 Ljubljana, Slovenia

[2]Department of Biophysics, University of Rochester, School of Medicine and Dentistry, Rochester, NY 14642, USA

Abstract. An elastic resistance of closed bilayer membrane to changes in curvature is treated in terms of local and non-local bending contributions to the membrane elastic energy. The method for the measurement of the corresponding elastic moduli by tether formation from giant phospholipid vesicles is described. The theoretical analysis of the method reveals the importance of the non-local bending for the stability of the tether formation. For the lipid 1-stearoyl-2-oleoyl-phosphatidylcholine containing smaller amounts of 1-palmitoyl-2-oleoyl-phosphatidylserine the mean value of the bending constant obtained is $1.2 \times 10^{-19} J$ and an estimate for the mean value of the non-local bending modulus is $4 \times 10^{-19} J$.

Phospholipid membranes exhibit specific elastic properties which are the consequence of their lamellarity. The significance of this for the behavior of closed membranes was first recognized by Evans [1,2] and Helfrich [3] who showed that there is a non-local contribution to the elastic energy of the membrane which is proportional to the integral of the mean curvature over the whole membrane. More recently, the corresponding relative stretching (relative expansivity) elastic energy term was explicitly derived from the elastic properties of the two monolayers constituting the bilayer and the geometrical constraint these two layers impose on each other [4]. The analysis revealed that the local and non-local bending moduli represent two independent elastic parameters of phospholipid bilayers. Experimental interest has in the past focused only on the determination of the local bending modulus ([5] and references therein). Recently a novel experimental approach involving the formation of thin bilayer cylinders (tethers) from giant phospholipid vesicles has been developed [6] and analyzed [7] to enable the determination [8] of both local and non-local bending moduli. The present report summarises this work. Firstly, the two bending moduli are defined for a closed symmetrical bilayer. Then the analysis of the tether experiment is presented emphasizing the stability of the system. Finally, the results of measurements are presented and discussed.

The two layers of a phospholipid bilayer are held in contact by hydrophobic forces but can be assumed to slide freely one by the other. Then the elastic

Springer Proceedings in Physics, Vol. 66 The Structure and Conformation of Amphiphilic Membranes
Editors: R. Lipowsky · D. Richter · K. Kremer

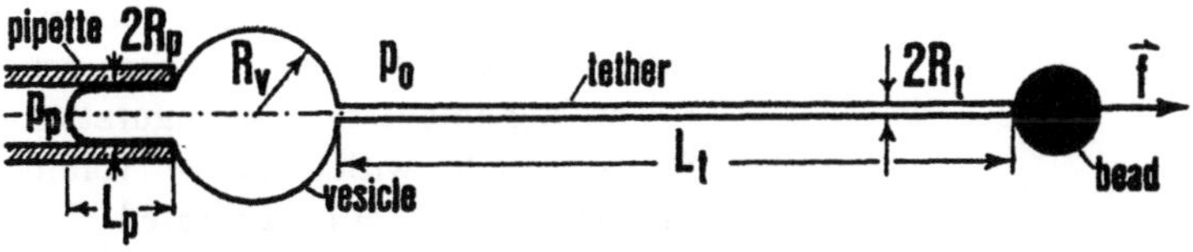

Figure 1: Schematic representation of the vesicle, which is aspirated in a pipette and pulled with constant force in the opposite direction. The shape is described by one parameter (pipette radius R_p) and four variables (pipette projection length L_p, vesicle radius R_v, tether length L_t, and tether radius R_t). The pressure surrounding the vesicle is p_0, the pressure in the pipette is p_p, and the force pulling the tether is f.

properties of the bilayer are determined by the elastic properties of the individual constituent layers subject to the constraint that their areas are interrelated by:

$$A^{ex} = A^{in} + h \int (c_1 + c_2) dA_0. \tag{1}$$

Here A^{ex} and A^{in} are the areas of the external and the internal layers, respectively, at their neutral planes, h is the distance between these neutral planes, c_1 and c_2 are the two principal curvatures, and integration is over the area of the neutral plane of the bilayer, A_0. Following a more general treatment [4], the elastic energy of a closed symmetric bilayer (W_{el}), i.e. a closed bilayer composed of layers of the same composition, is given by the sum of the area expansivity, bending and relative expansivity terms, respectively:

$$W_{el} = \frac{1}{2}K\frac{(A-A_0)^2}{A_0} + \frac{1}{2}k_c \int (c_1+c_2)^2 dA_0 + \frac{1}{2}\frac{k_r}{A_0}\left(\frac{\Delta A}{h} - \frac{\Delta A_0}{h}\right)^2, \tag{2}$$

where K, the area expansivity modulus of the bilayer, is equal to twice the area expansivity of a constituent monolayer, k_c, the membrane bending modulus, is equal to twice the bending modulus of a constituent monolayer, and k_r, the non-local bending modulus, is equal to $h^2K/4$. In Eq. 2, A is an expanded area of the neutral plane of the bilayer, ΔA is the difference between the areas of the two leaflets ($A^{ex} - A^{in}$), and ΔA_0 is the corresponding equilibrium difference. $\Delta A/h$ is equal to the integral of the curvature over the membrane area (see Eq. 1), and $\Delta A_0/h$ to the corresponding equilibrium value.

In the experimental procedure to be analyzed [6], a vesicle is aspirated into a pipette, forming a spherical portion outside the pipette and a projection within the pipette. Then the vesicle is placed into contact with a glass bead which sticks to it, i.e., a vesicle-bead pair is formed. After reducing the aspiration pressure the bead falls away from the vesicle, forming a tether between the bead and the body of the vesicle. The pressure can be adjusted to stop the bead movement, which means that equilibrium can be established. In this work the shape of the tethered vesicle is approximated by a simple geometrical model described in Fig. 1. The analysis involves the determination of the elastic energy of the membrane for a given configuration of the tethered vesicle, and

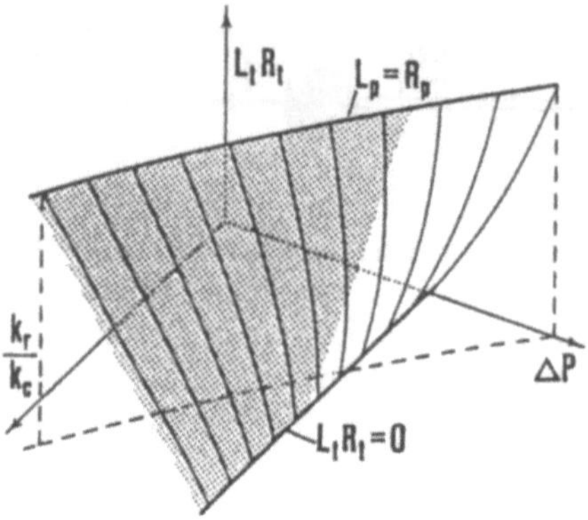

Figure 2: Points on the theoretically obtained displayed surface represent possible tether configurations for a vesicle with relative volume $3V_0/4\pi R_0^3 = 0.7$ aspirated into the pipette with the relative radius $R_p/R_0 = 0.4$ (R_0 is the radius of a sphere with the area A_0) for different values of the ratio k_r/k_c and dimensionless parameter $\Delta P = 4\pi^2 R_0 k_c \Delta p/(f+4\pi^2 k_r L_t^*/A_0)^2$. Tether configurations are stable within the shaded part of the surface.

the search for the configuration which corresponds to the minimum of the free energy of the system (F) comprising besides the elastic energy described above also the gravitational potential energy of the bead ($-fL_t$) and the contribution of the work done by the membrane due to hydrostatic pressure differences ($-\Delta p\pi R_p^2 L_p$ where $\Delta p = p_0 - p_p$). The significant contributions to the elastic energy of this system are [7] bending energy of the tether cylinder and the relative expansivity due to changes of the tether length, therefore the functional to be minimized with respect to the system variables reads

$$F = \pi k_c \frac{L_t}{R_t} + \frac{2\pi^2 k_r}{A_0}(L_t - L_t^*)^2 - fL_t - \Delta p\pi R_p^2 L_p. \tag{3}$$

With parameter L_t^* in Eq. 3 it is taken into consideration that the area difference of the constitutive vesicle is not necessarily the same as the area difference of the aspirated vesicle with the spherical external part, i.e. that the possibility exists that the areas of the layers of the aspirated vesicle may be relatively expanded (or compressed) before the tether is made.

To obtain the equilibrium tether configuration the first derivatives of the free energy, Eq. 3, are determined with respect to all the independent shape variables and set to zero. There are two such variables because the area expansivity contribution to the elastic energy (see Eq. 2) can be neglected and thus the behavior of the system can be studied under conditions of constant membrane area, and because the permeability of phospholipid membranes to water is low and thus the vesicle volume (V_0) can also be taken to be constant.

A convenient way to visualize the role of non-local bending in tether formation is represented in Fig. 2. The stability analysis reveals that only those tether configurations are stable for which at a given value of k_r/k_c the derivative of the product L_tR_t by pressure difference is negative. It can therefore be depicted from Fig. 2 that there is a certain critical value of the ratio k_r/k_c below which stable tether configurations can not exist.

The procedures for making vesicles and forming tethers are described elsewhere [6]. Experiments were performed [8] with the lipid 1-stearoyl-2-oleoyl-phosphatidylcholine plus smaller amounts of the charged lipid 1-palmitoyl-2-oleoyl-phosphatidylserine (2 % and 10 %). Essentially, the length of the projection in the pipette and the aspiration pressure were measured as functions of the tether length. From such data obtained from several vesicles the mean values obtained for the local and non-local bending moduli k_c and k_r are $1.2 \pm 0.2 \times 10^{-19} J$ and $4.1 \pm 2.4 \times 10^{-19} J$, respectively, and do not show any dependence on the percentage of charged lipid. The range of values obtained for k_r from different vesicles is unexpectedly broad. However, recent studies on the dynamic behavior of tethers reveal a possible effect of interlayer transport (flip-flop) of phospholipid molecules on the behavior of the system that could account for the variability of the present results.

The experimentally obtained ratio $k_r/k_c \approx 3.3$ can be [8] compared to a theoretical estimate. By using an approximative expression for the bending modulus, $k_c = Kh_m^2/12$ where h_m is the thickness of one monolayer, we get $k_r/k_c = 3h^2/h_m^2$. From the obtained mean value of k_r and a value of $0.2N/m$ for the membrane expansivity modulus we estimate that h is also about half the membrane thickness. The so predicted ratio $k_r/k_c \approx 3$ is thus in accordance with our observation.

References

[1] E.A. Evans: Biophys. J. **14**, 923 (1974)
[2] E.A. Evans: Biophys. J. **30**, 265 (1980)
[3] W. Helfrich: Z. Naturforsch. **29**, 510 (1974)
[4] S. Svetina, M. Brumen, B. Žekš: Stud. Biophys. **110**, 177 (1985)
[5] H.P. Duwe, J. Kaes, E. Sackmann: J. Phys. **51**: 945 (1990)
[6] L. Bo, R.E. Waugh.: Biophys. J. **55**, 509 (1989)
[7] B. Božič, S. Svetina, B. Žekš, R.E. Waugh: Biophys.J., submitted
[8] R.E. Waugh, J. Song, S. Svetina, B. Žekš: Biophys.J., submitted

Vesicle–Substrate Interaction Studied by Reflection Interference Contrast Microscopy

J. Rädler and E. Sackmann

Physik Department, Biophysics Group E22, Technische Universität München, James Franck-Str., W-8046 Garching, Fed. Rep. of Germany

Abstract. Giant DMPC-Vesicles interacting with a supported DMPC-bilayer are investigated by Reflection Interference Contrast Microscopy (RICM). Spherical vesicles whose shape fluctuations are suppressed by osmotic pressure can be observed to fluctuate like Brownian particles above the substrate. The interaction potential can be determined from the measured distribution function of distances. In the case of adhesion, the contact contour of deflated vesicles can be calculated. Contact roundings for of weak adhesion as well as contact angles for strong adhesion have been measured.

1 Introduction

The adhesion of vesicles to a wall has recently become of interest in the context of the variety of shape transformations that free vesicles exhibit [1]. Clearly, in the case of adhering vesicles the equilibrium shape and the dynamics of the fluid membrane is strongly dependent on the strength of the vesicle wall interaction [2]. We investigated repulsive vesicle wall interaction as well as vesicle adhesion.

2 Experimental setup and methods

The experimental setup is depicted in Fig.1. The vesicles are studied by reflection interference contrast microscopy. By this technique interference is observed between the object beam reflected at the membrane and the reference beam reflected at the glass-buffer interface. In the case of an osmotically swollen, spherical vesicle the interference pattern are known as Newton rings and the absolute sphere-substrate distance can be measured from the position of the fringes. Consequently, fast image processing allows monitoring distance fluctuations in real time [3].

Furthermore, the interference fringes at the edge of the contact zone of adhering vesicles of arbitrary shape can be used to calculate the vesicle contour close to the surface. Thus the nature of the equilibrium shape of adhering vesicles in the contact zone can be studied; i.e. the contact angle for strongly adhering vesicles or smooth contact curvatures in the case of weak adhesion.

Special thought must be given to the preparation of the surface of the substrate. Since optical techniques require glass as the underlying substrate, surface

Springer Proceedings in Physics, Vol. 66 The Structure and Conformation of Amphiphilic Membranes
Editors: R. Lipowsky · D. Richter · K. Kremer

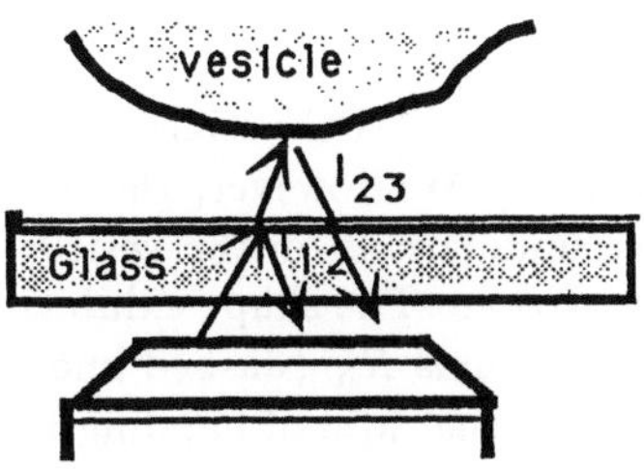

Figure 1: Principle of the reflection interference contrast microscopy. The interference between the object beam I_{23} and the reference beam I_{12} is observed.

roughness and unspecific electrostatic effects are considerable shortcomings in vesicle adhesion experiments. Most glass effects, however, are suppressed by silanizing the glass surface, i.e. chemically attaching C_{18} hydrocarbon chains to the surface and depositing a lecithin monolayer on top. Using this kind of substrate supported monolayer has the advantage of studying the symmetric lecithin-lecithin interaction that has already been measured by other techniques.

3 Repulsive interaction

Prieve et al. [4] showed recently that the interaction of a sphere which by Brownian motion fluctuates above a repulsive surface can be measured by sampling the distance distribution function of the random motion. The distribution function of distances contains the interaction potential simply by the Boltzmann relation. Giant DMPC vesicles are observed not to adhere to DMPC supported monolayers at low ionic strength ($< 20mM$) and osmotically swollen vesicles show distance fluctuations. The negative logarithm of the measured distance distribution is shown in Fig. 2. The solid line indicates a best fit to the theoretical interaction potential including electrostatic, Van der Waals and gravitational forces. The figure also shows that under the same conditions but increased ionic strength the theoretical interaction becomes attractive as experimentally observed (dashed line).

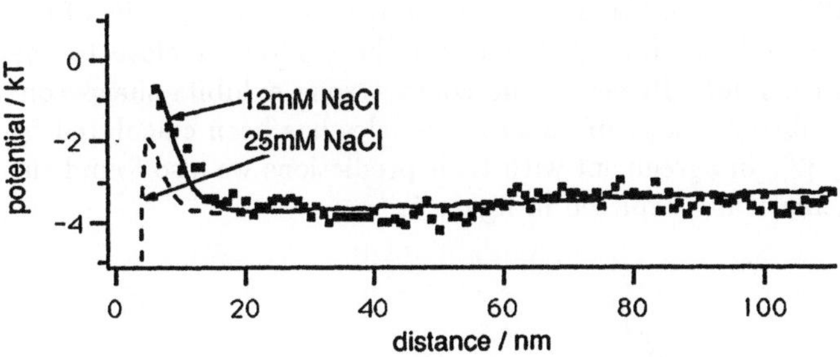

Figure 2: The interaction potential obtained from the distance fluctuations of a spherical DMPC vesicle above a DMPC supported monolayer in 100mM inositol plus 12mM NaCl. The solid line depicts the theoretical interaction potential for $\Psi = 1.8mV$. Increasing the ionic strength to 25mM leads to adhesion due to Van der Waals forces as indicated by the dashed line.

4 Weak adhesion

Adhesion is observed, if the ionic strength in a system of DMPC vesicles on a DMPC supported monolayer surface is increased to 20mM. However, the adhesion is weak which is obvious from visible surface undulations on the vesicle. Fig. 3 depicts the contour of the vesicle in the contact zone. The shape exhibits a contact rounding that is fitted by a circle of contact radius R_K (dashed line). The contact curvature has been predicted to depend on the ratio of bending to adhesion energy [6, 2] :

$$R_K = \sqrt{2K_C/\gamma} \tag{1}$$

Taking K_C to be $10^{-19}J$ the adhesion energy turns out to be on the order of $1.2 \cdot 10^{-8}J/m^2$. This value is small compared to $10^{-5}J/m^2$ measured by micropipette [6] and surface force technique [7]. However, in those cases tension is applied to the membranes, while in contrast at very low tension γ is reduced due to steric repulsion of the undulating membrane [8].

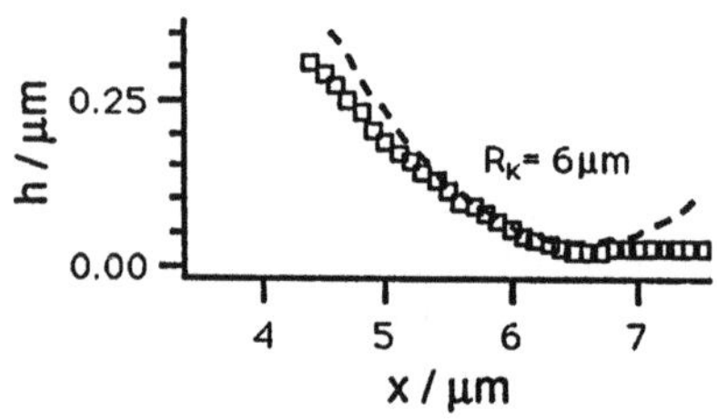

Figure 3: Calculated contour from the interference fringes of a DMPC-vesicle on a DMPC supported monolayer. The contour rounding gives an estimate for the adhesion energy. The dashed line depicts a circle of contact radius R_K (distorted scale !).

5 Strong adhesion

Strong intéraction is achieved by incorporating negative charges into the supported monolayer and positve charges into the vesicle. Fig. 4 shows an example of a vesicle (SOPC:Chol:DODAB) (49:49:2 mol%) adhered to a supported monolayer (DMPC:PS) (98:2 mol%). It depicts the reconstructed profile of the vesicle. The vesicle is only slightly deflated and the contour is close to the spherical shape (dashed line). However, the contact zone exhibits sharp contact angles. The equilibrium shape of adhering vesicles has been calculated by Seifert and Lipowsky [2]. In agreement with their predictions we also found the following shapes schematically depicted in figure 5.

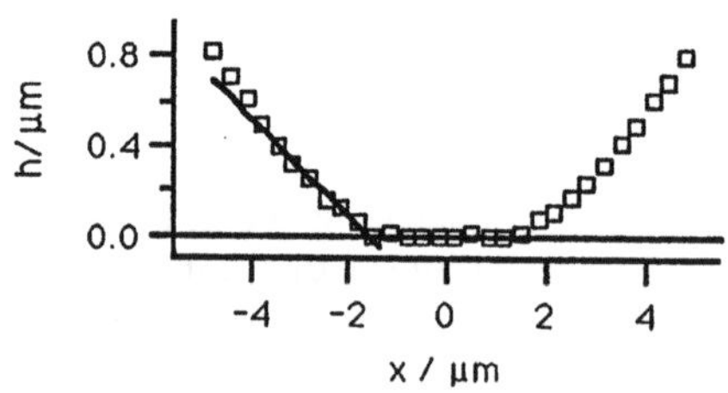

Figure 4: Reconstructed contour of a strongly adhering vesicle. The contact zone shows a contact angle.

Figure 5: Schematic representation of experimentally found shapes.

Acknowledgments

We like to thank A. Zilker for support with the image processing system. We also gratefully acknowledge the financial support by the DFG SA 246/20 and by the Fond der chemischen Industrie.

References

[1] Kaes,J. and Sackmann, E., Bio. Phys J. **60** (1991) 1

[2] Seifert,U. and Lipowsky,R., Phys.Rev A **42** (1990) 4768

[3] Raedler,J.and Sackmann,E. submitted

[4] Prieve,D.C. and Frey,N.A., Langmuir **6** (1990) 396

[5] Evans,E.A. and Parsegian,V.A., in Surface Phenomena in Hemorheology : Copley,A.L. and Seaman,G.V.F.,Eds., N.Y. Acad. Sci. (1983) 13

[6] Evans,E.A. and Metcalf,M., Biophys. J. **46** (1984) 423

[7] Marra,J. and Israelachvili.J., Biochemistry **24** (1985) 4608

[8] Servuss,R.M. and Helfrich,W. J.Phys. France **50** (1989) 809

Surface Induced Fusion of Vesicles into Planar Bilayers

G. Cevc[1], *W. Fenzl*[2], *and L. Sigl*[3]

[1]Urologische Klinik, Technische Universität München, W-8000 München 8, Fed.Rep.of Germany
[2]Institut für Festkörperforschung, Forschungszentrum Jülich, Postfach 1913, W-5170 Jülich, Fed. Rep. of Germany
[3]Sektion Physik, Ludwig-Maximilians-Universität München, Theresienstr. 37, W-8000 München 22, Fed.Rep.of Germany

Abstract: The spontaneous formation of planar bilayers at the surface of lipid vesicle suspensions has been detected using x-ray reflectivity as a surface sensitive technique.The initial reflectivity is found to be identical to that of a lipid monolayer. At a critical temperature (29°C for DMPC bilayers) a series of sharp peaks is observed in the reflectivity, indicating the formation of planar bilayers at the surface. It is proposed that critical fluctuations of the chain melting transition, gradually decreasing from 29°C at the surface to 23.5 °C in the bulk, promote fusion of surface aligned vesicles. Our finding provides a general method for the preparation of liquid-surface supported, well oriented membrane specimens.

Introduction

A monomolecular film of lipids is formed spontaneously at the surface of a vesicle suspension /1/. This technique is often used to prepare monolayers free of organic solvents /2/. The details of the molecular reorganisation of vesicles into a monolayer, however, are not well understood.
Several years ago N.Gershfeld has proposed, on the basis of radioactive tracer and surface tension measurements, that at some lipid-dependent critical temperature single bilayers should form spontaneously at the surface of a vesicle suspension /3/. The results of Gershfeld`s experiments have attracted little attention, perhaps because of the circumstantial experimental evidence. Today there exist well established x-ray or neutron methods to study liquid surfaces with molecular resolution /4/. We have therefore investigated the possible formation of bilayers at the surface of vesicle suspensions using x-ray reflectivity as a surface sensitive technique /5/

Springer Proceedings in Physics, Vol. 66 The Structure and Conformation of Amphiphilic Membranes
Editors: R. Lipowsky · D. Richter · K. Kremer

Results

The vesicle suspensions were prepared by mixing 100 mg of dimyristoylphosphatidylcholine (DMPC, Boehringer Mannheim) with 5 ml of ultrapure water and ultrasonication. They were filled into a round trough situated at the horizontal x-ray reflectivity instrument D4 at HASYLAB, DESY in Hamburg. In Fig.1 the reflectivity of the surface of a vesicle suspension is shown.

The data are almost identical to that of a monolayer spread from organic solvents /4/. A simplified model of the electron density distribution perpendicular to the surface consists of one box for the hydrocarbon tail and a second one for the polar headgroup. The parameters of the best fit, shown as a solid line in Fig.1, agree well with the corresponding dimensions of the DMPC molecule. We have therefore obtained for the first time direct structural evidence for a lipid monolayer at the surface of a vesicle suspension.

From Gershfeld's experiments it is known that for the phospholipid DMPC a change in the surface structure of the vesicle suspension should occur at a critical temperature T=29°C. In Fig.2 the result of a reflectivity experiment after annealing a sample 12 hours at this temperature is shown. The most striking effect is that several sharp peaks have evolved from the initially smooth monolayer curve. The experiment has been repeated several times at different temperatures. It is found that the new surface structure appears only at 29±0.5°C.

The sharp peaks observed in the reflectivity are a characteristic feature of a periodicity perpendicular to the surface. A model for the electron density near

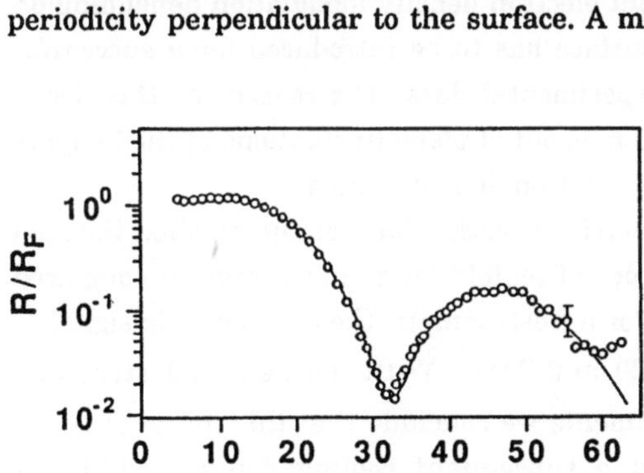

Fig.1: X-ray reflectivity of the surface of a suspension of DMPC vesicles in water at T=25°C divided by the Fresnel reflectivity vs. angle of incidence.The solid line is the result of a model calculation assuming a monolayer at the air-suspension interface.

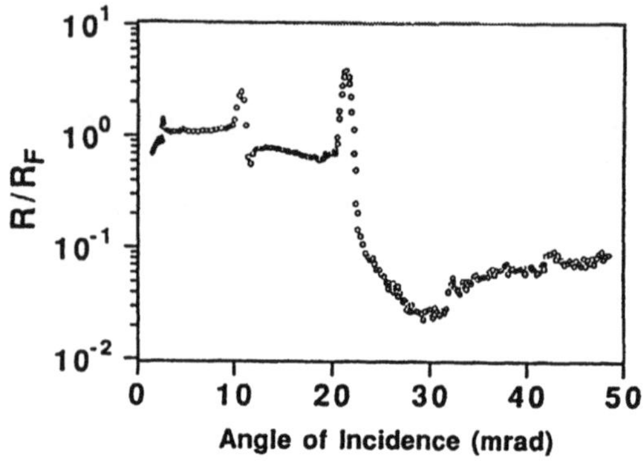

Fig.2: X-ray reflectivity of the surface of a suspension of DMPC vesicles in water after 12 h at T=29°C divided by the Fresnel reflectivity.

the surface is constructed where a certain number of planar bilayers is separated by a thin water layer from the monolayer at the liquid-air interface. The repeat distance of these membranes is obtained directly from the angular position of the reflectivity peaks. It is found to be equal to the intermembrane distance in the bulk in excess water. This observation strongly supports a planar multibilayer structure. Upon slow cooling the position of the peaks is shifted towards smaller wave-vectors between 26 and 25°C. The corresponding increase of the repeat distance indicates of a depth dependent chain-melting transition.

In addition, a decay in the apparent electron density modulation depending on the distance of a bilayer to the surface has to be introduced for a successful quantitative description of the experimental data. The reason for this decay could be quite interesting. For example, out of plane fluctuations of the bilayers lead to an apparent decrease of the electron density contrast.

We have also performed ellipsometric investigations of the surface induced multibilayer formation with samples of tenfold lower concentration compared to that of the x-ray experiments. For a fresh sample, the ellipsometric signal increases significantly between 30 and 31°C. With the detailled structural information from the x-ray experiments we conclude that this increase is due to the formation of bilayers. In a subsequent cooling run an additional increase of the ellipsometric signal is observed at 27°C. This indicates that the critical temperature for the surface induced vesicle fusion is shifted towards the bulk chain melting temperature with increasing number of bilayers.

Discussion

The following picture emerges based on our experimental observations. In a first step vesicles are attracted towards the liquid-air interface. Once close to the monolayer they tend to deform and align partly along the surface plane. In a third step the fusion of surface bound vesicles leads to planar bilayers. A simple phenomenological model for fusion induced by adhesion has been recently proposed by Lipowsky and Seifert /6/. We suggest that critical fluctuations of the chain melting transition, decreasing gradually from 29^{0}C at the surface to 23.5^{0}C in the bulk, promote fusion of surface aligned vesicles. The temperature dependence of the repeat distance of the membranes provides direct evidence for the proposed depth dependent chain-melting transition.

In practical terms our findings show that one can generate aligned lipid membranes supported by a liquid-air interface. These samples can prove useful for the investigation of model and native membranes, preferably in combination with the x-ray reflectivity technique.

We thank J. Als-Nielsen from Risø and HASYLAB Hamburg for offering us their technical facilities. This study has been supported by the Deutsche Forschungsgemeinschaft through SFB 266.

References

/1/ H. Schindler, Biochim. Biophys. Acta 555, 316 (1979)

/2/ S.P. Heyn, M. Egger and H.E. Gaub, J. Phys. Chem. 94, 5073 (1990)

/3/ N. L. Gershfeld, Biochim. Biophys. Acta 988, 335 (1989)

/4/ C. A. Helm, H. Möhwald, K. Kjaer and J. Als-Nielsen, Europhys. Lett. 4, 697 (1987)

/5/ G. Cevc, W. Fenzl and L. Sigl, Science 249, 1161 (1990)

/6/ R. Lipowsky and U. Seifert, Proceedings of the International Conference on Liquid Crystals, Vancouver 1990

The Shape of an Adhered Membrane Cylinder

M.M. Kozlov

Freie Universität Berlin, FB Physik,
Arnimallee 14, W-1000 Berlin 33, Fed. Rep. of Germany

We determined the shape of membrane cylinder adherent to a flat support and derived the relationship between the energy of membrane adhesion and the geometrical characteristics of membrane contour. The obtained theoretical results allow an alternative method of adhesion energy measurement in addition to the method which includes the measurement of contact angle. Two different cases are analytically analysed: zero transmembrane pressure differential at which the shape of the membrane is determined entirely by the bending effects; second, high transmembrane pressure differential the action of which surpasses the effect of bending elasticity. In the first case the adhesion energy is expressed only through the lengths of adsorbed and free parts of membrane contour which can be readily measured by optical methods. In the second case the expression for adhesion energy includes the additional geometrical characteristic - the curvature of membrane contour in the pole which can also be calculated from the results of optical experiments. The distribution of force factors along the membrane is analysed.

STATEMENT OF THE PROBLEM AND BASIC EQUATIONS

The mechanical properties of membranes are manifested in adhesion. Mutual adhesion of vesicles and their interaction with supports have been widely studied both experimentally and theoretically [1-3]. The purpose of this work is to express the energy of membrane adhesion in terms of the geometrical characteristics of membrane contour which seems facile to measure by optical methods.

Equation for the Local Equilibrium of a Membrane. Let us consider the contour of the membrane cylinder cross-section . Position of the point is characterized by the length of arc s counted off along the contour. The point of transition from the adherent segment to the free part shall be taken as the reference ($s = 0$). The coordinate of the membrane pole shall be designated as s_f. The length of the adherent part of the contour is equal s_a, the length of free part is equal s_f. The shape of the contour will be described by the function $\phi(s)$, where ϕ is the angle between the tangent to the contour and the horizontal axis. The angle is related to the curvature by the ratio $d\phi/ds = J$. At the vesicle pole $\phi = \pi$ and $dJ/ds = 0$. Using the angle ϕ, the shape of the membrane cylinder can be represented in Cartesian coordinates with axis x perpendicular to axis z. This representation has the form

$$z = \int_0^s ds \sin\phi, \qquad x = s_a + \int_0^s ds \cos\phi. \tag{1}$$

The shape of membrane cylinder contour is determined by the equations for local equilibrium of the membrane. The equation of equilibrium relative to the normal displacements, referred to as a generalized Laplace equation, can be written as

$$\gamma J - C_1 J^2 - \frac{dQ}{ds} = \Delta P, \tag{2}$$

Springer Proceedings in Physics, Vol. 66 The Structure and Conformation of Amphiphilic Membranes
Editors: R. Lipowsky · D. Richter · K. Kremer

where $\gamma[\frac{N}{m}]$ is the Gibbs surface tension, $C_1[N]$ is the bending moment, $Q[\frac{N}{m}]$ is the transverse shearing force, ΔP is the transmembrane pressure differential. We will use below also another force factor named zero moment $C_0[\frac{N}{m}]$ which is equal $C_0 = \gamma - JC_1$. The equation for lateral equilibrium appears to be

$$\frac{d\gamma}{ds} - C_1\frac{dJ}{ds} = 0. \tag{3}$$

The equation of the equilibrium of a surface segment relative to rotation is

$$Q = \frac{dC_1}{ds}. \tag{4}$$

We shall assume that the spontaneous curvature of the membrane equals zero and the bending moment C_1 is linked with the membrane curvature by a common linear relation $C_1 = E_{JJ}J$, where $E_{JJ}[Nm]$ is the membrane bending modulus. Two boundary conditions for equations (2)-(4) are a consequence of the symmetry of the cross -section shape relative to axis z. First, in the membrane cylinder pole angle $\phi = \pi$. Second, the transverse shearing force in the membrane cylinder pole is equal to zero, $Q = 0$. The third boundary condition is imposed in the point of transition of the flat segment of the membrane to the bent part, $s = 0$. Due to the membrane contour having no sharp points, in the point $s = 0$ the angle $\phi = 0$.

Energy of Membrane Adhesion; Equilibrium between the Membrane Cylinder and the Support . When the membrane element of area dA is brought into contact with the support, the adhesion forces do the work described by expression

$$dW_a = -g_a dA. \tag{5}$$

The value g_a is reffered to as adhesion energy and characterizes only the interaction between the membrane and the support.

Change in membrane cylinder free energy during the deformation with constant membrane area accompanying the adhision conform to the work of bending moment and is obtained using the expression

$$dF = C_1 AdJ. \tag{6}$$

Calculation of the energy change of the membrane element becoming flat and that of the free part of the membrane cylinder leads to the total change in the free energy of the membrane

$$dF = (\int_0^{C_{10}} J_0 dC_{10})dA \tag{7}$$

where C_{10} and J_0 are the bending moment and curvature of the membrane in the point of transition from the adherent to free segment ($s = 0$).

Based on expressions (6)-(7) we obtain that, in the membrane cylinder-support equilibrium, the bending moment and the curvature of the membrane are related as

$$\int_0^{C_{10}} J_0 dC_{10} = g_a. \tag{8}$$

Equation for Contour Shape. Within the framework of the above-assumed relationship between the bending moment and the curvature, equation (4) is represented as $Q = E_{JJ}\frac{dJ}{ds}$. Considering this relationship, from equations (2) - (3) we have

$$\frac{d}{ds}[\gamma - \frac{1}{2}E_{JJ}J^2] = 0 \tag{9}$$

where the surface tension is expressed as

$$\gamma = \frac{\Delta P}{J} + E_{JJ}\frac{1}{J}\frac{d^2 J}{ds^2} + E_{JJ}J^2 \tag{10}$$

VESICLE WITH LARGE PRESSURE DIFFERENTIAL

Let us consider the membrane cylinder with large pressure differential, $\Delta P \gg E_{JJ}J^3$, at which the surface tension significantly exceeds the product of the bending moment by the curvature. The final expressions for the curvature J of the membrane and angle ϕ depending on the arc length s have the following form :

$$J = \frac{\Delta P}{C_0} + \frac{(\pi - \frac{\Delta P}{C_0}s_f)}{\xi \sinh(\frac{s_f}{\xi})} \cosh\left[\frac{(s_f - s)}{\xi}\right], \tag{11}$$

$$\phi = \frac{\Delta P}{C_0}s + 2\frac{(\pi - \frac{\Delta P}{C_0}s_f)}{\sinh(\frac{s_f}{\xi})} \cosh\left[\frac{(2s_f - s)}{2\xi}\right] \sinh\left[\frac{s}{2\xi}\right]. \tag{12}$$

Expression (11) relates the value of the constant zero moment of the membrane with adhesion energy

$$\left(\frac{2g_a}{E_{JJ}}\right)^{\frac{1}{2}} = \frac{\Delta P}{C_0} + \frac{(\pi - \frac{\Delta P}{C_0}s_f)}{\xi} \coth\left[\frac{s_f}{\xi}\right]. \tag{13}$$

Experimentally Measurable Values. We obtained the exprcssions for the shape of the membrane cylinder cross-section and distribution of the force factors along the membrane that include the zero moment C_0 and the transmembrane pressure differential ΔP. Experimental conditions, as a rule, do not make it possible to measure these values. At the same time, the geometrical characteristics are readily measurable: the length of the contour of the membrane cylinder free segment s_f, the length of the contour of the adherent segment s_a, the curvature of the membrane in the pole J_∞. Based on these geometrical characteristics, one can calculate the energy of adhesion. It is sufficient to use expression (12) for the membrane cylinder shape and the relationship

$$s_a = -\int_0^{s_f} \cos\phi ds. \tag{14}$$

The problem can be solved numerically. The case of a small area of the adherent segment, $s_a \ll s_f$, gives the analytical results

$$C_0 = \pi^3 \frac{E_{JJ}}{s_f^3} \frac{s_a}{(\pi - J_\infty s_f)}, \tag{15}$$

$$\Delta P = \pi^3 \frac{E_{JJ}}{s_f^3} \frac{s_a J_\infty}{(\pi - J_\infty s_f)}. \tag{16}$$

Besides, expressions (15) and (16) relate the measured geometrical characteristics of the membrane cylinder contour and the adhesion energy

$$\left[\frac{2g_a}{E_{JJ}}\right]^{\frac{1}{2}} = J_\infty + \left[\frac{(\pi - J_\infty s_f)s_a\pi^3}{s_f^3}\right]^{\frac{1}{2}} \coth\left[\frac{\pi^3 s_a}{s_f(\pi - J_\infty s_f)}\right]^{\frac{1}{2}}. \tag{17}$$

VESICLE WITHOUT PRESSURE DIFFERENTIAL

Let us consider a membrane cylinder with zero pressure differential $\Delta P = 0$. In this case the shape of the membrane cylinder is determined by the bending moment and the transverse shearing force. The shape of the membrane cylinder at a small area of adhesion can be expressed in terms of the zero moment $C_{0\infty}$

$$J = \frac{\pi}{s_f}[1 + \frac{C_{0\infty}s_f^2}{\pi^2 E_{JJ}}\cos(\frac{\pi s}{s_f})], \tag{18}$$

$$\phi = \frac{\pi}{s_f}s + \frac{C_{0\infty}s_f^2}{\pi^2 E_{JJ}}\sin(\frac{\pi s}{s_f}). \tag{19}$$

The equation for adhesion energy has the form

$$(\frac{2g_a}{E_{JJ}})^{\frac{1}{2}} = \frac{\pi}{s_f} + \frac{C_{0\infty}s_f}{\pi E_{JJ}}. \tag{20}$$

Experimentally Measuranble Values. The expression of $C_{0\infty}$ in terms of length s_a of the adherent part of the membrane has the form

$$C_{0\infty} = 2\pi^2\frac{E_{JJ}s_a}{s_f^3} \tag{21}$$

Using this relationship one can easily express the force factors of the membrane through its geometrical characteristics. Besides, making use of expressions (20) and (21) one can easily relate the adhesion energy of the membrane with the opticaly measurable values: the length of the adherent part of the membrane s_a and the length of the free part s_f :

$$(\frac{2g_a}{E_{JJ}})^{\frac{1}{2}} = \frac{\pi}{s_f} + \frac{s_a}{\pi s_f^2} \tag{22}$$

Financial support by Alexander von Humboldt Foundation is gratefully acknowledged.

References.

1 Evans E. and Metcalfe M., Biophys.J. 45 (1984) 715-720.

2 Servuss R.M. and Helfrich W., J.Phys.France 50 (1989) 809-827.

3 Seifert U. Phys.Rev.A 43 (1989) 6803-6814.

Deformation of Giant Lipid Vesicles in an Electric Field

M. Kummrow and W. Helfrich

Freie Universität Berlin, FB Physik,
Arnimallee 14, W-1000 Berlin 33, Fed. Rep. of Germany

We deformed unilamellar giant vesicles of spherical shape into prolate ellipsoids of revolution by means of an alternating electric field. From the field distribution the electric forces acting on the vesicle are derived by use of the Maxwell stress tensor. The deformation is associated with a flattening of the undulations, thus increasing the apparent membrane area. A linear fit to the plot of the apparent area versus the logarithm of the lateral tension gives the bending rigidity as well as the initial lateral tension.

1. Field distribution around the lipid vesicle

The field outside the vesicle is a superposition of the homogeneous external field $\vec{E}_0$ and the (negative) dipole field of the vesicle. As the conductivity of the lipid membrane is much less than that of water the electric field on the surface of the vesicle is everywhere tangential to it. There is no field inside the vesicle but a strong screening field exists inside the membrane which is larger than E_0 by a factor of R/d, R and d being the vesicle radius and the membrane thickness respectively [1].

2. Electric forces

From the Maxwell stress tensor

$$T_{ik} = \frac{\epsilon_0}{2}\left(\rho\frac{d\epsilon}{d\rho} - \epsilon\right) E^2\delta_{ik} + \epsilon_0\epsilon E_i E_k \tag{1}$$

the electric forces acting on the membrane can be calculated. Here ρ is the mass density and ϵ the dielectric constant. The tensor (1) includes the isotropic forces of electrostriction which may be neglected as they hardly change the mass density. The remaining stresses are calculated via an imaginary box tightly enclosing a small piece of the vesicle membrane (ϑ is the polar angle with respect to the external homogeneous field). There are

a) stresses originating from the screening field:

$$T_{\vartheta\vartheta} \propto \epsilon_0 \left(E_0\frac{R}{d}\right)^2 \cos^2\vartheta \tag{2}$$

Springer Proceedings in Physics, Vol. 66 The Structure and Conformation of Amphiphilic Membranes
Editors: R. Lipowsky · D. Richter · K. Kremer

and

$$T_{rr}^{inside} \propto \epsilon_0 (2\epsilon_L - 1) \left(E_0 \frac{R}{d} \right)^2 \sin^2 \vartheta \tag{3}$$

(ϵ_L is the dielectric constant of the lipid). $T_{\vartheta\vartheta}$ is locally balanced by an inhomogeneous lateral tension σ_i, having no effect on the undulations but stretching the real membrane area, while T_{rr}^{inside} is just squeezing the membrane.

b) stresses originating from the external field. There is only a surface force density which consists of the polarization pressure and a vacuum part

$$T_{rr}^{ex} = -\frac{9}{8} \epsilon_0 \epsilon_W E_0^2 \sin^2 \vartheta \tag{4}$$

(ϵ_W is the dielectric constant of water). This stress gives rise to the ellipsoidal deformation.

The electric double layers associated with the transmembrane potential are of no importance as long as $\lambda_D / d \ll \epsilon_W / \epsilon_L$, λ_D being the Debye screening length.

3. Equilibrium of forces

In the deformed state the stress (4) from the outside water is balanced by an extra pressure Δp inside the vesicle and the homogeneous lateral tension σ_h such that for all ϑ

$$\Delta p = H \cdot \sigma_h - T_{rr}^{ex} \tag{5}$$

where H denotes the sum of principal curvatures. Taking eq.(5) at the equator and the pole one can calculate σ_h from

$$H_{equator} \cdot \sigma_h - T_{rr}^{ex} \left(\vartheta = \frac{\pi}{2} \right) = H_{pole} \cdot \sigma_h \tag{6}$$

4. Theory of undulations

An increase of lateral tension from an initial value $\sigma_0 > 0$ to $\sigma_h > \sigma_0$ gives rise to an relative increase of apparent area through the flattening of undulations

$$\frac{\Delta A}{A} = \frac{k_B T}{8 \pi k_c} \ln \frac{\sigma_h}{\sigma_0} \tag{7}$$

where $k_B T$ is the thermal energy and k_c the bending rigidity [2]. A linear fit to a plot of $\Delta A / A$ as measured versus the logarithm of σ_h as calculated will yield the bending rigidity from its slope as well as the initial lateral tension.

5. Some corrections

a) The real membrane area is also increased through the lateral tensions σ_i and σ_h and through the thinning of the membrane, so that $\Delta A / A$ was corrected assuming a reasonable value for the stretching modulus [3].

b) The polarization pressure is weakened as the vesicle is deformed, so that T_{rr}^{ex} must be corrected.

c) The volume of a given vesicle as calculated from the principal axes shows some scatter which does not depend on field strength or time. To improve accuracy the principal axes were corrected assuming a constant volume.

The effect of all corrections on the bending rigidity was 5 % or less.

6. Experiments

The experiments were made with giant flaccid vesicles of neutral lipids (EYPC, DLPC, DGDG, and POPC) swollen in 50-200 μM NaN_3 by a method similar to that of Reeves and Dowben [4,5]. They were watched normally to the field direction using a phase contrast microscope. Only vesicles of more than 20 μm in diameter and deforming reversibly were selected. The field strength ranged up to 100 V/cm with frequencies of 1 to 3 kHz. Sometimes the frequency was varied from 0.3 to 20 kHz without any effect on the deformation of the vesicle. Also the Debye screening length was checked to be small enough by adding 100 μM NaCl, it could also be estimated from the conductivity of the water in the sample cell. The latter, being of the order of $10^{-2}\Omega^{-1}\,m^{-1}$, suggests a Debye length of only about 10nm.

7. Results

The results are given in table 1. The statistical error of the result of a particular vesicle was about 10 % while the scatter among vesicles of the same lipid was up to 20 %. The initial lateral tensions were in the range from $1 \cdot 10^{-6}$ to $5 \cdot 10^{-4}$ mN/m.

With POPC only six out of nine vesicles gave a bending rigidity which was in the expected range. The other three as well as one EYPC vesicle gave increases of apparent area, again reversible, of up to 17 % which is much more than can be supplied by undulations only.

Table 1. Mean bending rigidities for various lipids as obtained in the present study.

lipid material	bending rigidity k_c in units of 10^{-20}J	number of vesicles
EYPC	2.47	7
DLPC	3.37	5
POPC	2.32	6
DGDG	1.01	6

A more detailed representation of the experiments and calculations is being published elsewhere [6].

8. References

[1] M. Winterhalter, W. Helfrich (1988), J. Coll. Interface Sci. **122**, 583
[2] W. Helfrich, R.M. Servuss (1984), Il Nuovo Cimento **3D**, 137
[3] R. Kwok, E. Evans (1981), Biophys. J. **35**, 637
[4] J.P. Reeves, R.M. Dowben (1969), J. Cell. Physiol. **73**, 49
[5] D. Needham, E. Evans (1988), Biochem. **27**, 8261
[6] M. Kummrow, W. Helfrich, Phys. Rev. A (submitted)

The Effect of the Electric Field on the Shapes of Phospholipid Vesicles

B. Žekš and S. Svetina

Institute of Biophysics, Medical Faculty, Lipičeva 2,
61105 Ljubljana, Slovenia and
J. Stefan Institute, Jamova 39, 61000 Ljubljana, Slovenia

Abstract. The shapes of phospholipid vesicles in an external electric field are evaluated using the bilayer couple model of the phospholipid membrane. Nearly spherical vesicles have for small fields a shape of prolate ellipsoids, which get more elongated with increasing electric field. For larger fields the elongation does not increase simply, but a bulge starts to grow on the equator. The shapes which minimize the membrane bending energy at given membrane area are calculated in dependence of the electric field and the corresponding volumes are determined.

The bilayer couple model of a phospholipid membrane [1] is based on the structure of the membrane, taking into account that the membrane consists of two layers in contact. It has been shown that the equilibrium shapes of vesicles can be obtained by minimizing the membrane bending energy with the restrictions that the area of the membrane is constant, the difference of the areas of the two layers is constant and the volume of the vesicle is constant [1]. Such a minimization procedure leads to a variety of different shapes which agree well with experimentally observed shapes and their sequences [2]. For axially symmetric vesicles the shapes can be obtained by solving the differential equation [1]

$$\frac{x}{2(1-x^2c_p^2)}\left[xc_p\frac{dc_p}{dx}\left(2c_p+x\frac{dc_p}{dx}\right)+M+Lc_p+Nc_p^2\right]+\frac{d}{dx}\left(2c_p+x\frac{dc_p}{dx}\right)=0\ , \quad (1)$$

where c_p is the principal curvature along the parallels, x is the distance from the axis and M, L and N are the three Lagrange multipliers, which are determined from the conditions that the volume, area and area difference have constant values. Here c_p and x are dimensionless quantities and are expressed relatively to the sphere with the same area as the vesicle. Because the membrane bending energy is scale invariant, the shape depends only on the relative volume (v) and relative area difference (Δa). Once the shapes are obtained for given v and for different values of Δa, one can obtain the minimum at this volume without the restriction for Δa by solving the equation

$$\frac{\partial w_b(v,\Delta a)}{\partial \Delta a}=0\ , \quad (2)$$

where $w_b(v,\Delta a)$ is the bending energy expressed relative to the corresponding sphere. The solutions of equation (2) are equivalently obtained by taking the

Springer Proceedings in Physics, Vol. 66 The Structure and Conformation of Amphiphilic Membranes
Editors: R. Lipowsky · D. Richter · K. Kremer

Lagrange multiplier related to the area difference equal to zero, i.e. $N = 0$. In a similar way also the unrestricted minimum with respect to the volume conservation can be obtained by taking $L = 0$. Both conditions taken simultaneously, i.e. $N = 0$ and $L = 0$, lead of course to a spherical shape, because the sphere has a minimal bending energy at constant area.

An external static electric field deforms phospholipid vesicles because of the electric Maxwell stresses acting on the membrane. The field tends to compress spherical vesicles laterally and elongate them in the direction of the field [3,4,5] and the vesicles get elliptically deformed for small fields. For larger fields the electric field cannot be treated as perturbation and more complicated shapes appear as a result of the minimization of the membrane bending energy in the presence of the electric field [6,7]. The problem can be divided into two parts. The first is the electric one. For a given vesicle shape one must solve the Laplace equation for the electric potential, with the conditions that the electric field is equal to the external electric field (E_0) far away from the vesicle, while at the vesicle surface the normal component of the field equals zero, corresponding to a nonconductive membrane and to the condition that the current in the external ionic solution must be tangential at the surface. The mathematical procedure, which is based on the expansion in Legendre polynomials, gives as a result the tangential electric field on the surface of the vesicle. The second part of the problem consists of finding the equilibrium shape. The minimization procedure, which will be presented in detail elsewhere, leads to a differential equation with the same structure as Eq. (1), except that the Lagrange multiplier M is modified and includes the electric field term

$$M \rightarrow M - \frac{\epsilon\epsilon_0 E_0^2 R^3}{k_c} \frac{2}{x^2} \int_0^x e^2(x')x'dx' \qquad (3)$$

Here $e(x)$ is the tangential component of the electric field relative to E_0. The external electric field influences the shape only in the combination

$$\tilde{E}_0^2 = \frac{\epsilon\epsilon_0 E_0^2 R^3}{k_c} \quad , \qquad (4)$$

which is proportional to the ratio of the electric and bending energies of the vesicle. Here ϵ is the permittivity of water, k_c is the membrane bending constant and R is the radius of the corresponding sphere of the vesicle. It can be seen from expression (4) that the problem is no longer scale invariant and that the effect of the electric field is stronger for larger vesicles and for smaller bending constants.

Numerical analysis shows that nearly spherical vesicles in small external electric fields have shapes of prolate ellipsoids, as expected from theoretical predictions based on treating the electric field as a small perturbation [3,4]. For larger fields a bulge with larger curvature develops on the equator. This characteristic shape (Fig. 1) is not related to the requirements for constant volume and constant area difference, as can be seen in Fig. 1 where the shape is shown for $N = 0$ and $L = 0$, which means that this is the equilibrium

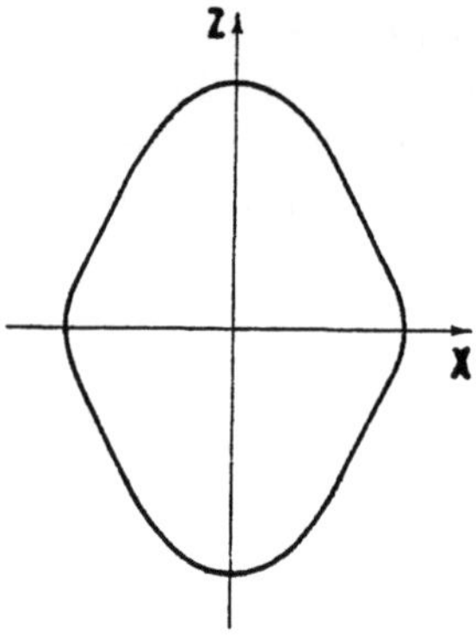

Figure 1: The vesicle shape corresponding to $\tilde{E}_0 = 4.04$ and $N = L = 0$ ($v = 0.95$, $\Delta a = 1.025$).

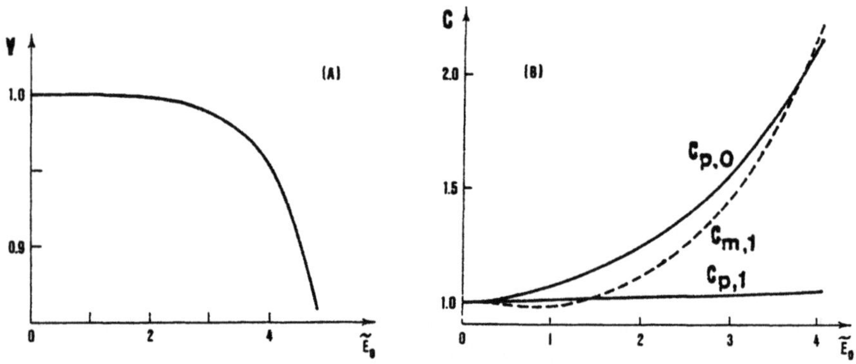

Figure 2: (A) The dependence of the relative volume v and (B) the dependences of the curvature on the pole ($c_{p,0}$) and of the two curvatures on the equator ($c_{p,1}$ - along the parallels, $c_{m,1}$ - along the meridians) on the normalized electric field $\tilde{E}_0$ for $N = L = 0$.

shape corresponding to given membrane area and given electric field without restrictions for v and Δa, which are allowed to adjust freely. In Fig. 2A the dependence of the obtained relative volume on the normalized electric field $\tilde{E}_0$ is shown. It can be observed that for small electric fields ($\tilde{E}_0 \leq 3$) the electric field does not affect the equilibrium volume very much, but for larger fields the dependence $v(\tilde{E}_0)$ becomes very steep. The character of the corresponding shapes can be estimated from Fig. 2B, where the curvature on the pole and the two curvatures on the equator are shown. For small fields, the curvature on the pole increases with the field, while on the equator the radius of the ellipsoid (equal to the inverse $c_{p,1}$) decreases and its curvature $c_{m,1}$ decreases. The prolate ellipsoid is therefore getting longer and thinner with the increasing field. At larger fields ($\tilde{E}_0 \geq 1$) the curvature on the equator ($c_{m,1}$) increases and becomes at $\tilde{E}_0 \approx 1.5$ larger than the inverse radius. At even larger fields ($\tilde{E}_0 \geq 4$) the curvature on the equator is larger than the curvature on the pole, which means that the shape has a pronounced bulge on the equator. We have therefore ellipsoidal shapes for small fields, the shapes with a bulge on the equator for large fields and a transition region for intermediate fields.

The observed shapes of vesicles in the external electric field are usually ellipsoidal [5]. It is hard to understand, why the large field regime with typical shapes as the one in Fig.1 is usually not reached, as the required electric fields are not very large. The normalized electric field $\tilde{E}_0 = 1$ corresponds for $R = 20\mu m$ and $k_c = 10^{-19} J$ to $E = 1.3 V/cm$. One possibility is that the electrode polarization diminishes the local field on the vesicle drastically. The other explanation would be that the electric field induced stresses in the membrane are so large that it cannot be considered as laterally incompressible [5]. In such a case the vesicle would adjust to the external field not only by changing its shape but also by changing the area of its membrane. These effects will be analysed in more detail elsewhere.

References

[1] S. Svetina and B. Žekš, *Eur. Biophys. J.* **17**, 101 (1989) and the references therein.

[2] K. Berndl, J. Käs, R. Lipowsky, E. Sackmann, and U. Seifert, *Europhys. Lett.* **13**, 659 (1990).

[3] W. Helfrich, *Z. Naturforsch. C* **29**, 182 (1974).

[4] M. Winterhalter and W. Helfrich, *J. Colloid Interface Sci.* **122**, 583 (1988).

[5] M. Kummrow and W. Helfrich, this workshop, to be published.

[6] B. Žekš, S. Svetina, and V. Pastushenko, *Stud. Biophys.* **138**, 137 (1990).

[7] B. Žekš and S. Svetina, in *Biologically Inspired Physics*, editor L. Peliti, Plenum, N.Y., 1991.

AC Field Controlled Formation of Giant Fluctuating Vesicles and Bending Elasticity Measurements

M.I. Angelova[2], *S. Soléau*[1], *P. Méléard*[1], *J.F. Faucon*[1], and *P. Bothorel*[1]

[1]Centre de Recherche Paul Pascal, CNRS, Avenue A. Schweitzer, F-33600 Pessac, France

[2]Central Laboratory of Biophysics, Bulgarian Academy of Sciences, 1113 Sofia, Bulgaria

Abstract. It is shown that the formation of giant fluctuating unilamellar vesicles can be considerably improved by applying an AC electric field during the swelling of the lipid. The analysis of the thermal fluctuations shows that the bending modulus of egg-yolk phosphatidylcholine decreases versus time (probably, due to the lipid chemical degradation) and that the incorporation of 37 mol% cholesterol in the bilayer leads to a 60% increase of its rigidity.

1 Introduction

Bending elasticity has been a subject of increasing interest during the past decade, due to its implications both in the phase stability of microemulsion and in biological processes involving shape transformations of membranes. Several methods have been used to measure the bending modulus k_c of lipid bilayers, most of them being based on the analysis of their shape fluctuations [1, 2, 3]. Following the works of Schneider et al. [4] and Sackmann et al. [5], an exact mathematical model has been recently proposed to account for the thermal fluctuations of quasi-spherical vesicles [6, 7]: the Legendre decomposition of the angular autocorrelation function of the fluctuations allows the determination of the mean-squared amplitudes of the spherical harmonics, and the effect of membrane tension can then be explicitly taken into account. Furthermore, it has been shown that the use of the autocorrelation function considerably decreases the noise contribution, and that the effect of the integration time of the video camera on harmonics having low correlation times can be corrected using an appropriate numerical factor. So, one of the last remaining problem for a more systematic investigation of the bending elasticity of lipid membranes is to obtain vesicles satisfying both theoretical and experimental requirements. Indeed, these vesicles must be large enough ($> 10\mu$m) so that their contour coordinates can be determined accurately by optical microscopy. They must also be unilamellar, and perfectly isolated, i.e. free of any neighbouring vesicles, filament or other kind of material. And, obviously, they must perform thermal fluctuations (large enough for the image analysis to be carried out). That implies very low membrane tensions ($< 10^{-5}$ mN/m). However, vesicles must be "quasi-spherical", which means that their time-averaged shape must be

Springer Proceedings in Physics, Vol. 66 The Structure and Conformation of Amphiphilic Membranes
Editors: R. Lipowsky · D. Richter · K. Kremer

spherical, and that the amplitudes of fluctuations must remain small compared to the vesicle radius.

2 Preparation of Vesicles

Some experimental procedures leading (in principle) to a good yield of giant unilamellar vesicles have already been described. These methods are based on the dialysis (or dilution) of lipid solutions against aqueous buffers [8], on the effect of chaotropic salts such as trichloroacetate [9], on the fusion of small vesicles by freezing and thawing [10], or on a double emulsion process using chloroform and an aqueous solution [11]. However, in all these methods, the vesicle formation takes place in the presence of an osmotic stress, and the resulting vesicles are generally not well-isolated, and exhibit relatively large membrane tensions able to hinder thermal fluctuations. In addition, traces of the organic solvent always rest and could modify the properties of the membranes. So, to our knowledge, all the groups interested in the mechanical properties of the lipid bilayer have used, up to now, more gentle procedures based on that early proposed by Reeves and Dowben [12]. Typically, a thin lipid film is formed on a solid substrate (glass, teflon,...) by evaporation of lipid solution under vacuum. Water is added, and giant vesicles are allowed to form spontaneously. Unfortunately, this method suffers of, at least, two severe limitations: (i) the kinetics of vesicle formation is very slow and usually requires several days, which can lead to chemical degradation of the lipid, and (ii) the yield of "ideal" vesicles is very low, so it takes long time (it could be even practically impossible) to find one of them in the sample.

The method suggested in this study is based on the early works of Angelova and Dimitrov on the electro-formation of vesicles [13]. The cell for preparation of vesicles consisted of two ITO coated glasses separated by a silicone spacer of 0.6 mm. The lipids (L-α phosphatidylcholine from frozen egg yolks (EPC); cholesterol (Chol), Serva 17090) were dissolved in chloroform/methanol 9:1 solvent. A drop of 2.5 μl lipid solution was deposited and dried in nitrogen stream on one of the electrodes. AC field of 10 Hz 1 V was applied to the cell and milliQ water (R$\sim$ 15$\div$18 MΩ/cm) added. Usually, the time of swelling under AC field was 2 h. The AC field was switched off and vesicles left to relax during several hours. The video records were taken and treated using phase-contrast optical microscopy, video-enhanced contrast technics and image analysis according to the procedure of Faucon at al. [7]. The applied AC field induced formation and contuneous grow of vesicles up to $\sim 10 \div 15 \mu$m, stepwise increase of the size to $\sim 50 \div 70 \mu$m (as a result of lateral fusions of smaller ones), detachment and diffusion of vesicles from the electrode surface.

Figs.1-2 present vesicles obtained from the mixture Chol/EPC 37:63 mol/mol, lipid amount corresponding to 5$\div$10 bilayers, after 2 h 30 min swelling under AC of 10 Hz, 1.75 V. It has to be stressed that *the spontaneous* formation of vesicles in that case was not possible. Most of the vesicles: a) were probably unilamellar: (i) their contours had the weakest possible contrast, and (ii) the

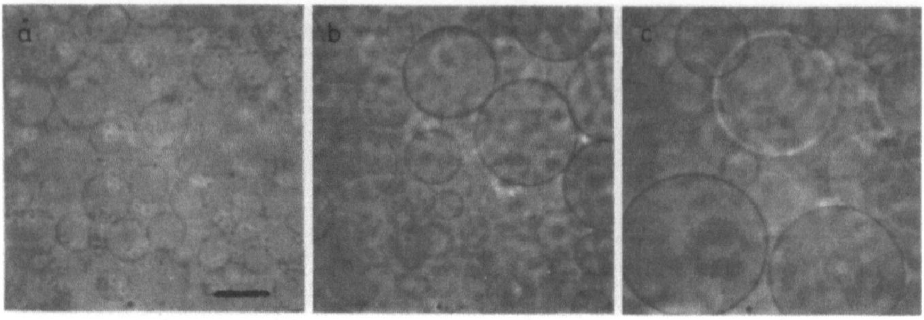

Fig. 1 Vesicle populations (Chol/EPC 37:63 mol/mol, after 2 h 30 min swelling under AC field of 10 Hz, 1.75 V) at different distances from the electrode surface: a) at 5 μm; b) 15 μm; c) 30 μm. Bar = 25 μm.

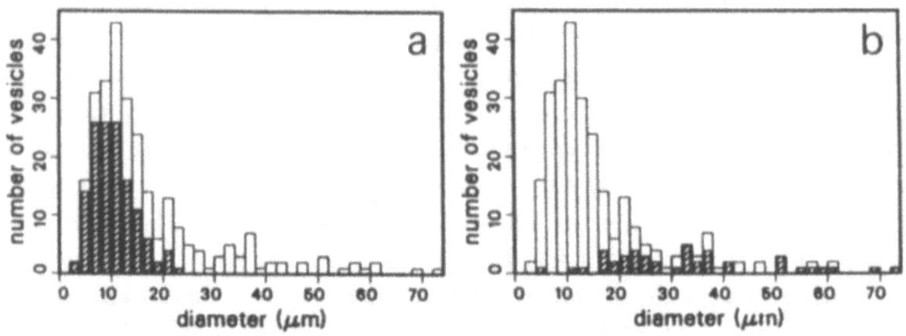

Fig. 2 Size distribution for the vesicles (for the sample shown in Fig. 1) found between: a) 0 and 10 μm ▨ , 0 and 75 μm ▯ ; b) 30 and 75 μm ▨ , 0 and 75 μm ▯ .

k_c values given by the image analysis supported that idea as well; b) they were quite homogeneous in size at a given distance from the electrode surface with the average diameter increasing with that distance and the time of action of AC field; c) large number of vesicles got rid of tension and performed suitable thermal fluctuations after relaxation time of a few hours; d) lots of them were well isolated from the others.

3 Bending Elasticity

Bending elasticity measurements were performed as previously described [7]. The time course of k_c for EPC vesicles formed under AC field is shown in **Fig.3** We have found a monotonous decrease of k_c from $(0.66 \pm 0.06) \cdot 10^{-19}$ J for one day to $(0.45 \pm 0.05) \cdot 10^{-19}$ J for 13 days after the vesicle formation. This result can be explained by a chemical degradation of the lipid versus time, which leads to formation of lyso-lipids, fatty acids and oxidation products. Thus, the

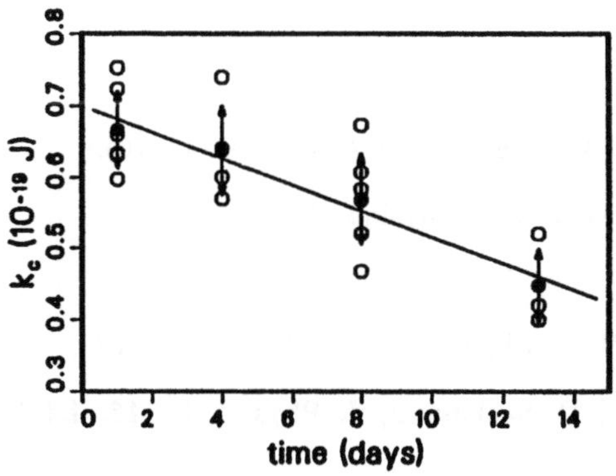

Fig. 3 Time course of k_c for EPC vesicles formed under AC field.

bending modulus proved to be very sensitive to the chemical composition of the bilayer and to the presence of degradation products.
It is noteworthy that a previous study performed using the same mathematical analysis but the *conventional method* of vesicle preparation led to a mean value of $k_c = (0.45 \pm 0.11) \cdot 10^{-19}$ J on about 40 vesicles [7]. It is clear from the above results that such a low value, as well as the corresponding relatively large standard deviation, can be due to the long time needed for the vesicle formation in that case.

It can also be mentionned that the reduced membrane tension $\bar{\sigma} = \frac{\sigma R^2}{k_c}$, with σ = actual membrane tension and R = vesicle radius, is rather low for all the analysed vesicles and ranges between about -3 and 25. Furthermore, as shown in **Fig. 3**, the dispersion on k_c values obtained at a given time is relatively small, which seems to rule out the existence of oligolamellar vesicles. This demonstrates the usefulness of the electro-formation process to obtain tension-free unilamellar vesicles.

Some determinations of the bending modulus have also been performed on mixtures of EPC with cholesterol. In this case too, flaccid giant vesicles were easily obtained, with diameters ranging between 14 and 18 μm. As expected, an increase of the bending modulus was observed, with a mean value of $k_c = (1.06 \pm 0.15) \cdot 10^{-19}$ J when the molar fraction of cholesterol is equal to 0.37. This corresponds to an increase of about 60% compared to pure EPC, which is much smaller than the three or four fold increase previously reported [2, 3]. The main problem in this case, besides the inevitable question of the number of bilayers constituting the contour, lies in the homogeneity of the dried lipid film, and of the membranes of the resulting vesicles.

Acknowledgment: We thank M. Moineau for his technical assistance.

References

[1] R.M. Servuss , W. Harbich and W. Helfrich, BBA **436**, 900 (1976).

[2] H.P. Duwe, J. Kaes and E. Sackmann, J. Physique **51**, 945 (1990).

[3] E. Evans and W. Rawicz, Phys. Rev. Lett. **64**, 2094 (1990).

[4] M. Schneider, J. Jenkins and W. Webb, J. Physique **45**, 1457 (1984).

[5] H. Engelhardt, H.P. Duwe and E. Sackmann, J. Phys. Lett. **46**, L395 (1985).

[6] S.T. Milner and S.A. Safran, Phys. Rev. A **36** 4371 (1987).

[7] J.F. Faucon, M.D. Mitov, P. Méléard, I. Bivas and P. Bothorel, J. Physique **50**, 2389 (1989).

[8] N. Oku, J.F. Scheerer, R.C. MacDonald, BBA **692**, 384 (1982).

[9] N. Oku and R.C. MacDonald, J. Biol. Chem. **258**, 8733 (1983).

[10] N. Oku and R.C. MacDonald, Biochemistry **22**, 855 (1983).

[11] S. Kim and G.M. Martin, BBA **646**, 1 (1981).

[12] J.P. Reeves and R.M. Dowben, J. Cell Physiol. **73**, 49 (1969).

[13] M.I.Angelova and D.S.Dimitrov, Faraday Discuss. Chem. Soc. No.**81**, 303, 345 (1986); D.S.Dimitrov and M.I.Angelova, Progr. Coll. Pol. Sci. **73**, 48 (1987); M.I.Angelova and D.S.Dimitrov, Mol. Cryst. Liq. Cryst. **152**, 89 (1987); M.I.Angelova and D.S.Dimitrov, Progr. Coll. Pol. Sci. **76**, 59 (1988).

Part III

Membranes in Complex Fluids

Micelles and Vesicles of Gangliosides

L. Cantù[1], *M. Corti*[2], *and M. Musolino*[1]

[1]Department of Chemistry and Biochemistry, University of Milan, via Saldini 50, I-20133 Milan, Italy
[2]Department of Electronics, University of Pavia, via Abbiategrasso 209, I-27100 Pavia, Italy

Abstract. Aggregation properties of various gangliosides into micelles and vesicles are described. Vesicles form spontaneously in solution and are in thermodynamic equilibrium with a small concentration of a lamellar phase. Vesicles seem to have small bending elasticity, of the order of kT.

Self association of amphiphiles into aggregated structures in water is determined by the balance of hydrophobic and hydrophilic interactions, together with geometrical considerations regarding the amphiphile itself [1]. In this paper we shall present a class of biological amphiphiles, called gangliosides, which give a clear evidence of the effect of the molecular geometry on the final structure of the aggregate. Gangliosides are sialic acid containing glycosphingolipids that are normally present in the outer leaflet of the plasma membranes of vertebrate cells [2]. They have a marked amphiphilic character with an oligosaccharide chain headgroup and a double tailed hydrophobic part. In nature a large variety of gangliosides is found, which differ in the length and conformation of both the oligosaccharide and hydrocarbon chains. We shall consider the sequence of gangliosides GM4, GM3, GM2, GM1, GD1a, GT1b all with the same hydrophobic portion, mainly C18 fatty acid and C20 sphigosine, but with increasing complexity of the oligosaccharide chain, from 2 to 7 sugar groups. Differently from phospholipids, gangliosides may form micelles since the large sugar headgroup can supply enough curvature to form globular structures rather than vesicles or bilayers.Light scattering and neutron scattering measurements of the aggregation properties of gangliosides have already been reported [3,4] and recently reviewed [5].

The gangliosides GT1b, GD1a, GM1 and GM2 form non-spherical micelles with increasing aggregation number and dimension . Deviation from sphericity increases with size, as expected from geometrical considerations [3] .The smallest gangliosides GM3 and GM4 form vesicles [6,7,8]. The important finding is that these vesicles, made of a single component, form spontaneously without supply of external energy and therefore in thermodynamic equilibrium. This is an ideal vesicle system which usual theories of self assembly, based on statistical mechanics, can be correctly applied to. After a preliminary observation by electron microscopy [4], the GM3 vesicle distribution has been extensively studied [6,7] by the laser light scattering technique, which is not intrusive and, therefore, quite suitable to study equilibrium properties. The main finding is that in solution vesicles are in equilibrium with a small amount of large aggregates, which are of lamellar type. Temperature affects both the vesicle distribution and the percentage of large aggregates.GM3 prepared as sodium salt was dissolved in a 30 mM NaCl water solution at a concentration of 0.1 mM at room temperature, 25°C. Solutions clarify without agitation. Vesicles form spontaneously also in pure water or at different salt concentrations. Aggregates in solution were studied as a function of temperature, between 7°C and 50°C. In the experimental work reported in [6] and [7], care was taken to use the light scattering technique at best, since complex aggregate distributions cannot be easily obtained with standard deconvolution of dynamic data only. On the other hand, combination of absolute measurements of the scattered intensity and of the correlation function of intensity fluctuations, due to Brownian motions, is very effective to eliminate arbitrariness and give confidence to the distribution parameters.The result of the fit of the static and dynamic laser light scattering data obtained as a function of temperature and scattering angle is that the aggregate distribution can be properly represented by an Helfrich distribution [9] of unilamellar vesicles and a small amount of large flat objects schematized as oblate ellipsoids. At 50°C the average aggregation number <N> and radius R of the vesicle distribution are <N>=14300 and R=245Å respectively. At 7°C, <N>=17100 and R=265Å. Out of the total concentration of GM3, 98.5% goes into vesicles at 50°C and 96.5% at 7°C. The remaining material goes into oblate ellipsoids with semimajor axis equal to 4500Å (±10%) and axial ratio 80 (±20%). Equilibration times of the aggregate distribution are found to be small, since temperature scans are fully reversible and quite independent of scan rate. The bimodal distribution given by the coexistence of vesicles and large aggregates seems an intrinsic feature of this system which form vesicles spontaneously, since large aggregates reform rather rapidly after removal from the solution by microporous filtration.The coexistence of vesicles with a lamellar phase in

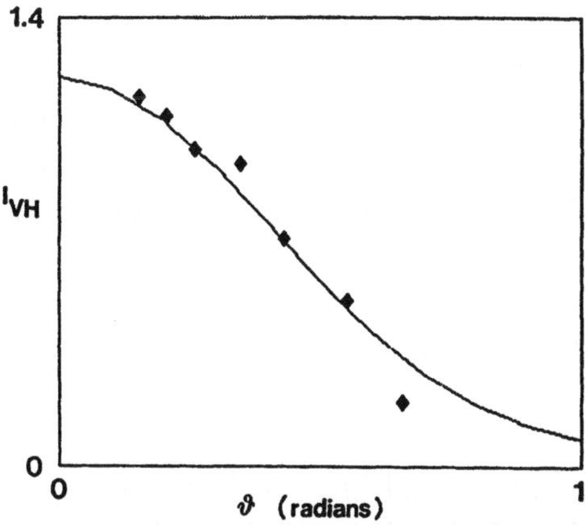

Fig.1. Depolarized light from the GM3 solution as a function of the scattering angle. The full line is the calculated intensity for thin discs.

thermodynamic equilibrium is an important question, which justifies the effort of finding an independent experimental check by measuring the depolarized scattered light as a function of angle. Since GM3 aggregates are made up of individual GM3 molecules which can be optically anisotropic, in principle depolarized scattering can be possible. Depolarized scattering theory [10] predicts that for thin spherical vesicles of any size, made up of anisotropic molecules, no depolarization can occur, due to internal cancellation effects. Therefore, if the solution is probed with vertically polarized light and horizontally polarized scattered light is observed, aggregates with a non-symmetrical shape should be present in solution. The dots of Fig.1 represent the measured depolarized light as a function of angle for a solution of GM3 at a temperature of 25°C. The intensity scale is normalized to the polarized light scattered by pure water. Leakage of the vertically polarized component of the scattered intensity is less than a few percents of the measured depolarized intensity at low angles and has been subtracted. Depolarized light intensity has been calculated by Picot et al. [11] for thin anisotropic discs and long rods. The full line in Fig.1 represents the angular intensity distribution for thin discs with a radius of 3500Å. Fitting with rods is not acceptable. From the ratio of the horizontal and vertical component of the scattered light the intrinsic anisotropy δ is equal to 0.028 (notation of [11]). This is a strong independent confirmation of the results presented in [7] and [8] obtained with vertically polarized light. The horizontal polarization selects only the discs of the bimodal distribution.The very fact that GM3 vesicles form spontaneously in solution is a strong indication that the bending elasticity of the bilayer is low, of the order of kT, the thermal energy. Since these vesicles are made of a single component amphiphile, their spontaneous curvature cannot be different from zero; thus any energetic stabilization effect is forbidden [12]. Therefore GM3 vesicles are stabilized entropically which indeed can be possible only for very low bending elasticity.

The authors would like to thank the group of Prof.Tettamanti for the biochemical support. L.Cantù thanks Fidia Spa for funding her work.

1) Physics of Amphiphiles:Micelles,Vesicles and Microemulsions, M Corti and V. Degiorgio Eds.(North Holland, Amsterdam, 1985).
2) G. Tettamanti in [1], p. 607.
3) L. Cantù, M. Corti, S. Sonnino and G. Tettamanti, Chem. Physics of Lipids, 41, 315 (1986).
4) S.Sonnino, L. Cantù, M. Corti, D. Acquotti and G. Tettamanti, Chem. Physics of Lipids, 52, 231 (1990).
5) M. Corti, L. Cantù and P. Salina, Adv. Coll. Interface Sci. 36, 153 (1991).
6) L.Cantù, M. Corti, M. Musolino and P. Salina, Europhysics Lett., 13, 561 (1990).
7) L. Cantù, M. Corti, P. Lago and M. Musolino, in SPIE vol 1430 Photon Correlation Spectroscopy: Multicomponent Systems, 144 (1991).
8) L. Cantù, M. Corti, M. Musolino and P. Salina, Progress Colloid and Polymer Sci. 84, 21 (1991).
9) W. Helfrich, J. Phys. (Paris), 47, 321 (1986)..
10) S.R. Aragòn and M. Elwenspoek, J. Chem. Phys., 77, 3406 (1982).
11) C. Picot, G. Weill and H. Benoit, J. Colloid and Interface Sci., 27, 360 (1968).
12) S. Safran, P. Pincus and D. Andelman, Science, 248, 354 (1990).

Self-Assembly of Bipolar Lipids

A. Relini, F. Cavagnetto, and A. Gliozzi
Dipartimento di Fisica, Università di Genova,
via Dodecaneso 33, I-16146 Genova, Italy

Abstract. Several experimental techniques such as ^{1}H-NMR, light scattering and electron microscopy have been employed to study the self-assembly of aqueous dispersions of bipolar lipids and of their mixtures with monopolar ones. These experiments have shown that the theory of lipid self-assembly introduced by Israelachvili can still be applied to these compounds.

1. Introduction

The membrane of *Sulfolobus solfataricus*, a thermophilic archaeobacterium living at $T = 90^0$C and $pH \simeq 2$ has unusual thermal stability, probably due to the peculiar structure of its lipids [1]. The polar lipid extract (PLE) of the membrane of this microorganism is composed of several bipolar fractions. Glycerol dialkyl glycerol tetraether (GDGT) and glycerol dialkyl nonitol tetraether (GDNT) constitute the backbone of all these compounds, which are comprised of a polar headgroup at each end of a double C_{40} hydrocarbon chain. Aim of this work is to study the physical principles that govern the molecular organization of these lipids.

2. Materials and Methods

Bipolar lipids were extracted as described previously [2]. Electron microscopy was performed on lipid aqueous dispersions and on large unilamellar vesicles (~ 100 nm in diameter), prepared at room temperature by controlled detergent removal using a buffer (10 mM Tris, 150 mM NaCl, 0.1 mM EDTA, titrated to pH 7 with HCl). The sample was deposited on Formvar coated grids and stained with 1% phosphotungstic acid solution at pH 7. Small unilamellar vesicles were prepared in distilled water by sonication at 60°C and their radius R_s was obtained by light scattering, using a laser at 514 nm connected to a Brookhaven Instrument BI 2030 digital correlator. When mixtures of bipolar lipids with egg phosphatidylcholine (PC) were considered, the formation of closed vesicles was checked by the splitting of the choline peak induced by the paramagnetic ion Pr^{3+} in ^{1}H-NMR spectra [3]. Light scattering and NMR measurements were performed at 60°C.

3. Theory

It has been shown [4] that once the optimal surface area per molecule a_0, hydrocarbon chain volume v and maximum chains length l_c are specified for a given lipid, the ratio

$$P_c = \frac{v}{a_0 l_c}$$

determines the shape and size of the structure that the lipids can pack into. Lipid molecules with P_c between 0.5 and 1 can form lamellar phases or vesicles; on the contrary, when P_c is greater than one, no vesicles are obtained. In the former case it is also possible to calculate the lower limit to the vesicle radius R_c. The extension of this theory to two-component vesicles [5] has been adapted to binary mixtures of monopolar and bipolar lipids.

4. Results

Electron micrographs of lipid aqueous dispersions are shown in Fig.1. For any mixture lamellar phases are obtained only above a critical molar ratio, typical of the bipolar-monopolar lipid system. The above requirement is fulfilled only in cases 2 and 4 of Fig.1. A quantitative comparison between theory and experiment is given in Table 1. In the case of GDGT, GDNT and PLE, P_c is greater than 1 and no vesicle formation occurs. Details about the calculations and the assumptions used are given elsewhere [6]. It can be observed that increasing the bipolar lipid content the vesi-

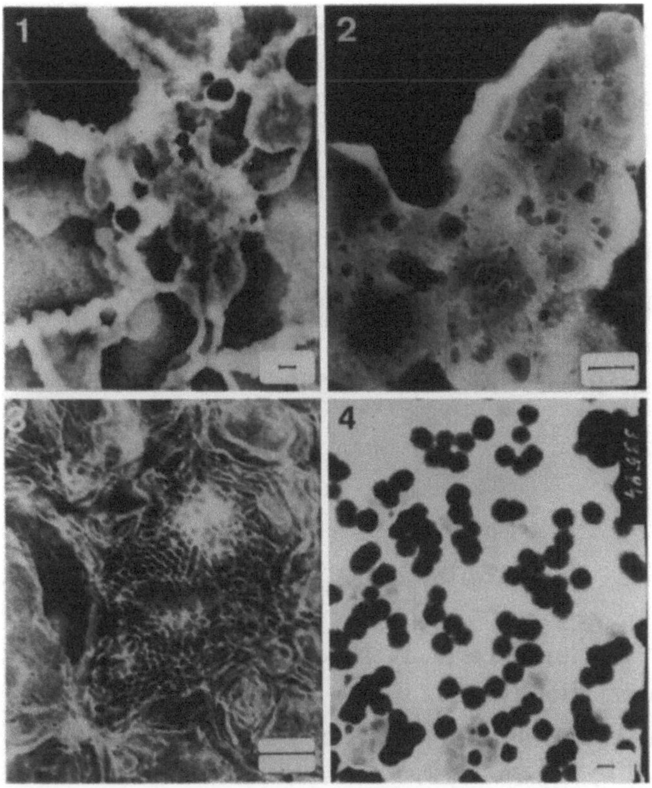

Fig.1. Negative stain electron micrographs of lipid aqueous dispersions. 1) GDGT/PC (1:4 molar ratio); 2) GDNT/PC (1:4 molar ratio); 3) PLE; 4) PLE/PC (1:2 molar ratio) vesicles obtained by controlled detergent removal. The bars correspond to 100 nm.

Table 1

Comparison between theoretical and experimental vesicle radii.

lipid	P_c	$R_c(nm)$	$R_s(nm)$
GDGT/PC 1:6	0.94	28 ± 2	25 ± 4
GDNT/PC 1:4	0.96	40 ± 2	38 ± 4
GDNT/PC 1:9	0.92	23 ± 2	24 ± 5
PLE/PC 1:4	-	-	57 ± 6

cle radius increases. Since PLE/PC is more than a binary system, the theory cannot be applied in this case. In all the other cases the agreement between measurements and theoretical predictions is satisfactory.

Acknowledgements

This work has been supported by MURST 60% and 40% grants, by the CNR grant "Progetto Finalizzato Chimica Fine II" and by CEE Esprit Basic Research Action N. 3200.

References

[1] M. De Rosa, A. Gambacorta, and A. Gliozzi, Microbiol. Rev. **50**, 70 (1986).
[2] M. De Rosa, A. Gambacorta, B. Nicolaus, B. Chappe, and P. Albrecht, Biochim. Biophys. Acta **753**, 249 (1983).
[3] Z. Mirghani, D. Bertoia, A. Gliozzi, M. De Rosa, and A. Gambacorta, Chem. Phys. Lipids **55**, 85 (1990).
[4] J. N. Israelachvili, S. Marcelja, and R. G. Horn, Q. Rev. Biophys. **13**, 121 (1980).
[5] J. N. Israelachvili, D. J. Mitchell, and B.W. Ninham, Biochim. Biophys. Acta **470**, 185 (1977).
[6] F. Cavagnetto, A. Relini, Z. Mirghani, A. Gliozzi, D. Bertoia, and A. Gambacorta, (submitted).

Determination of Size and Structure of Lipid IV_A Vesicles by Quasi-Elastic Light Scattering and Small-Angle X-Ray Scattering

N. Maurer and O. Glatter

Institute of Physical Chemistry, University of Graz, Heinrichstr. 28, A-8010 Graz, Austria

ABSTRACT. Heavy sonication of lipid IV_A in different buffer solutions at pH values around 7.4 produces mostly unilamellar vesicles, which are stable over several weeks. For short sonication times these vesicles are very large, about 250 nanometers in terms of an average hydrodynamic radius with a slight polydispersity, as determined by quasi-elastic light scattering. Therefore small-angle X-ray scattering curves obtained with a conventional instrument do not contain enough information to determine the overall size of the lipid IV_A aggregates. However, the pair distance distribution function (PDDF) of the scattering curve is in agreement with the assumption of large vesicular particles. The thickness of the double layer is very small compared to the overall dimension. It can be evaluated assuming a lamellar particle. The corresponding PDDF provides a thickness of about 5 nm. The electron density distribution within the double layer can be calculated directly from the PDDF by a convolution square root operation. Head groups and hydrocarbon chains are represented by two regions of different electron densities.

1. INTRODUCTION

Lipopolysaccharide (LPS or endotoxin) is a major component of the outer surface of the outer membrane of Gram-negative bacteria [1]. Essentially all of the physiological activities of LPS in both animal and isolated cells can be attributed to the lipid A

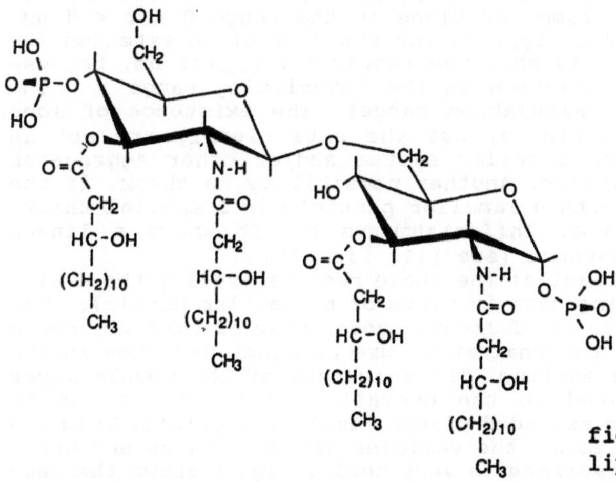

fig. 1. Structure of lipid IV_A.

Springer Proceedings in Physics, Vol. 66 The Structure and Conformation of Amphiphilic Membranes
Editors: R. Lipowsky · D. Richter · K. Kremer

portion of this molecule. Lipid IV_A is a bioactive precursor of lipid A. Both, lipid A as well as lipid IV_A, are used as model compounds to investigate the responses of biological systems to LPS. These substances can be isolated from mutants of Gram-negative bacteria in milligram amounts of high purity, whereas LPS usually is a heterogeneous product. Chemically, lipid IV_A is an amphiphile consisting of a polar head group, two β 1'→6 linked glucoseamine phosphates, and four hydrophobic β-hydroxymyristoyl side chains (fig. 1).

2. METHODS

Methods and evaluation techniques are described in detail elsewhere [2,3]. Briefly, the SAXS data were evaluated with the indirect transformation method ITP [4]. The electron density distribution $\rho(r)$ was computed by the convolution square root technique DECON [5-7] from the PDDF calculated with ITP [8].

3. RESULTS AND CONCLUSION

Size of the aggregates: The overall size of the lipid IV_A aggregates was determined by quasi-elastic light scattering. These experiments show that the lipid IV_A aggregates are very large. The formation of vesicles under the same conditions was also evidenced by electronmicrographs [9]. In our sample the average size of the vesicles in terms of an hydrodynamic radius determined by the cumulant method was 250 nm with a polydispersity of 16%. The size distribution calculated by the inversion of the field correlation function by an inverse Laplace transform shows a peak at the same size with a somewhat higher polydispersity. The mean size of 250 nm is too large to be accounted for by micelles and indicates formation of vesicles.

SAXS curves do not contain enough information to determine the overall size and shape. Nevertheless, the overall PDDF (fig. 2) is in agreement with the assumption of large vesicular particles. A PDDF of a cylindrical micelle, which would be the other physically possible structure, is not in agreement with our data (see [10] figs. 3 and 5). The PDDF of the particles shows the inhomogeneities across the lamellar plane in the range $0 < r < 5$ nm. The shape of the PDDF is typical for the PDDF of an extended lamellar particle. We only show the result for D_{max}=12 nm, because it gives the best resolution in the interisting range $r < 5$nm (D_{max} determines the calculation range). The existence of some micelles cannot be excluded, but the data clearly are not in agreement with a pure micellar system and/or other aggregates like cylindrical micelles. Another possibility to check, if the data are consistent with a lamellar particle is a special thickness-Guinier plot, i.e. $\ln(I(h)*h^2)$ vs h^2. It shows a linear course over a wide h-range (see [11] fig. 3).

Thickness of the lamella: The above results justify the evaluation of the scattering data in terms of a lamellar particle. The scattering curve (fig. 3), desmeared under this assumption, shows the influence of the internal structure at great h-values in the form of a broad side maximum. The thickness of the double layer is very small compared to the overall dimensions. It can be evaluated assuming an extended planar lamellar particle. This assumption is adequate since the vesicles are very large and hence the curvature of the surface is very small. Fig. 4 shows the mea-

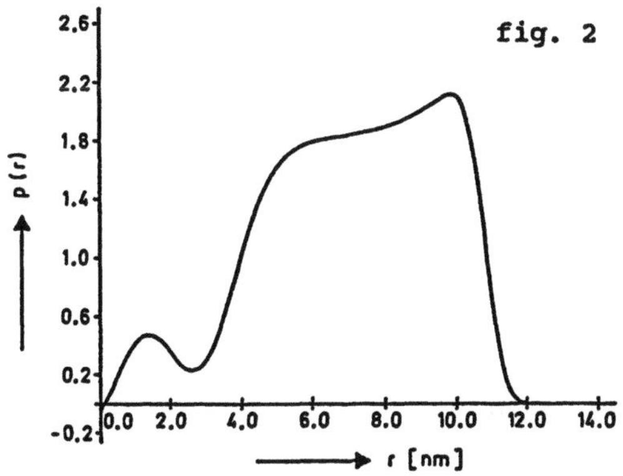

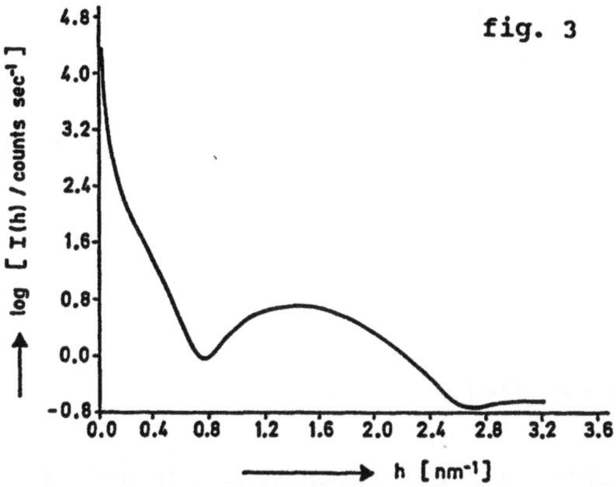

figs. 2,3. (2) PDDF of the whole aggregate and (3) desmeared scattering curve.

sured scattering function (open circles) together with the approximation function (full line) under the assumption of a lamellar structure. The corresponding PDDF of the thickness ($p_t(r)$) is depicted in fig. 5. The maximum thickness was limited to 6 and 8 nm, respectively. From the fact that this function differs from zero significantly only for $r < 5$ nm, one can directly deduce that it cannot be a multilamellar system. The slight oscillations for $r > 5$ nm are due to the inaccuracy of the experimental data. The thickness of the unilamellar layer is about 5 nm. However, the existence of very few multilamellar vesicles cannot be excluded. It should be noted that the calculations for a PDDF of a cross-section assuming cylindrical symmetry did not give any physically meaningful results.

Internal structure of the lamella: Starting with the $p_t(r)$ calculated from the experimental data (open circles in fig. 6, full line in fig. 5) the best possible centrosymmetrical electron density distribution was calculated as a sum of equidistant step functions of constant width (full line, fig. 7). The fit to the given $p_t(r)$ is shown in fig. 6 as full line. The profile with seven equidistant step functions suggests the computation of a two step model. The corresponding profile for optimized width and height is shown as a dashed line in fig. 7.

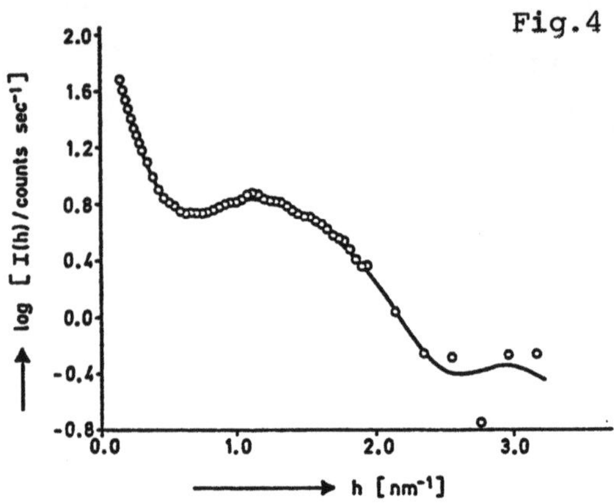

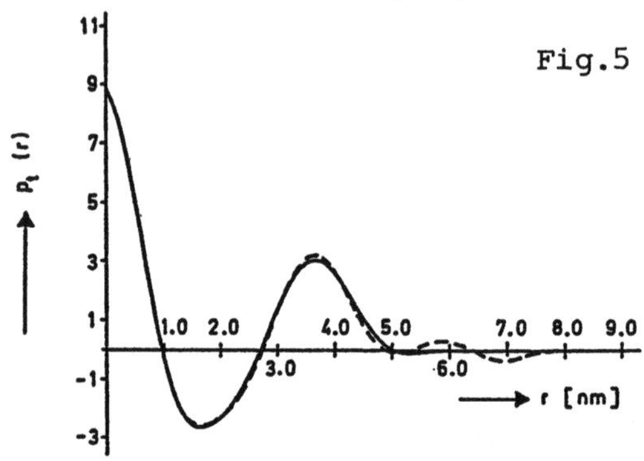

figs. 4,5. (4) Measured scattering curve (open circles) and its approximation function (full line). (5) PDDF of the thickness.

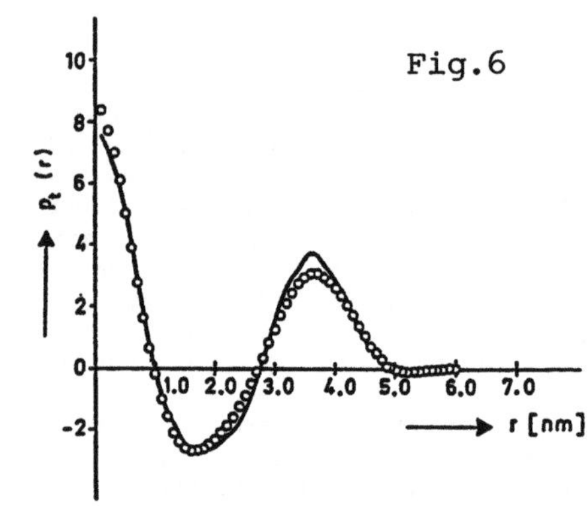

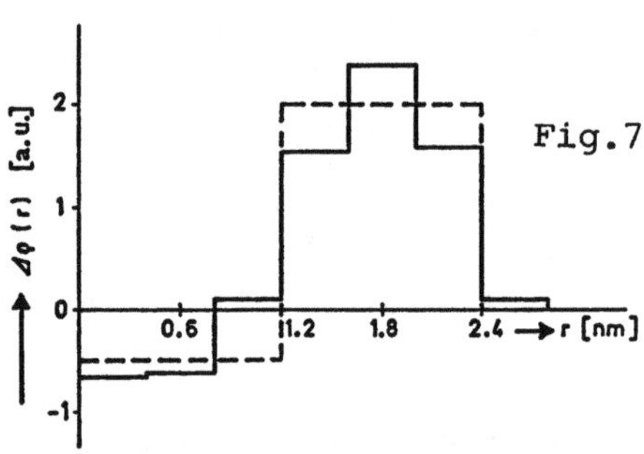

figs. 6,7. (6) PDDF of the thickness calculated from the experimental data (open circles) and the fit (full line) from the electron density distribution with seven step functions.
(7) electron density distribution from a sum of seven equidistant step functions of constant width (full line) or as a two step function of adjustable width and height (dashed line).

Conclusion: Our results lead to the conclusion that lipid IV_A forms unilamellar vesicles, whose lamellae consist of two regions of different electron densities corresponding to the head groups and the hydrocarbon chains. The lamella has a thickness of 4.8 nm with a hydrocarbon chain length of 1.2 nm.

4. REFERENCES

1. E.Th. Rietschel: *Chemistry of Endotoxin*, Vol. 1, Ser. *Handbook of Endotoxin*, ed. R.A. Proctor (Elsevier, 1984)
2. R. Pecora: *Dynamic light scattering* (Plenum Press, N.Y. 1985)
3. O. Glatter, O. Kratky: *Small-angle X-ray scattering* (Academic Press, London 1982)
4. O. Glatter: J. Appl. Cryst. **10**, 415-421 (1977)
5. O. Glatter: J. Appl. Cryst. **14**, 101-108 (1981)
6. O. Glatter, B. Hainish: J. Appl. Cryst. **17**, 435-441 (1984)
7. O. Glatter: J. Appl. Cryst. **21**, 886-890 (1988)
8. O. Glatter: J. Appl. Cryst. **13**, 577-584 (1980)
9. M. Hofer, R.Y. Hampton, C.R.H. Raetz, H. Yu submitted to: Chem. Phys. Lipids (1991)
10. O. Glatter: Progr. Colloid. Polym. Sci. **84**, in press (1991)
11. N. Maurer, O.Glatter: J. Appl. Cryst. **24**, in press (1991)

Multicomponent Vesicular Aggregates (MCVA): Spontaneous Vesiculation of Perfluorinated Single-Chain Surfactant Mixtures

S. Szönyi[1;[1]], *A. Cambon*[1], *H.J. Watzke*[2;[1]], *P. Schurtenberger*[2], and *E. Wehrli*[3]

[1]Laboratoire de Chimie Organique du Fluor, Université de Nice, Parc Valrose, F-06034 Nice Cedex, France
[2]Institut für Polymere, ETH Zürich, CH-8092 Zürich, Switzerland
[3]Institut für Zellbiologie, ETH Zürich, CH-8092 Zürich, Switzerland

1.0 Spontaneous Vesiculation and Membrane Organization

We report on spontaneous vesiculation of single-chain perfluorinated surfactant mixtures. The surfactants bearing oppositely charged headgroups form stable unilamellar vesicular aggregates upon gently mixing non-equimolar solutions. We have synthesized and employed a variety of perfluoro alkane single-chain surfactants (**Figure 1**). The obtained transparent solutions contained polydispersed aggregates. The unilamellar vesicles were characterized and investigated by quasi-elastic light-scattering, electronmicroscopy and nuclear magnetic resonance spectroscopy.

Vesicles are self-assembled membraneous structures, which consist of a closed surfactant bilayer shell. Recently a new way was found to produce vesicles by mixing cationic and anionic single-chain surfactants, which form under proper conditions stable, unilamellar aggregates [**2**]. The electrostatic interaction produces a "double-chain" ion-pair surfactant, which forms lamellar sheets and eventually closed-shell and single-wall surfactant aggregates [**3**].

We were interested in finding single-chain surfactants that can spontaneously form vesicles and produce stable membranes [**4**]. We included perfluoro alkane surfactants in our studies, because they have lower CMCs for the same number of chain carbon atoms and exhibit a strong tendency to form lamellar bodies at higher surfactant concentrations [**5**] .

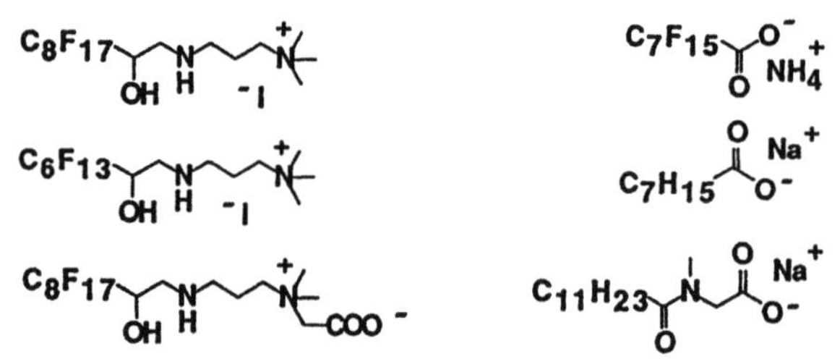
1 2 3 4 5 6

Figure 1: Cationic, anionic, and zwitterionic perfluoro alkane and hydrocarbon surfactants used in the experiments. (Perfluoro octanoic acid ammonium salt **4**, sodium octanoate **5** and sodium lauroyl sarcosinate were purchased from Fluka, Switzerland. Surfactant **1**, **2**, and **3** were synthesized following the procedure of Szönyi and coworker [**6**]).

Table 1: Hydrodynamic radii of vesicles determined by quasi-elastic light-scattering at 24.8°C. The light-scattering was performed on a Malvern Zetasizer 3, under an angle of 90°, using a Coherent INOVA 200 argon ion laser (λ=488nm).

mixture	Ψ	R(h)/nm [7]	N†
1/4	0.77	34.1 ± 10.9	1.00
1/4	0.46	13.4 ± 3.4	0.36
		37.8 ± 18.0	0.64
1/5	0.46	6.9 ± 2.8	0.15
		51.9 ± 38.8	0.85
1/6	0.46	44.8 ± 15.3	1.00
2/4	0.73	51.3 ± 20.8	1.00
3/4	0.60	54.8 ± 25,5	1.00

†N is the intensity weighted fraction of aggregates with radius R(h)

Our surfactants easily form vesicles from both the monomolecular and micellar state. **Table 1** and **Figure 2** summarize the microscopic appearance and the hydrodynamic radii of the different preparations. The surfactants were mixed in different molar ratios. Ψ indicates the molar ratio of the negatively charged component.

The appearance of the samples in electronmicroscopy is similar to that of "normal" double-chain surfactant vesicles and phospholipid liposomes. The dynamic light-scattering data confirm aggregates, which contain an inner aqueous volume. By varying the composition in mixtures of surfactants **1** and **4** a strong effect is observed on the sizes and the size distributions (**Table 1**). Vesicles formed under a high excess of the negatively charged and shorter surfactant **4** exhibited a fairly broad monomodal size distribution (see also **Figure 2B**). Vesicles formed under an excess of the positively charged and longer surfactant **1** exhibited a bimodal size distribution (see also **Figure 2A**). Exchanging the perfluorinated surfactant **4** for a hydrocarbon analog of the same carbon number, surfactant **5**, results in a similar bimodal distribution. All other mixtures exhibited unimodal size distributions (**Table 1**). (Stock solutions of sodium octanoate **5** and ammonium perfluoro octanoate **4** were prepared in concentrations below CMC. Stock solutions of surfactants **1**, **2**, **3**, and **6** were above CMC).

The membrane organization is strongly affected by the compositional variations. Information on the mobility of membrane molecules can be gained from the determination of T_2*, which was calculated from the line width of the headgroup proton NMR spectra of surfactant **1**. **Figure 3** shows the ^{1}H-NMR spectra of the mixtures of **1** and **4** having the compositions Ψ=0.77 (spectrum A) and Ψ=0.46 (spectrum B), respectively. While spectrum A resembles that of the micellar stock solution (T_2*=27ms), spectrum B indicates drastical changes in the molecular mobility (T_2*=6ms) and modifications of the magnetic environment (chemical shift) of molecule **1**. Since dis-

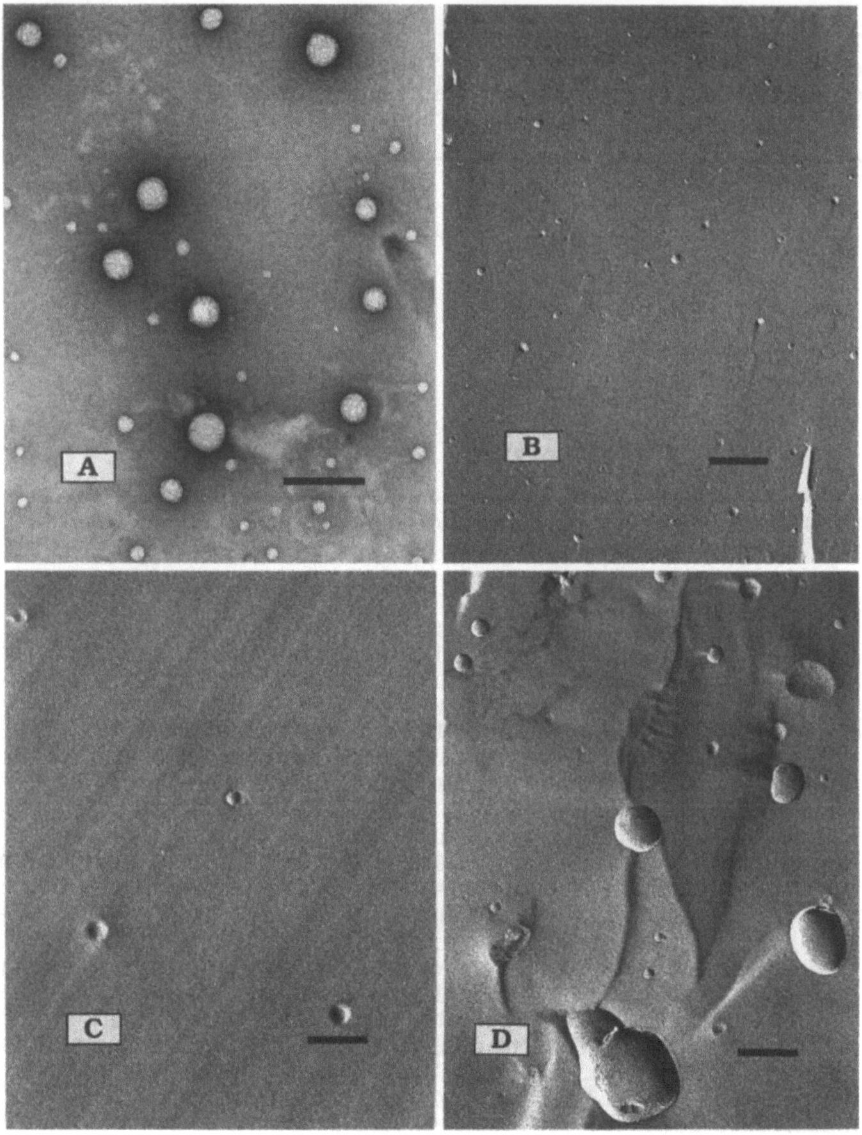

Figure 2: Negative staining (**A**) and freeze fracture electronmicrographs (**B-D**). **A**: mixture of **1** and **4**, Ψ=0.25, [S]=6.5mM; **B**: mixture of **1** and **4**, Ψ=0.60, [S]=6.5mM; **C**: mixture of **2** and **4**, Ψ=0.43, [S]=4.3mM; **D**: mixture of **3** and **4**, Ψ=0.44, [S]=22.3mM. The inserted bar indicates 0.2 μm. (The sample were stained by uranyl acetate. Samples for freeze fracture were prepared by propane jet vitrification following the procedure of Müller and coworker [**8**].The replicas were studied in a Philips EM 301 electronmicroscope).

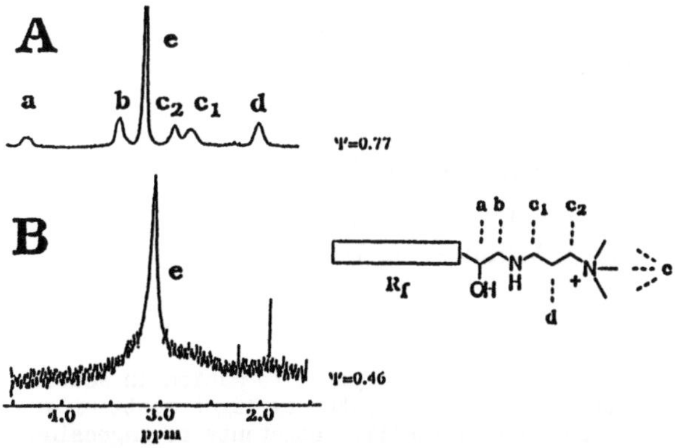

Figure 3: 300 MHz Proton nuclear magnetic resonance spectra of mixtures of **1** and **4** in D_2O with molar ratios of Ψ=0.77 and Ψ=0.46, respectively. Surfactant concentrations were 6.5mM and 3.4mM, respectively. (The samples were prepared in D_2O. The proton nuclear resonance spectra were recorded on a Bruker Spectrometer AM 300 WB. D_2O was taken as internal standard).

tinct vesicular aggregates can be found for both compositional ranges (**Figure 2A** and **2B**) membrane organization of the vesicles must differ appreciably. Our results show that single-chain perfluoro alkane amphiphiles are a promising group of spontaneous vesiculating surfactants with interesting self-assembling features.

Acknowledgement: The authors acknowledge gratefully the help of Dr. P. Skrabal, F. Bangeter, S. Egelhaaf, and M. Ambühl.

2.0 References:

[1] to whom correspondence should be adressed.
[2] E. W. Kaler, A. Kamalakara Murthy, B. E. Rodriguez, and J. A. N. Zasadzinski: Science (Washington) **245**, 1371 (1989).
[3] S. A. Safran, P. Pincus, and D. Andelman: Science (Washington) **248**, 354 (1990).
[4] M. Ambühl, P. L. Luisi, and H. J. Watzke: to be published, (1991).
[5] K. Fontell and B. Lindman: J. Phys. Chem. **87**, 3289 (1983) (and references therein).
[6] S. Szönyi and A. Cambon: to be published, (1991).
[7] Data analysis by an exponential sampling inverse Laplace transform program as described by N. Ostrowsky, D. Sornett, P. Parker, and R. Pike: Opt. Acta **28**, 1059 (1981).
[8] M. Müller, N. Meister, and H. Moor: Mikroskopie (Wien) **36**, 129 (1980).

Formation of Colloid and Liquid Crystal Phases of Magnesium Dodecylbenzenesulfonate: Interpretation by Fractals

D. Težak, I. Fischer-Palković, S. Heimer, and F. Strajnar

Laboratory of Physical Chemistry, Faculty of Science,
University of Zagreb, 41001 Zagreb, Marulićev trg 19,
P.O. Box 163, Croatia

Abstract. The processes of colloid and liquid crystal formation in aqueous amphiphilic - electrolyte solutions were investigated by light scattering and polarizing microscopy. The apparent solubility constants of magnesium dodecylbenzenesulfonate were calculated. The fractal dimensions ranging between $1<D<2.6$ and $3<D<4.3$ explained a different cluster mass distribution in colloidal and lamellar phases.

1. Introduction

It was found earlier that dodecylbenzenesulfonic acid (HDBS) - water mixtures exhibit nematic, lamellar and inverse cubic phases |1|. The lamellar phases were also found by mixing HDBS with electrolyte solutions |2,3|. In this work the formation of different phases in aqueous solutions of $Mg(NO_3)_2$ and HDBS were recognized by fractal approach applying light scattering technique. The experiments performed at different temperatures showed a considerable influence of the kinetics of nucleation on the formation and structure of mesophases or colloidal phases.

The behaviour of small amphiphilic molecules which form the lyotropic liquid crystalline phases with interacting metal ions can be considered as a kinetic model approach to the amphiphilic membranes. The structure of membranes can be observed as a basic bilayer or a multilayer.

2. Experimental

2.1 Materials. All chemicals of p.a. grade were used as a commercially distributed: $Mg(NO_3)_2$ from "Kemika" Zagreb, HDBS from "Prva Iskra" Barić. The aqueous solutions were standardized and prepared as described previously |3|. HDBS was not isomerically pure, and therefore all equilibrium constants were taken as apparent values and denoted by an asteriks.

2.2 Techniques. Almost all phenomena were observed in heterogeneous systems using Zeiss tyndallometer and Brice Phoenix universal light scattering photometer DU 2000. Zeiss polarizing microscope was used for determination of phases. Investigation of ionic species in solutions were performed by conductometry.

2.3 Treatment of data. The ionic equilibria in solutions of $Mg(NO_3)_2$ and HDBS were considered as follows:

$$\text{(i)} \quad Mg^{2+} + 2DBS^- \rightleftharpoons Mg(DBS)_2 \tag{1}$$

with the equilibrium constant:

$$k_s^* = a(Mg^{2+}) \cdot a^2(DBS^-) \; ; \tag{2}$$

$$\text{(ii)} \quad MgHCO_3^+ + DBS^- \rightleftharpoons MgHCO_3DBS. \tag{3}$$

Springer Proceedings in Physics, Vol. 66 The Structure and Conformation of Amphiphilic Membranes
Editors: R. Lipowsky · D. Richter · K. Kremer

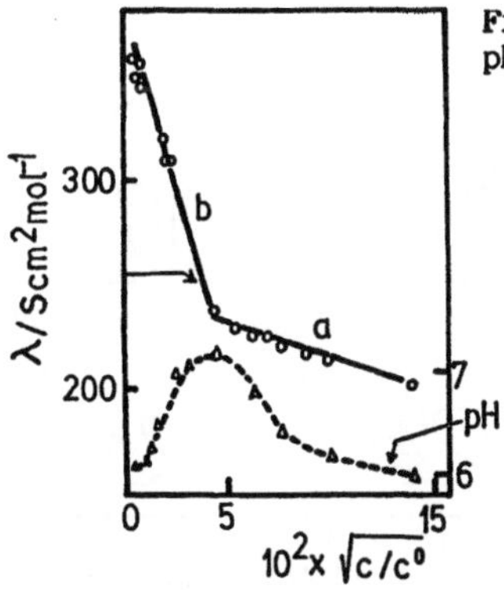

Figure 1. Conductivity and pH data for $Mg(NO_3)_2$ solution.

The mean activity coefficient was calculated applying Debye-Hückel law. Taking into account the formation of species according to the expression (3) |4|, the following experiment was done in order to prove the presence of Mg^{2+} or $MgHCO_3^+$ - ions in solutions: (i) the conductivity measurements showed that practically 100% of Mg^{2+}- ions were present in the aqueous solutions of $Mg(NO_3)_2$ above $3x10^{-3}$ mol dm^{-3} (λ_∞ = 247.97 S cm^2mol^{-1} vs. $\lambda_{\infty theoretical}$ = 249 S cm^2mol^{-1}, slope a in Figure 1; (ii) corresponding pH-behaviour related to sudden change of the slope in the conductivity line indicated the presence of $MgHCO_3^+$ species at ionic strength lower than $3x10^{-3}$ mol dm^{-3} $Mg(NO_3)_2$ (slope b).

3. Results

The formation of new phases of magnesium dodecylbenzenesulfonate from aqueous solutions is characterized by the processes of dimerization, premicellization and micellization of amphiphile |2|. The micellar complexes with Mg^{2+}-ions are formed when c.m.c. is exceeded. It is obvious from Figure 2 that the stoichiometry of magnesium dodecylbenzenesulfonate is not simple. The activity presentation in the "solubility product domain" shows two slopes of the solubility limit line: slope 0.5 for $|Mg(NO_3)_2|$/mol dm^{-3} between $4x10^{-2}$ and $3x10^{-3}$, and slope 0.77 (mean value) between $3x10^{-3}$ and $1.5x10^{-4}$. Therefore, it can be assumed that the 50% of magnesium in the form of Mg^{2+} and 50% in the form of $MgHCO_3^+$ is present in solution in the region II. Calculating the solubility product this fact was taken into account by multiplying k_s^* with 0.5. The solubility products presented in Table 1 promote the corrected values of those published earlier |2|.

The formation of crystalline and liquid crystal phases is presented in the precipitation diagrams in Figure 3 by the precipitation domains of liquid crystals (LC) mostly lamellar |3|, the solid crystals (SC), mixed LC and SC domains. The disappearance of LC domains at higher temperatures is obvious in the systems, which have been prepared at the temperatures presented in Figure 3. There is a different behaviour of the samples which had been prepared at room temperature and after that kept in the microscopic field under elevated temperature; these samples showed the lamellar

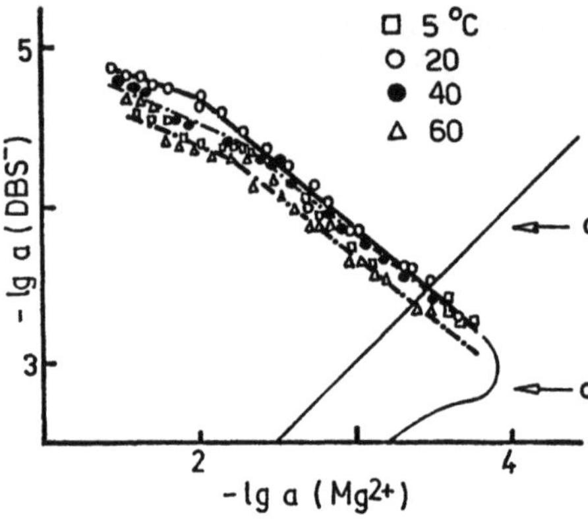

Figure 2. Activity plot of the precipitation limits in "solubility product" region.

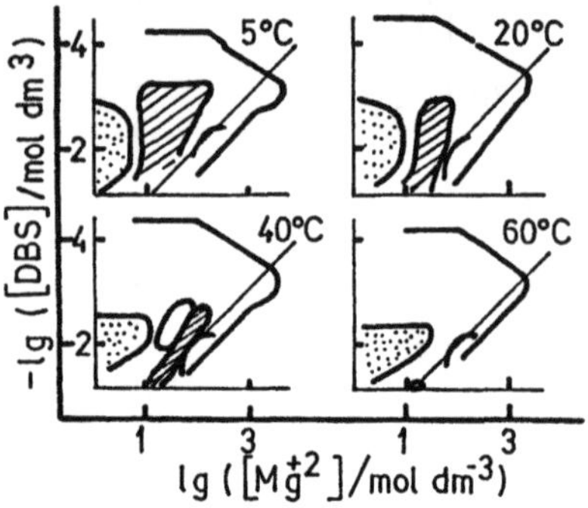

Figure 3. Phase diagrams: LC region denoted by lines, SC region denoted by points, mixed LC+SC regions in between.

Table 1. Solubility product of magnesium dodecylbenzenesulfonate

t °C	slope	pk_s^*	$\Delta_r G^*$/kJ mol^{-1}
5	0.68	0.866±0.059	52.52
20	0.83	10.801±0.070	60.60
40	0.83	10.918±0.046	65.43
60	0.74	9.449±0.089	60.25

phase covering the whole LC area presented in the 5 °C diagram. Even the SC phases created at room temperature were transformed into lamellar at elevated temperature. The almost regular effect was the transformation of usual focal conics into the birefringent tubular texture at high temperature.

The dominant role is played by the changes in solubility and micellar properties at elevated temperatures, as well as slow self-organization of amphiphile molecules. The intrinsic efficacy of the temperature and the increased concentrations of amphiphile and electrolyte on the growth and polydispersity was shown by van Os et al. |5|, and can be also considered as a possible effect of alkylbenzenesulfonates.

The formation of phases depends on diffrent nucleation rates in different concentration domains. The crystalline phase in the high concentration excess of electrolyte is characterized by fast nucleation of a great number of crystal nuclei. Therefore, the supernatant had been exhausted, and no appearance of new phase was possible. At the equivalency region the formation of liquid crystals occurs as a result of slow self-organizing of amphiphile molecules into bilayers and multilayers.

The preliminary results of the fractal approach to the mass distribution in colloidal aggregates bellow c.m.c., and LC phase above c.m.c., are presented in Table 2. The power law of intensity of scattered light I(Q) plotted vs. scattering vector ($Q=4\pi n/\lambda \cdot \sin\Theta/2$) gives a linear plot with surface fractal dimensions (D_s) for a smooth object |6| of high mass density which stand $3<D_s<4.3$. For colloidal aggregates the mass fractal dimensions (D_m) indicate $1.6<D_m<2.6$. The very fast nucleation of primary

Table 2. Fractal dimensions calculated in various regions.

	$Mg(NO_3)_2$	HDBS	D	R^2
(a) colloid aggregation (10 min.)	$1x10^{-2}$	$6x10^{-4}$	1.62	0.997
	$1x10^{-2}$	$4x10^{-4}$	1.98	0.999
	$1x10^{-2}$	$2x10^{-4}$	2.65	0.999
(b) LC formation (1 day)	$6x10^{-2}$	$6x10^{-2}$	3.85	0.998
	$4x10^{-2}$	$4x10^{-2}$	3.68	0.996

All concns. in mol dm^{-3}; R = correlation coefficient for a linear function.

particles of colloidal aggregates exhibited the fractal dimensions constant after 10 minutes, while the proper values for fractal dimensions in the case of liquid crystal formation were obtained after several hours (some representative values in Table 2).

4. Conclusion

The difference has been shown between slow processes of the formation of lamellar mesophase and fast nucleation of crystalline phase. The preliminary investigations have shown that there are well expressed differences in fractal behaviour of colloidal aggregation and surface fractals characterizing the processes of liquid crystal formation. Obviously, well defined focal conics |2| present the smooth objects with very coherent mass distribution inside self-organized liquid crystalline structure.

5. References

|1| Đ. Težak, G. Hertel and H. Hoffmann, *Liq. Cryst.* **10**, 15 (1991).

|2| Đ. Težak, F. Strajnar, D. Šarčević, O. Milat and M. Stubičar, *Croat. Chem. Acta* **57**, 93 (1984).

|3| Đ. Težak, F. Strajnar, O. Milat and M. Stubičar, *Progr. Colloid & Polymer Sci.* **69**, 100 (1984).

|4| F.S. Nakayama, *J. Chem. Eng. Data* **16**, 178 (1971).

|5| Binana-Limbelé, N.M. van Os, L.A. Rupert and R. Zana, *J. Colloid Interface Sci.* **141**, 157 (1991).

|6| D.W. Schaefer, *Science* **243**, 1024 (1989).

Ginzburg–Landau Theory of Bulk and Interfacial Properties of Amphiphilic Systems

G. Gompper and S. Zschocke

Sektion Physik, Ludwig-Maximilians-Universität München, Theresienstr. 37, W-8000 München 2, Fed.Rep.of Germany

Abstract. We have studied ordered phases, and their interfacial behavior, in a simple Ginzburg-Landau model of oil-water-surfactant mixtures. Many of the phases observed in experiment are recovered. A wetting transition of the lamellar phase near a wall is predicted. We also show how the elastic properties of amphiphilic monolayers are related to bulk properties of microemulsions in our model.

1. Introduction

Monolayers of amphiphiles or surfactants form spontanously at the interface between a polar and a non-polar liquid, typically water and oil, whereby the interfacial tension can be reduced several orders of magnitude. Due to this enormous reduction of the interfacial free energy, phases with an extensive amount of internal interface become possible, like the lamellar phase, where the monolayers form a one-dimensional stack, or the microemulsion, where the monolayers form a random, spatially isotropic array [1]. In all these phases, the monolayers are believed to behave as a two-dimensional, incompressible liquid. Since their area is therefore (almost) constant, their fluctuations are controlled by elastic *bending* energy, given by the famous expression of Helfrich [2],

$$\mathcal{H} = \int dS[\sigma + \Delta H + 2\kappa H^2 + \bar{\kappa} K] \tag{1}$$

where $H = (1/2)(1/R_1 + 1/R_2)$ is the mean curvature (with the local radii of curvature R_1 and R_2) and $K = 1/(R_1 R_2)$ is the Gaussian curvature. Here, σ is the (bare) surface tension, κ the bending modulus, $\bar{\kappa}$ the saddle-splay modulus, and Δ is related to the spontanous curvature [3].

We want to show that an entirely different approach, a phenomenological Ginzburg-Landau theory of oil-water-surfactant mixtures [4-6,7,8], can be used both to calculate the structure of the ordered phases, and to study their interfacial properties. In particular, we can calculate the elastic coefficients in (1), and thereby obtain a relation between interfacial properties of the monolayers and the bulk properties of microemulsions.

Springer Proceedings in Physics, Vol. 66 The Structure and Conformation of Amphiphilic Membranes
Editors: R. Lipowsky · D. Richter · K. Kremer

2. Ginzburg-Landau Model

The starting point of our analysis is the Ginzburg-Landau free energy functional [4]

$$F\{\phi(\mathbf{r})\} = \int d^3r[c(\nabla^2\phi)^2 + g(\phi)(\nabla\phi)^2 + f(\phi)], \tag{2}$$

for a scalar order parameter ϕ, which is proportional to the local difference of the oil- and water-concentrations. The functional (2) can be considered as a simple density functional. As in other density functionals, the theory is defined by the requirement that the free energy density, f, of the ***homogeneous*** phases is known, as well as the direct correlation function, or equivalently the scattering intensity, again in the homogeneous phases [9]. Now, there is little detailed information available about these quantities. The general structure of f and g, on the other hand, is quite easily accessible, either from a microscopic model, or from experiment. The free energy density $f(\phi)$ must have three local minima, corresponding to the three homogeneous phases oil, water and microemulsion. At coexistence, two or more of these minima, ϕ_1, ϕ_2, .., have the same value $f(\phi_1) = f(\phi_2) = ...$ Furthermore, it is well known that the scattering intensity of the microemulsion has a pronounced peak at non-zero wave-vector q, whereas there is only a peak at $q = 0$ in the oil-rich and water-rich phases [10]. This implies that g is negative in the microemulsion, and positive in the other two homogeneous phases [4,5].

For numerical calculations, we employ a piecewise parabolic form [4,11] of f,

$$f(\phi) = \begin{cases} \omega_+(\phi - \phi_+)^2 + \frac{1}{2}\mu\,\phi & \text{for } \phi_{0+} < \phi \\ \omega_0\phi^2 + f_0 & \text{for } \phi_{0-} < \phi < \phi_{0+} \\ \omega_-(\phi - \phi_-)^2 - \frac{1}{2}\mu\,\phi & \text{for } \phi < \phi_{0-}, \end{cases} \tag{3a}$$

where ϕ_{0+} and ϕ_{0-} are defined such that f is continuous, and a piecewise constant form of g,

$$g(\phi) = \begin{cases} g_+ & \text{for } \phi_{0+} < \phi \\ g_0 & \text{for } \phi_{0-} < \phi < \phi_{0+} \\ g_- & \text{for } \phi < \phi_{0-}. \end{cases} \tag{3b}$$

3. Phase Diagram

The phase diagram [6] of our Landau model (2) contains three homogeneous phases by construction, the oil-rich phase, the water-rich phase, and the microemulsion. However, for large negative g_0 it contains also spatially modulated phases, see Fig.1. These modulated phases are the lamellar phase, a 1-dimensional stack of monolayers, separated by oil and water, a 2-dimensional, hexagonal array of water cylinders in oil, and a 3-dimensional FCC crystal of swollen micelles [12]. The hexagonal phase is shown in Fig.2. Order parameter profiles which vary in more than one spatial direction cannot occur in a simple

Fig.1

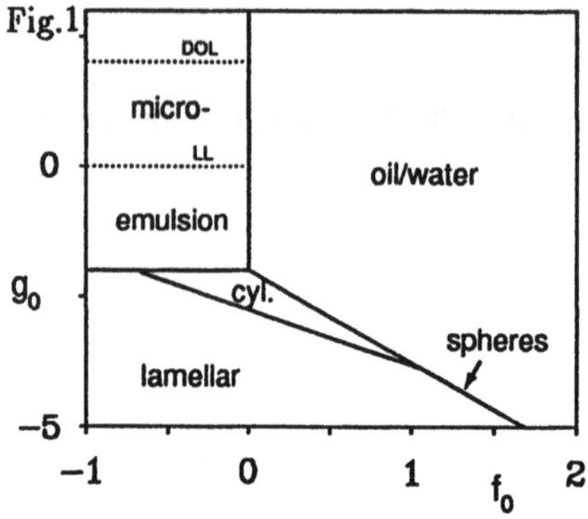

Fig. 2

Fig. 1. Phase diagram for $c = 1$, $g_+ = g_- = 4.6$, $\omega_+ = \omega_- = 4$, $\omega_0 = 1$, $\phi_+ = 2$, $\phi_- = -1$, $\mu = 0$, as a function of f_0 and g_0. All transitions are first-order. The correlation function changes its behavior at the disorder line (DOL) from oscillatory to monotonic. The peak of the scattering intensity moves away from $q = 0$ at the Lifshitz line (LL).

Fig. 2. Contour plot of the order parameter density in the hexagonal phase of cylinders. The parameters of (3) are the same as in Fig.1, and $f0 = 0.808$, $g_0 = -3.5$.

square gradient theory (with $c = 0$, $g(\phi)$ =const.) due to Derrick's theorem [13]. The existence of modulated phases is therefore due to the Laplacian term in (2).

In order to observe the hexagonal and the FCC phases, the system must have 'spontanous curvature', i.e. the monolayers must bend preferentially to one side. In the Landau model, this implies that the $\phi \to -\phi$ symmetry must be broken. Also, the strength of the interaction between monolayers must be small, and the saddle-splay modulus must be negative. For our phase diagram, all these requirements have been met, see Section 5 below.

4. Wetting of a Wall by the Lamellar Phase

In experiments with oil-water-surfactant mixtures, walls of the container are always present. A wall is a planar defect. It seems therefore natural to expect that as the first-order transition from oil/water coexistence to the lamellar phase is approached, a wall will nucleate the lamellar phase: a thin film should form, which grows continuously until the whole sample is in the lamellar state.

In order to study this question, we add a surface free energy contribution

$$F_s\{\phi(\mathbf{r})\} = \int d^3r\ \delta(z)\ [f_s(\phi) + g_s(\nabla\phi)^2], \tag{4a}$$

where

$$f_s(\phi) = \mu_s\phi + \omega_s\phi^2, \tag{4b}$$

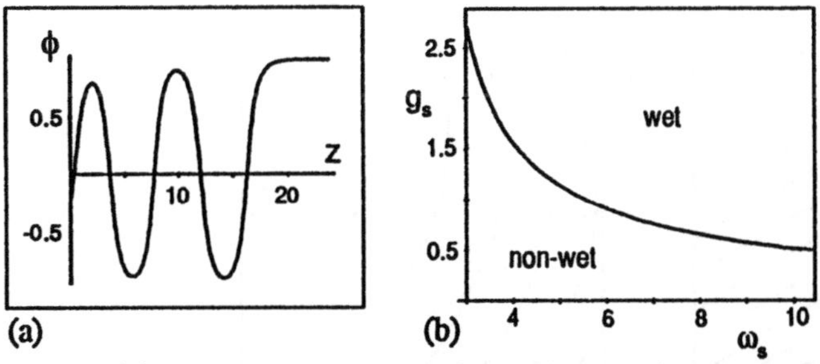

Fig. 3. (a) Order parameter profile in the vicinity of a wall, when the system is close to coexistence of oil/water and the lamellar phase, with $g_s = 1.5$, $\omega_s = 6.0$, and $\mu_s = 0$. (b) Surface phase diagram at 3-phase coexistence between oil, water and lamellar phase, with $\mu_s = 0$. The parameters in (3) are $c = 1$, $g_+ = g_- = 4.4$, $g_0 = -4.5$, $\omega_+ = \omega_- = 4$, $\omega_0 = 1$, $\phi_+ = 1$, $\phi_- = -1$, $f_0 = 0.7023$, $\mu = 0$.

to our Landau model (2). Here, μ_s describes the interaction of oil and water molecules with the wall, ω_s the changed pair interactions of molecules near the wall, and g_s the modified amphiphilicity. With this model, we find that the wall can indeed nucleate the lamellar phase. The order parameter profile for such a case, slightly off coexistence, so that the film is still microscopic, is shown in Fig.3a. However, this is by no means a generic property; only when the wall favors neither oil nor water ($\mu_s \simeq 0$), and when both $g_s > 0$ and $\omega_1 > 0$, does a wetting transition occur. The full surface phase diagram is shown in Fig.3b for $\mu_s = 0$. The order of the wetting transition is hard to determine, because after the first layers have formed, the free energy differences of profiles with a different number of layers are extremely small.

5. Elastic Properties of a Monolayer

The elastic constants in the Helfrich Hamiltonian (1) can be obtained from a calculation of the Landau free energy of oil-water interfaces of various shapes at two-phase *coexistence* [6]. The different geometries considered are a plane, and a cylinder and a sphere, both of radius R. For large R, the interfacial profile of cylinder and sphere approach the planar profile, $\bar{\phi}$. This can be used to expand the free energy of both cylinders and spheres in powers of R^{-1}, and identification of the coefficients of this expansion with the corresponding coefficients in (1), with the result [6]

$$\sigma = \int_{-\infty}^{+\infty} dz\ s(z), \qquad \Delta = 2\int_{-\infty}^{+\infty} dz\ z\ s(z), \qquad (5a,b)$$

$$\kappa = \int_{-\infty}^{+\infty} dz\ [2c(\bar{\phi}')^2], \qquad \bar{\kappa} + 2\kappa = \int_{-\infty}^{+\infty} dz\ z^2\ s(z), \qquad (5c,d)$$

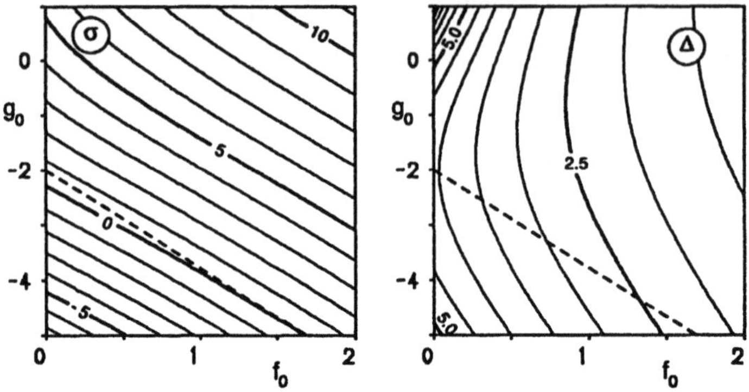

Fig. 4. Contour plot of surface tension σ and 'spontanous curvature modulus' Δ, as a function of f_0 and g_0. The other parameters of the free energy functional (2), (3) are the same as in Fig.1. The dashed line marks the position of the phase transition between oil/water phases and the spacially modulated phases.

where

$$s(z) = [2g(\bar{\phi})(\bar{\phi}')^2 + 4c(\bar{\phi}'')^2]. \tag{5e}$$

The numerical results [6,14] for model (2), (3) are shown in Fig.4. Our result (5) agrees with the results obtained for the standard square gradient theories, i.e. for $c = 0$, $g(\phi) = \text{const.} > 0$, in a low-temperature expansion [15,16]. In this case, $\kappa \equiv 0$, and $\bar{\kappa} > 0$. The bending rigidity is therefore due to the Laplacian in the Landau functional (2). Also, a negative saddle splay modulus is only possible for $c > 0$. The purely mechanical derivation of Δ and $\bar{\kappa}$ given in Ref.17 (for $\kappa = 0$) leads to the identification of $s(z)$ with the stress profile through the monolayer. Finally, we want to emphasize that the solutions with spherical (or cylindrical) symmetry in our model *at oil/water coexistence minimize* the free energy, while the metastable droplet solutions found in the square gradient theories correspond to a local maximum in the free energy [18,19].

The results for σ, Δ, κ and $\bar{\kappa}$ can be used as input in the interfacial models [20-22] of microemulsions.

6. Summary

We have studied ordered phases, and their interfacial behavior, in a simple Ginzburg-Landau theory. Many of the phases observed in experiment are recovered. A wetting transition of the lamellar phase near a wall is predicted, which is waiting for an experimental observation.

Acknowledgements: Stimulating discussions with M. Schick, K. Mecke, and H. Wagner are greatfully acknowledged.

REFERENCES:

[1] See, e.g. *Modern Ideas and Problems in Amphiphilic Science*, edited by W.M. Gelbart, D. Roux, and A. Ben-Shaul, to be published.

[2] W. Helfrich, Z. Naturforsch. **28a**, 693 (1973).

[3] Δ is related to the spontanous radius of curvature, R_0, by $\Delta = -4\kappa/R_0$. Our expression (1) agrees with the usual form of the Helfrich Hamiltonian for *tensionless* monolayers, $\mathcal{H} = \int dS[2\kappa(H-H_0)^2+\bar{\kappa}K]$ with $H_0 = 1/R_0$, in the special case $\sigma = 2\kappa R_0^{-2}$.

[4] G. Gompper and M. Schick, Phys. Rev. Lett. **65**, 1116 (1990).

[5] G. Gompper, R. Hołyst, and M. Schick, Phys. Rev.A**43**, 3157 (1990).

[6] G. Gompper and S. Zschocke, Europhys. Lett. (in press).

[7] In the vicinity of the tricritical point, the results of a Landau theory with g =const>0 and $c = 0$ agree very well with the experimental data, see H. Kleinert, J. Chem. Phys. **84**, 964 (1986).

[8] A Landau model with two order parameters has been considered by K. Chen, C. Jayaprakash, R. Pandit, and W. Wenzel, Phys. Rev. Lett. **65**, 2736 (1990).

[9] See, e.g. M. Baus, J. Phys. Condens. Matter **2**, 2111 (1990), and references therein.

[10] M. Teubner and R. Strey, J. Chem. Phys. **87**, 3195 (1987).

[11] The parabolic approximation has been introduced in the context of wetting by R. Lipowsky, Z. Phys. B**55**, 345 (1984).

[12] We have not considered the stability of ordered bicontinuous phases, like the single and the double diamond structures discussed by Z.-G. Wang and S. Safran, Europhys. Lett. **11**, 425 (1990).

[13] G.H. Derrick, J. Math. Phys. **5**, 1252 (1964).

[14] For the numerical calculations we use the equimolar radius as the interface position.

[15] H.-W. Diehl, D.M. Kroll, and H. Wagner, Z. Phys. B**36**, 329 (1980).

[16] S.C. Lin and M.J. Lowe, J. Phys. A**16**, 347 (1983). R.K.P. Zia, Nucl. Phys. B**251**[FS13], 676 (1985).

[17] W. Helfrich, in *Physics of Defects*, edited by R. Balian, M. Kléman, J.-P. Poirier (North-Holland, 1981).

[18] J.S. Rowlinson and B. Widom, *Molecular Theory of Capillarity* (Clarendon, Oxford, 1982).

[19] M.P.A. Fisher and M. Wortis, Phys. Rev. B**29**, 6252 (1984).

[20] S.A. Safran, L.A. Turkevich, and P.A. Pincus, J. Phys. (Paris) Lett. **45**, L69 (1984).

[21] D. Andelman, M.E. Cates, D. Roux, and S.A. Safran, J. Chem. Phys. **87**, 7229 (1987).

[22] D. Huse and S. Leibler, J. Phys. France **49**, 605 (1988).

Shape and Size Fluctuation of Microemulsion Droplets

B. Farago

Institut Laue-Langevin, 156X, F-38042 Grenoble Cedex, France

Abstract: It has been suggested that the properties of amphiphilic layers as they exist in microemulsions or lamellar systems, are dominated by their curvature elasticity. This physical picture has been used to understand the microemulsion phase diagrams [1,2] in the bicontinous regime. For simple droplet systems the shape fluctuations of the microemulsion globules have also been described in terms of the curvature elasticity theory [3,4].
We present a direct quantitative test of this theory taking into account both the predicted static and dynamic fluctuations. We find that both the splay modulus, K, as well as the saddle-splay modulus, $\overline{K}$, must be considered to provide a consistent fit of both the statics and the dynamics.

1.Introduction

The analysis of the experiments considers the thermal shape fluctuations of the microemulsion droplets, which distort them from their average spherical form. These fluctuation give an extra contribution to the quasielastic neutron scattering which is detected by Neutron Spin Echo (NSE), whereas Small Angle Neutron Scattering (SANS) data yields information on the time averaged shape and on the size distribution. To optimize the information content, we used deuterated solvents for the outer continuum as well as for the core and normal hydrogenated surfactants (which form the film interface). The coherent scattering length densities of the oil and the water are nearly the same while the surfactant film has a large contrast compared to them (shell contrast). Since the scattering arises from this contrast, the detected signal comes solely from the surfactant layer, which exhibits the shape fluctuations.
Two series of samples will be discussed. 1. AOT/decane/water water in oil droplets as a function of size 2. the same system at a given size, investigating the effect of buthanol as added cosurfactant.
The scattering form factor of a spherical shell of radius R with an infinitesimaly thin wall thickness is given by $P(q) \approx (j_0(qR))^2$ where q is the momentum transfer, and $j_0(x)$ is the zero-th order spherical Bessel function, which has its first zero at $qR=\pi$. In reality, the polydispersity smears out this zero to a shallow minimum, nevertheless its position accurately determines the average size and its depth contains information about the size distribution. For fitting purposes we consider the Schultz distribution [5,6] which has two parameters the average size R_0, and a width parameter Z . The mean square deviation is given by :
$\langle(R-R_0)^2\rangle = R_0^2/(Z+1)$. For a microemulsion Z is typically ≈ 20.
The SANS data were taken at the small angle cameras D11 and D17 of the Institute Laue Langevin in Grenoble. The NSE experiments were carried out at the spin echo spectrometer IN11 also at the ILL where we studied the same samples as in the SANS experiments. As explained elsewhere [7], NSE measures directly the normalized intermediate scattering function $I(q,t)/I(q,0)$. In most cases we employed neutron wavelength of $\lambda=8.5Å$ and $\delta\lambda/\lambda=18\%$ resulting in a time window of 0-18 nsec. This allows the observation of motions which have

Springer Proceedings in Physics, Vol. 66 The Structure and Conformation of Amphiphilic Membranes
Editors: R. Lipowsky · D. Richter · K. Kremer

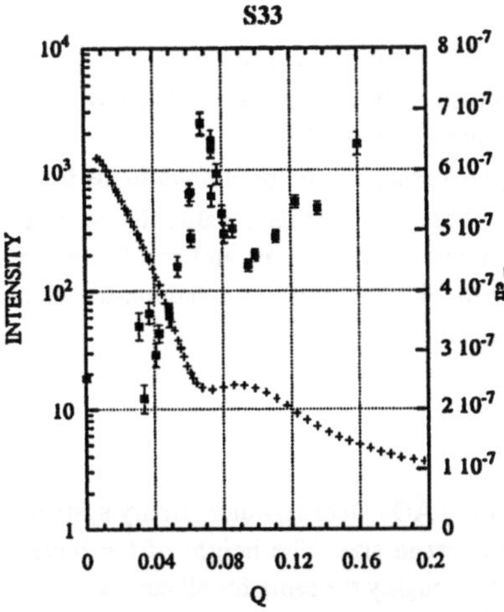

Fig.1. NSE and SANS data for S33 plotted on the same q scale. It shows that the maximum in $D_{eff}(q)$ coincides with the local minimum of the static form factor.

comparable characteristic times, and detectable displacements corresponding to our wave vector range (.02Å^{-1}-.12Å^{-1}); like the center of the mass diffusion and, as we will show, the shape fluctuations. In the case of small fluctuations around a time averaged spherical shell this can be written up to $o(u^2)$ as:

$$I(q,t)=\left\langle \exp(-D_{tr}(R)q^2t)V_s(Dr)^2\left[f_0(qR)+\sum_{\ell\geq 2}\frac{2\ell+1}{4\pi}f_\ell(qR)\left\langle|u_\ell|^2\right\rangle\exp(-\frac{t}{\tau_\ell})\right]\right\rangle_R \tag{1}$$

where

$$f_0(x)=j_0(x)^2+j_0(x)\sum_{\ell\geq 2}\frac{2\ell+1}{4\pi}\left\langle|u_\ell|^2\right\rangle\left[\left\{2-x^2+\ell(\ell+1)\right\}j_0(x)-2xj_1(x)\right]$$

$$f_\ell(x)=\left[(\ell+2)j_\ell(x)-xj_{\ell+1}(x)\right]^2$$

where $D_{tr}(R)$ is the tracer diffusion coefficient of the droplets, V_S is the volume of the surfactant shell u_ℓ is the dimensionless fluctuation amplitude of the ℓ-th mode corresponding to the ℓ-th sphericàl harmonićs, τ_ℓ is the corresponding relaxation time [3,8] and $<>_R$ stands for the average over he size distribution. Eq.1 is model independent ; the description of the dynamics enters through $<|u_\ell|^2>$ and τ_ℓ. Most of the measured curves did not show strong deviation from a single exponential lineshape. This means that in the q region, where the prefactor of $\exp(-t/\tau_\ell)$ has a detectable amplitude in Eq.1., $D_{tr}q^2$ and $1/\tau_\ell$ have comparable values. Therefore, for easy parametrisation one may fit $\exp(-q^2D_{eff}(q)t)$ to the experimental points and define thereby $D_{eff}(q)$.

$D_{eff}(q)$ is shown in Fig1. and Fig.4. as a function of q. In order to understand these NSE data we have to distinguish two q regions. At small q we average over motions on a scale of 1/q so up to $q\approx 1/R$ we can not detect any internal motion or shape fluctuation. There we have $D_{eff}=D_{tr}(R)/S(q)$. From the SANS data we found that $S(q)\approx 1$ for q>.03Å^{-1}. As it is seen from

Eq 1. I(q,t) is composed of a sum of exponentials with q dependent prefactors. The peak structure of $D_{eff}(q)$ can be understood as follows: as we go from small q toward higher q, the form factor of the $\ell>0$ modes starts to contribute significantly to the scattered intensity as compared to the time and polydispersity averaged form factor ($<j_0(qR)^2>$).

This results in an increase of $D_{eff}(q)$. At $qR=\pi$, $<j_0(qR)^2>$ reaches a minimum while the form factor of the $\ell=2$ mode is near to its first maximum. Therefore at this q value the relative contribution of the shape fluctuation component reaches a maximum, and results in the observed peak. This implies, and this is confirmed by the detailed fitting procedure, that this peak region is dominated by the $\ell=2$ mode contribution.

2. Statics

Fig 2. shows the SANS pattern as a function of size for the AOT/decane/water ternary system. The shift of the minimum shows the change in the average size. The height of the bump (corresponding to the first maximum of the form factor) is roughly the same for all curves which means that the polydispersity is similar independently of the average radius.

When trying to elucidate the role of butanol, to minimize experimental errors and to make the data directly comparable, we have chosen the compositions such that the average size was roughly the same.

To keep the size constant we had to chose a increasing water/AOT ratio with increasing butanol concentration. This indicates that at least part of the alcohol is located in the interface, making

Fig.2.

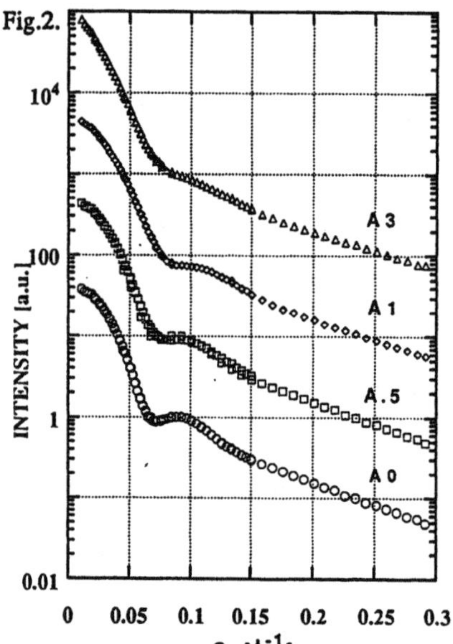

AOT size serie

Fig.3.

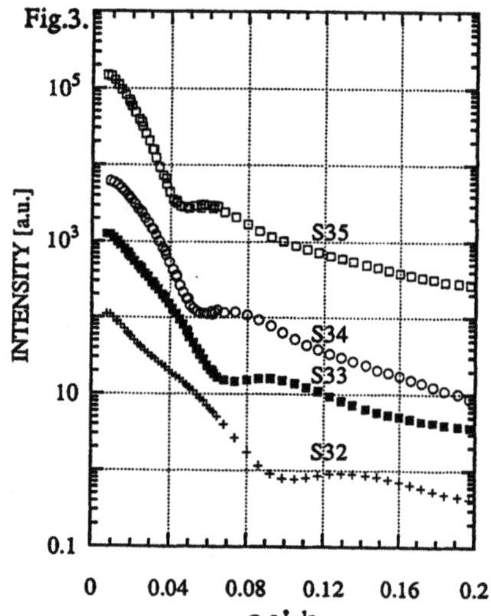

Fig.2. SANS data of the AOT/d-decane/D_2O samples. S32 stands for 3:2 AOT/water weight ratio.

Fig.3. SANS data with different butanol concentration (A0 no buthanol; A3 the most buthanol. For composition see [11]) The data are shifted with a factor 10 for better visibility.

more surface available, only a small fraction, if any, is in the water phase, and the rest is in the oil. Fig 3. displays the properly corrected and normalized scattering patterns for different butanol concentrations. We find that with increasing butanol concentration the minimum of the form factor becomes shallower and broader, indicating an increasing polydispersity. We further note that the intensity scattered in the forward direction increases also. This is a signature of increasing attractive interaction.

3. Dynamics

For the different size AOT system the maximum scales with the minimum of the SANS curves. From Fig.4 it is clear that the peak height of D_{eff} increases with the butanol content in spite of the increasing polydispersity which tends to smear out any structure. Consequently we can immediately draw the qualitative conclusion that the floppiness of the interface increases in the presence of the cosurfactant. To extract dynamical parameters we have to choose a theory to fit the data.

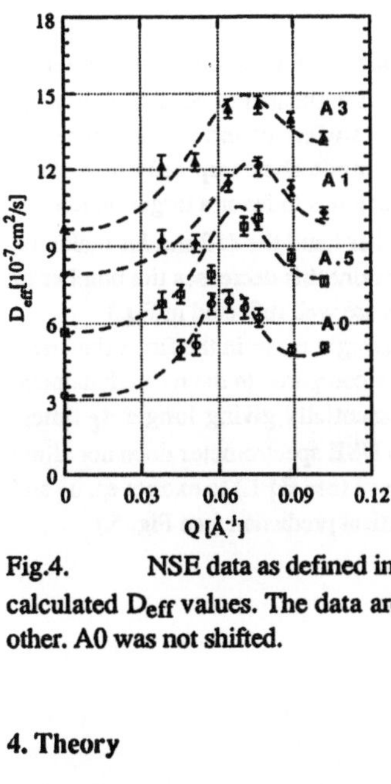

Fig.4. NSE data as defined in the text. Empty symbols are the measured, full symbols the calculated D_{eff} values. The data are shifted with a constant $2*10^{-7}cm^2/s$ with respect to each other. A0 was not shifted.

4. Theory

For the quantitative interpretation of the data we start with the theory of Milner and Safran [3] which is based on the following premise: The main restoring force for the thermal ondulations of the drops is the bending energy ; the total interfacial area is fixed by the quantity of surfactant in the system. The constant surface condition is taken into account in the free energy (E) via a Lagrange multiplier γ: $E'=E-\gamma A_t$, where A_t is the total surface of the droplets to be kept constant. γ has a physical meaning, it describes the effective tension (or 2D pressure) present in the interface (This should not be confused with a macroscopic surface tension). We also include the saddle-splay energy of the Helfrich expression [9] , which comes from the Gaussian curvature:

$$E_{bend}=\frac{K}{2}\int dS\left(\frac{1}{R_1}+\frac{1}{R_2}-\frac{2}{R_s}\right)+\bar{K}\int dS\frac{1}{R_1R_2} \qquad (2)$$

R_1 and R_2 being the two radii of curvature and R_s the spontaneous curvature. The Gauss-Bonnet theorem implies that the saddle-splay contribution is a topological constant and in the case of droplets it gives $n*4\pi*\bar{K}$, where n is the number of droplets. However with the constant surface, constant volume condition, varying the polydispersity varies the number of droplets, consequently $\bar{K}$ can not be neglected. Therefore augmenting the energy term with this expression in ref.3. and ref.4., the fluctuation amplitudes and time constants can be readily recalculated [10]. Here we are interested in the polydispersity (ℓ=0 mode), and the ℓ=2 modes, which dominate the NSE data. We find [11]:

$$<|u_0|^2>=\frac{kT}{2K(6-A)} \qquad <|u_2|^2>=\frac{kT}{4KA} \qquad \tau_2=\frac{\eta R^3}{KA}\frac{55}{24} \quad (3)$$

where
$$A=4\frac{R}{R_s}-3\frac{\bar{K}}{K}$$

and
$$\gamma=\frac{K}{R_s^2}\left[6-8\frac{R}{R_s}+2\left(\frac{R}{R_s}\right)^2+3\frac{\bar{K}}{K}\right] \quad (4)$$

where η is the vicosity. u_0 can be converted to the Z parameter in the Schultz distribution by $Z=(4\pi/u_0^2)-1$.The characteristic time of the size fluctuations is estimated to be in the μs range (as the water can be considered as incompressible, it involves surfactant and water exchange) well out of our observation window (up to 18 ns), so this mode shows up only as a static property. Physically $\bar{K}$ couples to the number of droplets. $\bar{K}<0$ tends to favor a larger number of drops and hence increases the polydispersity (u_0). On the other hand the $\ell>2$ modes make the droplet surface rippled. With the constant total surface constraint this decreases the number of droplets. Thus $\bar{K}<0$ tends to inhibit these modes. These trends are well reflected in Eq.3

At the present state we neglect any viscous dissipation which might occur in the finite thickness of the surfactant interface. If it has a considerable contribution compared to the hydrodynamical dissipation, the relation between u_ℓ and τ_ℓ will change, essentially giving longer τ_ℓ times. Unfortunately the present time window and accuracy of the NSE spectrometer does not allow the decomposition of the signal into multiple exponential decays (see Eq.1.) to extract eg. u_2 and τ_2 separately.However the data are compatible with the theoretical prediction (see Fig. 5.)

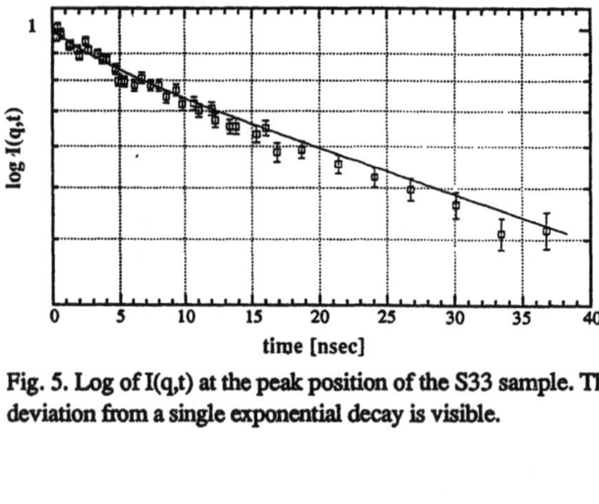

Fig. 5. Log of I(q,t) at the peak position of the S33 sample. The solid line is the fitted curve. The deviation from a single exponential decay is visible.

5. Discussion

A qualitative analysis of the data already allows us to draw two very important conclusions:
1.) $\overline{K}/K$ is close to -2: We know this because in the case of the decane/AOT/water system the polydispersity is practically independent of the droplet size and it is around 18%-20% [12]. In Eq.4. we can get such a high polydispersity either by choosing K to be small or by taking A close to six. The first solution turns out to be incompatible with the NSE data (since it gives values for u_2 which are too large) so we take $A \approx 6$. On the other hand A should be only weakly dependent on R, which means $R_s >> R$ and $\overline{K}/K$ must be close to -2. Negative $\overline{K}$ was also suggested by Binks at al. [13].
2.) The fit is insensitive to the spontaneous curvature, R_s: It is known experimentally that the maximum droplet size is ≈ 100Å. For $\overline{K}=0$ this limit occurs at $R_{max} \approx R_s$. When $\overline{K} \neq 0$ this limit is $R_{max} \approx R_s(1+\overline{K}/2K)$. With the known value of $R_{max} \approx 100$Å and $\overline{K}/K \approx -1.9$ this gives $R_s \approx 2000$Å. Indeed if $R_s >> R$,it has little influence on the $\ell=2$ mode, and the polydispersity should change rapidly only when R is close to $R_s(1+\overline{K}/2K)$.
Fig 3 shows the results of the fit on the SANS data. The dashed lines represent what would be the scattered intensity without the $\ell>2$ modes. On fig.1 and 4. $D_{eff}(q)$ was determined for the fitted points in the same way as for the measured points. The size dependent data can be fitted with the same K value and $\overline{K}/K$ ratio as the A0 butanol free sample. With the addition of butanol we observe a strong decrease of K with a nearly constant $\overline{K}/K$ ratio. The quantitative description gives a reduction of K from 3.8kT down to 2kT and $\overline{K}/K \approx -1.89$. We note that this negative $\overline{K}$ is absolutely essential to describe simultaneously the relatively large polydispersity and the moderate shape fluctuations. Also in the framework of the theory [3,10] we can calculate γ, the tension present in the interface (Eq.4.). Note that the R/R_s terms are small, the $6+3\overline{K}/K$ nearly cancels as $\overline{K}/K$ is close to -2, so γ automatically becomes small.

This work is based on the results of several collaborations between the author and J.S. Huang, S.A.Safran, S.T.Milner (Exxon), D. Richter (KFA Jülich) L.Magid (Tenessee),R.Strey (Göttingen).

References

1. S.A.Safran J.Chem.Phys 78 2073 (1983) and D.Andelman,M.E.Cates,D.Roux and S.A.Safran J.Chem.Phys. 87 7229 (1987)
2. S.A Safran , L.A. Turkevich Phys. Rev. Lett. 50 1930 (1982)
3. S.T. Milner, S.A. Safran Phys. Rev. A36 4371 (1987)
4. M.Borkovec J.Chem.Phys.91 6268 (1989)
5. M.Kotlarchyk, R.B. Stephens, J.S.Huang J.Phys.Chem 92 1533 (1988)
6. M. Kotlarchyk, S.H.Chen J.Chem.Phys 79 2461 (1983)
7. Neutron Spin Echo, edited by F.Mezei, Lecture Notes in Physics Vol. 128. (Springer-Verlag, Heidelberg, 1980)
8. J.S.Huang, S.T. Milner, B. Farago, D. Richter Phys. Rev. Lett. 59 2600 (1987)
9. W. Helfrich, Z. Naturforsch., Teil C 28, 693 (1973)
10. S.A.Safran to be published
11. B. Farago, D. Richter, J.S.Huang, S.A.Safran, S.T. Milner Phys. Rev. Lett. 59 2600 (1987)
12. Our unpublished results
13. B.P.Binks, J.Meunier, O.Abillon, D.Langevin Langmuir 5 415 (1989)

Microemulsions in Technical Processes

K. Stickdorn and M.J. Schwuger
Institut für Festkörperforschung, Forschungszentrum Jülich,
Postfach 1913, W-5170 Jülich, Fed. Rep. of Germany

Abstract. Microemulsions and reverse micellar systems are of growing interest in research and industry. They are, in contrast to emulsions, thermodynamically stable and optically transparent mixtures of water, unpolar solvent, surfactant and in most cases electrolyte and cosurfactant. A general survey of different already established or potential applications of microemulsions and reverse micellar systems is given. The major characteristics of the multicomponent systems are pointed out and related to the technical processes.

1. Properties

There are some characteristic features of microemulsions which are of special interest for the technical application of these systems:

extremely low interfacial tension. The interfacial tension between the different phases of the microemulsions reaches a minimum when a so called three-phase microemulsion is obtained. This microemulsion contains similar amounts of oil and water and exhibits the highest mutual solubility of these two components. The interfacial tension between oil and water is about two or three orders of magnitude higher than in microemulsion [1].

monodisperse particle-size distribution. There is experimental evidence that on a microscopic scale microemulsions are structured into water-rich and oil-rich domains which are separated by an amphiphilic layer. The microstructure depends on the composition of the system: at low oil concentrations the microstructure is oil swollen globules in a continuous aqueous matrix; at high oil concentrations reverse water-swollen micelles are formed; when the system contains similar amounts of water and oil, bicontinous microemulsions are formed [2]. The domain size of the dispersion is determined by the interfacial tension between the aqueous and the oil-rich phase in the presence of a saturated amphiphile monolayer.

good solubilisation properties. As basic components the microemulsions contain water, surfactant and oil; therefore they are good solvents for both nonpolar organic molecules and polar molecules and ions.

thermodynamical stability. In contrast to emulsions, microemulsions are thermodynamically stable colloidal dispersions.

phase separation can easily be achieved. The phase diagrams of many different microemulsions are well known and general patterns are described [3].

Springer Proceedings in Physics, Vol. 66 The Structure and Conformation of Amphiphilic Membranes
Editors: R. Lipowsky · D. Richter · K. Kremer

With this knowledge specific phase separations can be achieved by changing temperature, pressure, proton- or salt-concentration as well as chemical composition.

Besides these major properties there are some further special characteristics of microemulsions:

optical transparency. The dimensions of the droplets or domains in the microemulsions are so small that the mixtures are macroscopically homogeneous and therefore transparent.

low viscosity. Compared to liquid crystals, which are very often formed in ternary mixtures, microemulsions exhibit a low viscosity.

large interfacial area. The microstructure of the systems implies a large interfacial area between the oil-rich and the water-rich domains.

2. Applications

Microemulsions and reverse micellar systems are already used in some industrial fields but there exist also a lot of potential applications which are not well established until now. We do not claim to give a complete survey of all processes which involve microemulsions, but with this contribution we want to point out the importance of this field.

Enhanced oil recovery

About 30 % of the total content of an oil reservoir can usually be recovered by primary tapping and water flooding. By means of enhanced recovery further 20 % can be produced. One of the techniques for enhanced recovery is flooding with microemulsions and is based on its very low interfacial tension. This enables the mobilisation of oil-droplets or oil-ganglia which are trapped in the reservoir pores. The interfacial tensions for crude oil/water are in general about 10^{-1}-10^{-2} mN/m whereas 10^{-5} to 10^{-6} mN/m can be achieved when an appropriate microemulsion system is chosen.

Metal extraction

This process is used to extract minerals from low-grade ores. Two steps are necessary: first of all the mineral is dissolved in acidic or basic solution; second the metal is extracted using a specific extractant. The extractants are generally large organic molecules which are forming metal-complexes. These organic molecules can be dissolved in the hydrophobic domains of the microemulsion whereas the metal-ions are located in the aqueous domains. The extraction velocity in microemulsion systems is about 10 to 100 times higher than in conventional two-phase systems [4].

Lubricants and cutting oils

These compounds very often contain surfactants and water as additives, which causes two effects:

- corrosion-inhibition; because some surfactants can solubilise corrisive and oxidative agents;
- increase in heat capacity of the oil because of the water content.

Pharmaceuticals and cosmetics

Until now mostly emulsions and liquid crystals are used in pharmacy and cosmetics. Compared to conventional organic solvents, the mentioned systems show a better solubilisation and also a better stabilisation of the pharmaceuticals. Using microemulsions as medium offers further important advantages [5,6,7]:
- cheap preparation of the formulations because of the facile preparation of the microemulsions;
- better storage because of the thermodynamic stability;
- no immun reaction of the organism or fat embolism because of the small droplet size;
- formulations can be used as injectants if a composition with a large one phase region is chosen;
- higher diffusion rates and skin penetration fluxes of the pharmaceuticals compared to conventional formulations are observed;
- the denaturation of proteins caused by protein-surfactant interactions is avoided by the water domains.

Washing

Detergents are mixtures of water and surfactant mixed with different additives. The application of microemulsions as washing agents improves the oil soil removal from solid surfaces and the capillaries of the fabrics because of the low interfacial tension. Furthermore the soil particles can better be kept in suspension with these agents. Raise of washing efficiency can be achieved by adding the agents which are increasing the viscosity of the solution.

Reaction medium for organic and biochemical reactions

The composition of microemulsions allows dissolving polar and nonpolar organic molecules, ions and proteins or enzymes. Therefore these media are very suitable for the performance of chemical reactions. Compared to reactions in two-phase media, the reaction rate is higher [8]. This is caused by two effects: the large internal interface and an accumulation of the reactants at the interface. Dependent on the reactants, phase separation can sometimes be used for the isolation of the products and recovery of catalysts. For the performance of biochemical reactions in microemulsions the function of the water droplets as microreactor is most important, which causes the enzymes not to lose their activity in the aqueous domains [9].

Particle preparation

Microemulsions can be used to prepare ultrafine particles with a very small particle-size distribution. Usually two microemulsions of the same composition, only containing different reactants, are mixed and the precipitate can be isolated. Again the droplets can be regarded as microreactors. They determine the size of the particles, which is not necessarily the one of the droplets. Although the mechanism is not yet totally elucidated, this technique is already used for the production of e.g. catalysts, semiconductors, photographic suspensions and superconductors [10,11,12].

References

1. D. Langevin: Reverse Micelles, Proceedings of the European Science Foundation Workshop (eds.: P.L. Luisi, B.E. Straub, Plenum Press, New York 1984) pp. 287-303

2. W. Jahn, R. Strey: J. Phys. Chem. 92, 2294-2301 (1988)

3. M. Kahlweit, R. Strey: J. Phys. Chem. 94, 3881-3894 (1990)

4. P. Fourre, D. Bauer, J. Lemerle: Anal. Chem. 55, 662 (1983)

5. S. Keipert, I. Siebenbrodt, F. Lüders, M. Bornschein: Pharmazie 44, 433 (1989)

6. D.W. Osborne, A.J.I.Ward, K.J. O'Neill: Drugs and the Pharm. Sci. 42, 349 (1990)

7. D.W. Osborne, A.J.I. Ward, K.J. O'Neill: Drug Development and Industrial Pharmacy 14, 1203 (1988)

8. R. Schomäcker, K. Stickdorn, W. Knoche: J. Chem. Soc. Faraday Trans. 87, 847 (1991)

9. P.D.I. Fletcher, B.H. Robinson, R.B. Freedman, C. Oldfield: J. Chem. Soc. Faraday Trans. 1 81, 2667 (1985)

10. T.F. Towey, A. Khan-Lodhi, B.H. Robinson: J. Chem. Soc. Faraday Trans. 86, 3757 (1990)

11. A. Kan-Lodhi, B.H. Robinson, T. Towey, C. Herrmann, W. Knoche, U. Thesing: The Structure, Dynamics and Equilibrium Properties of Colloidal Systems (eds.: D.M. Bloor, E. Wyn-Jones, Kluwer Academic Press 1990) pp. 373-383

12. P. Ayyub, A.N. Maitra, D.O. Shah: Physica C 168, 571 (1990)

The Effect of Additives on Surfactant Sheets in Microemulsions

M.A. Lopez-Quintela, W. Korneta, and L. Liz*

University of Santiago de Compostela,
E-15706 Santiago de Compostela, Spain
*Permanent address: Technical University, Radom, Poland

Microemulsions are thermodynamically stable, topologically microstructured phases containing at least surfactant, oil and water. The special features of microemulsions are due to properties of a surfactant which preferes to be located between oil and water regions. Nowadays an important trend in the new materials technology is to obtain finer magnetic powders of better particle size distribution . In our group the method of making such powders consisting of particles having very well controlled size using microemulsions was developed (USA patent no.4.983.217-1991). The most important parameters of micro-emulsions for this kind of applications are the self-diffusion coefficient of droplets existing in the microemulsion (because chemical reactions which take place are controlled by diffusion of droplets) and the size of these droplets. In order to obtain experimentally these parameters, the dynamic light scattering technique (DLS) is usually applied. This technique posses the advantage that it does not disturb the microemulsion.
The correlation function of the scattered light intensity $g(Q,\tau)$ is measured, where $\overline{Q}$ is the scattering vector (magnitude Q) and τ is the correlation delay time. We performed DLS experiments in the high-Q and short-τ limits. At large Q the spatial scale probed is much smaller than the spacing between droplets. At short times the droplet starts to diffuse without taking notice of its neighbours. The first cumulant $\Gamma(Q)$ of $g(Q,\tau)$ which is extracted from the experimental data in the high-Q and short-τ limit, and taking into account that droplets are identical can be written as a sum of two terms [1]. The first term is $D_s^sQ^2$, where D_s^s denotes a short-time centre-of-mass self-diffusion coefficient. The second term describes the contribution of shape fluctuations of the droplet to $\Gamma(Q)$. The light scattering apparatus we used is equipped with an argon ion laser operating on the 488 nm and digital correlator. The microemulsion which

Springer Proceedings in Physics, Vol. 66 The Structure and Conformation of Amphiphilic Membranes
Editors: R. Lipowsky · D. Richter · K. Kremer

we study consists of igepal CA 520 (40% vol.) and equal volumes of water and decane. This microemulsion we now apply to make particles. We added the salt KCl and butanol to obtain desired properties for our applications. The structure of microemulsions with equal volumes of water and oil has recently been studied by measuring diffusivities of each component [2]. These studies show that at low salinities the oil and surfactant diffuse together. We measured electrical conductivity of our microemulsion and its relatively high value suggests that small oil droplets stabilized by a protecting layer of surfactant molecules are dispersed in the water. In order to decrease the contribution of shape fluctuations to $\Gamma(Q)$, we first performed DLS experiments close to the lamellar phase. Some results are given in the table below.

Table.-*Values for $\Gamma(Q)/Q^2$ and the hydrodynamic radii R_H obtained from DLS measurements at the scattering angle 90° in the microemulsion consisting of igepal (40%vol.) and equal volumes of water and decane.*

Additive	None	KCL [5% wgh. in water]	butanol [6% vol.]
T [^oC]	45.2	35.0	25.5
$\Gamma(Q)/Q^2$ [cm^2/s]	$4.5 \cdot 10^{-7}$	$3.5 \cdot 10^{-7}$	$5.8 \cdot 10^{-7}$
R_H [nm]	8.73	8.86	4.30

The size of droplets close to the lamellar phase depends on additives, what should be connected with properties of lamellar phase. Interfaces in a lamellar phase close to the droplet phase are undulated. In nonionic systems the dependence of the frequency of the undulation mode on the square of the scattering vector $\bar{\kappa}$, which lies inside the lamellae planes, has the minimum for $\kappa^2 \neq 0$ [2]. Lamellar phase can be viewed as a stack of surfactant layers separating oil and water regions. The surfactant head-group carry a permanent electrostatic dipol with a preferential orientation perpendicular to the interface. In order to study the stability of the lamellar phase, one adds deformations to the average order of layers and looks for their amplitude. Let's describe small deformation of a layer by a continuous local displacement $\Psi(\bar{r})$ from a reference plane (xy).

The displacement $\Psi(\bar{r})$ is in the z-direction, whereas the vector $\bar{r}$ is in the reference plane. The free-energy functional of deformations we write in the following form:

$$F = \int_V \left[\gamma\left(\nabla\Psi\right)^2 + K\left(\nabla^2\Psi\right)^2 + \delta\left(1 - E_z P_z\right) \right] d\bar{r} \quad , \tag{1}$$

where $\gamma>0$ is the interfacial tension determined by the hydrophobic interactions between surfactant tails [1], K denotes bending elastic constant, whereas $\delta>0$ expresses the strength of the electrostatic head-group repulsion interactions. γ, K and δ are in units of $k_B T$. The constant K is due predominantly to surfactant tails which oppose distortion and it is positive, i.e.opposes bending right from the start. Head-group repulsive forces favour bending because deformations decrease the energy of dipolar interactions, what is described by the last term in the expression (1). In our approach the energy cost of bending can be compensated by the energy gain for the decrease of dipolar repulsion. For small deformations dipolar energy is caused mainly by z-components of dipoles. In the expression (1) we denoted by $P_z(\bar{r})$ and $E_z(\bar{r})$ the z-components of the distribution of dipolar moments and fields respectively. They are connected by Maxwell equations in electrostatic approximation i.e. $\mathrm{rot}\bar{E}(\bar{r})=0$ and $\mathrm{div}(\bar{E}(\bar{r})+\bar{P}(\bar{r}))=0$. Moreover one can write the relation : $E_z^2(\bar{r}) \simeq 1-(\nabla\Psi(\bar{r}))^2$. Using this relation, Maxwell equations and decomposing $\Psi(\bar{r})$ into its spatial modes, one can write the free energy in the form $F \simeq \sum \Delta(\kappa^2)\Psi(\bar{\kappa})\Psi(-\bar{\kappa})$, where the summation is over $\bar{\kappa}$. $\Delta(\kappa^2)=\gamma\kappa^2+K\kappa^4+\delta k^4/(k^2+\kappa^2)$, where $k=\pi/a$ and a is the average distance between layers. The probability of undulation described by $\bar{\kappa}$ is $\exp(-\Delta(\kappa^2))$. The biggest probability has the undulation with κ_c^2 which minimizes $\Delta(\kappa^2)$. κ_c^2 can generally be 0, π^2 or it should satisfy the following conditions: $(\gamma+2K\kappa^2)(k^2+\kappa^2)^2-\delta k^4=0$, $\kappa_c^2 \geq 0$, $\Delta(\kappa_c^2) \leq \Delta(0)$ and $\Delta(\kappa_c^2) \leq \Delta(\pi^2)$. The period of undulation having the biggest probability is $\lambda_c=2\Pi/\kappa_c$. The addition of short-chain alcohols like butanol decreases K, because $K \simeq n^{2.5}$, where n is the average length of a surfactant tail [2]. If K decreases λ_c decreases. The big value of K suppresses all undulations. Adding more and more surfactant decreases γ and λ_c decreases. The addition of a salt has the effect of screening electrostatic interactions and δ decreases, so λ_c increases. The changes in the size of droplets near lamellar phase induced by additives are correlated with changes of λ_c described above. We performed also DLS at the angle 90° for the same samples at temperatures near the phase separation of the microemulsion. For $\Gamma(Q)/Q^2$ we obtained the following values: $3.42 \cdot 10^{-7}$ cm^2/s - the sample without additives at T=50.0°C and $2.66 \cdot 10^{-7}$ cm^2/s - the sample with KCl at T=40.0°C. Near the phase separation the damping of shape fluctuations is rather low and this mainly causes the decrease of $\Gamma(Q)$. Finally we would like to report the last data obtained from DLS at the angle 90° for the microemulsion which consists of ethyl oleate (7.5% vol.), water, labrasol-saturated polyglycolysed glycerides (surfactant) and isostearic plurol (cosurfactant). The volume proportion of surfactant to cosurfactant equals 3. This microemulsion has potential applications in the pharmaceutic industry (it is not toxic), so we performed measurements at T=32°C (the temperature of the human body). The phase diagram is presented during this Workshop. We obtained for $\Gamma(Q)/Q^2$ the following values: $2.19 \cdot 10^{-8}$ cm^2/s - for 12.5% vol. of the water, $4.49 \cdot 10^{-8}$ cm^2/s - for 32.5% vol.of the water and $8.54 \cdot 10^{-8}$ cm^2/s - for 67.5% vol. of the water. The increase of $\Gamma(Q)/Q^2$ with the decrease of the amount of a surfactant and cosurfactant is associated mainly with the increase of the size of oil droplets. Because the volume fraction of oil is small it is difficult to accomodate all of the surfactant on interfaces. The head-group area A for high surfactant concentrations is small, $K \simeq 1/A$ [2] is big, and sizes of droplets are big.

ACKNOWLEDGMENTS

W.K. would like to thank the Ministry of Education in Spain for support.

R E F E R E N C E S

[1] "Physics of Amphiphiles:Micelles, Vesicles and Microemulsions" (North-Holland 1985) p.793,24.

[2] "Physics of Amphiphilic Layers" (Springer-Verlag 1987) p.357,153,12.

Membrane Curvature and Structural Transitions for Charged/Uncharged Phospholipid Mixtures

D. Lerche[1], *N.L. Fuller*[2], *and R.P. Rand*[2]

[1]Institute of Medical Physics and Biophysics, Charité, Humboldt University, O-1000 Berlin, Fed.Rep.of Germany
[2]Brock University, St. Catharines, Canada

Abstract. We have measured the effect of charged phospholipids on the phase structure and the curvature of phospholipid monolayers. 12 to 15 mole% dioleoylphosphatidylserine (DOPS) converts the reverse hexagonal phase of dioleoylphosphatidylethanolamine (DOPE), to a multilamellar phase. The intrinsic monolayer curvature decreases systematically with DOPS content up to at least 20 mole% DOPS.

There is increasing evidence that electrostatic characteristics contribute significantly to the mechanics of lipid and biological membranes. Theoretical estimations of the contribution of electrostatic energy to the spontaneous curvature of a membrane have been made [1,2]. We have recently measured the spontaneous curvature of neutral phospholipid monolayer membranes (3), including those that make up the reverse hexagonal (H_{II}) phase of the neutral lipid dioleoylphophatidylethanolamine (DOPE). The aim of this study therefore was to measure the effect of charged lipids on these phospholipid monolayers The structure of the H_{II} phase is shown in figure 1. This report describes preliminary results wherein DOPE was charged by systematic addition of dioleoylphosphatidylserine (DOPS), a charged lipid with identical hydrocarbon chains.

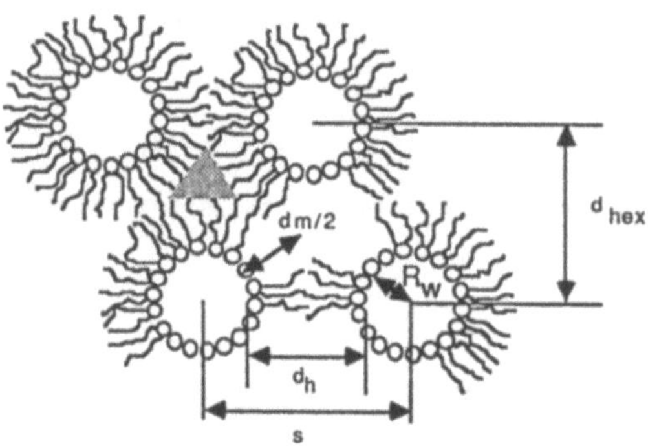

Figure 1. Structural parameters describing the reverse hexagonal H_{II} phase. These are derived from X-ray diffraction, sample composition, and component densities (3).

2. Materials and Methods

The experiments were performed as previously described in detail [3]. Briefly, synthetic DOPE and DOPS and tetradecane, purchased from Avanti Polar

Editors: R. Lipowsky · D. Richter · K. Kremer

Lipids, Inc., Birmingham, AL., were mixed in organic solvent in defined proportion and dried under vacuum. The dry lipid was hydrated to 70 wt % 2mM TES buffer (pH 7.3) by weighing into small sealed bottles, and equilibrating in the dark, at room temperature, for 48 hours. X-ray diffraction was used to characterize the structures, lamellar and/or hexagonal phases, formed by the various lipid mixtures, and to determine the dimensions of these structures.

3. Results

Figure 2 shows that from 0 to 12 mole% DOPS the lipid forms an H_{II} phase in excess water as does DOPE alone. In this low charge region there is a small but systematic increase in the H_{II} lattice spacings, from 64Å to 68Å at 12 mole % DOPS. However, above 15 mole% DOPS the lipids form single lamellar (L) phases with high water contents. This shows that the effect of between 12 and 15 mole% charged lipids is to induce the transition from the hexagonal to lamellar phase, a transition of the monolayers from highly curved to planar.

Figure 3 shows that if 20 wt% of the lipid is tetradecane (td), no L phases form, up to at least 20 mole% DOPS in excess water. The H_{II} lattice expands systematically to 78Å. We interpret the conversion of lamellar to hexagonal structure on addition of tetradecane as the result of the removal of hydrocarbon chain stress as formulated by Kirk and Gruner (4). We interpret the increase in spacing in excess water and tetradecane as a relaxation of the lipid to a stress free state, and particulary to a state that allows the monolayer in the H_{II} phase to reach its intrinsic radius of curvature (3).

We have estimated from these dimensions the intrinsic radius of curvature, R_o, of the monolayers within those hexagonal phases containing excess water and tetradecane. We have determined for DOPE that a pivotal position near the polar group-hydrocarbon chain interface remains constant in area as the monolayer is bent [3], a position that has been defined as the place where the radius of curvature should be measured. For all DOPS concentrations near excess water the "bilayer" dimension, d_l in figure 1, is approximately 37Å. The polar groups themselves are 6Å thick. Therefore we take

$R_o = 0.5 \times (s-d_h) - 6$ Å. as the value of the intrinsic radius of curvature for the tetradecane-containing phases of figure 3.

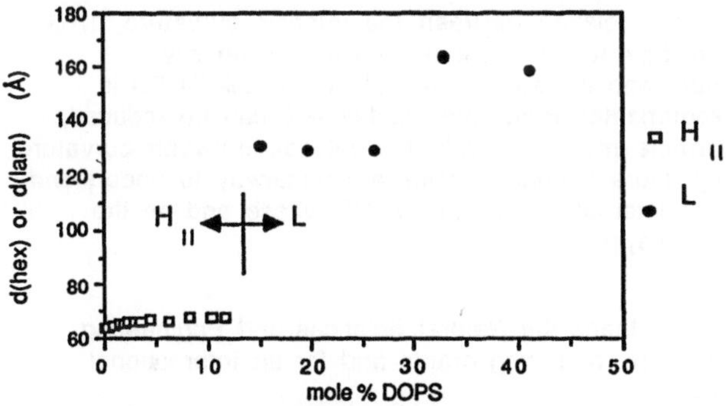

Figure 2. X-ray diffraction spacings of the H_{II} and lamellar phases of DOPE/DOPS mixtures in excess water (70 wt% water for DOPS above 30 mole%). d(hex) is the dimension shown on figure 1 and d(lam) is the total repeat spacing of the lamellar lattice

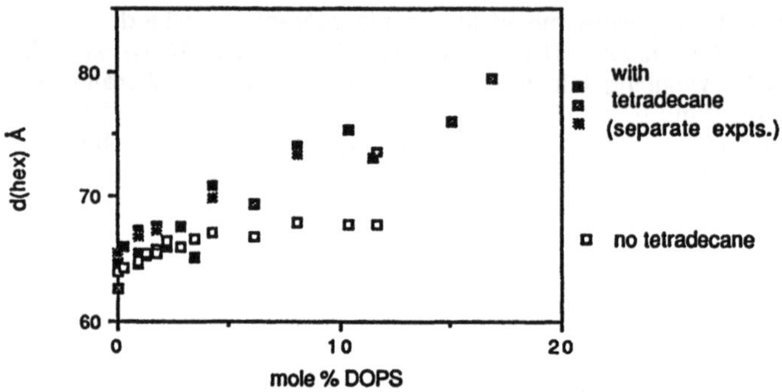

Figure 3. dimension of the hexagonal phase formed by DOPE/DOPS mixtures in excess water, with and without tetradecane. (tetradecane was 20wt.% of the total non-aqueous portion of the sample). Small systematic differences, equivalent to 2-3 mole % charge, between experiments was seen.

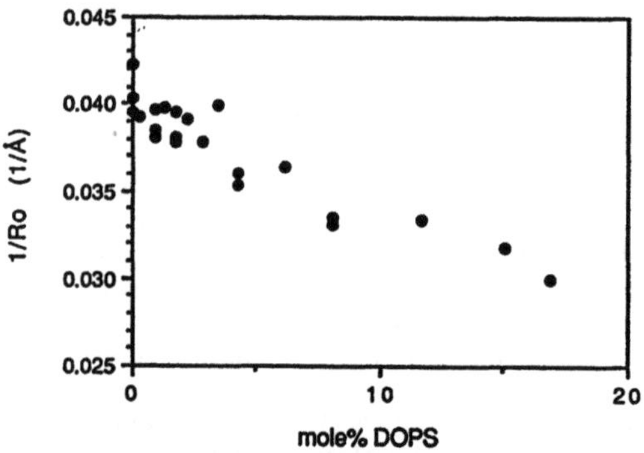

Figure 4. Dependence of the intrinsic monolayer curvature of the H_{II} phase on the mole% DOPS in DOPE.

In figure 4 we show the relation between the intrinsic curvature, $1/R_o$, and mole% DOPS which we take to represent surface charge density. Correlation between curvature and the square root of the mole% DOPS is equally good. Until the experimental errors seen in figure 3 can be reduced, these results cannot determine more precisely the relation between curvature and surface charge density. More detailed studies are underway to understand the effects of charge on the internal dimensions of this phase and on the bending modulus of the monolayer.

Acknowledgements - We thank the Natural Sciences and Engineering Research Council of Canada for operating grants and for an International Scientific Exchange Award.

REFERENCES

1. Lerche, D., Kozlov, M. M. and Markin, V. S. (1987) Electrostatic Free Energy and Spontaneous Curvature of Spherical Charged Layered Membrane. Biorheol. 24: 23-34.

2. Winterhalter, M. and Helfrich, W. (1988) Effect of Surface Charge on the Curvature Elasticity of Membranes. J. Phys. Chem. 92: 6865-6867.

3. Rand R. P., Fuller, N. L., Gruner, S. M. and Parsegian, V. A. (1990) Membrane Curvature, Lipid Segregation, and Structural Transitions for Phospholipids under Dual Solvent Stress. Biochemistry 29: 76-87.

4. Kirk, G. L., and Gruner, S.M. (1985) Lyotropic Effects of Alkanes and Headgroup Composition on the L_α - H_{II} Lipid Liquid Phase Transition: Hydrocarbon Chain Packing versus Intrinsic Curvature. J. Physique 46: 761-769.

Epitaxial Relationships Between Adjacent Phases in Hydrated Monoolein

R.H. Templer, N.A. Warrender, R. Meadows, and J.M. Seddon

Department of Chemistry, Imperial College, London SW7 2AY, UK

Abstract. We have observed a distinct epitaxial relationship between the L_α and H_{II} phase in monoolein. Our results on the epitaxy between the L_α and Ia3d cubic phase are less conclusive, but suggest that there are two preferred epitaxial orientations.

1. Introduction

An important question which has so far been little explored is that of the geometric relationships between the lattices of adjacent lyotropic mesophases. Such information is clearly liable to be of great relevance to understanding structural transformations in membranes, and from the point of view of physical chemistry, may shed light on the mechanisms underlying such transitions. Very precise, yet subtle epitaxial relationships have been found to exist between a variety of mesophase topologies in a number of normal topology systems [1,2]. In this paper we report our first efforts to determine epitaxial relationships in the type II system, monoolein in water.

2. Apparatus and Sample Preparation

We have built a high sensitivity two dimensional detector which allows us both to scan samples rapidly for regions of good alignment, as well as monitoring the effect of annealing procedures on the alignment process. The detector is a second generation version of the CCD based integrating detector designed and built at Princeton University [3].

The principle of the device's operation is straightforward. A phosphor converts incoming X-Rays into visible light. The optical image is then photo-amplified by the image intensifier (Thomson TH9312), whilst being reduced in size from 40 to 25mm in diameter. The intensified and reduced output image is then projected on to a slow scan scientific CCD camera (Thomson TH7882 CDA with $23 \times 23 \mu m^2$ pixels in a Star 1 camera head, Photometrics, USA). Noise and dark current in the system is reduced by cooling the CCD and the image tube. In operation this means we obtain a quantum limited record of the X-Ray diffraction, where the major noise contribution is from the shot noise in the X-Ray signal itself.

Using our rotating anode generator (GX20, Enraf Nonius, Netherlands) with a 100μm point focus and Franks optics the detector enabled us to record images with exposures as short as 30 seconds, which made it feasible to scan large sample vol-

 Springer Proceedings in Physics, Vol. 66 The Structure and Conformation of Amphiphilic Membranes
Editors: R. Lipowsky · D. Richter · K. Kremer

umes to find the best aligned regions, as well as observing the behaviour of such monodomains as we altered thermodynamic parameters.

Samples were deposited on to thin mylar sheet glued on to a cylindrical copper section. A central slot in the copper former allowed the X-Ray beam to pass freely through a tangential section of the curved mylar sheet. In this way the meridional parts of the reciprocal lattice were rotated on to the Ewald sphere and could be viewed simultaneously.

The sample holder fitted into a hermetically sealed cell which was maintained at 100% relative humidity. The cell temperature could be varied between -30 to 100°C with the temperature being controlled to ±0.03°C. The X-Ray windows of the cell were made from 0.001" mylar. For long exposures the mylar did give a small angle scattering signal, but this was easily subtracted away from the raw data.

Monoolein (Larodan, Sweden) with a stated purity greater than 99% was first diluted in a solvent to make a concentrated, but still relatively inviscid fluid. The solution was then loaded into a Pasteur pipette inserted through O-ring seals leading into the sample cell and positioned above the mylar sheet. We obtained our best results when the rate of solvent evaporation was slowest and the sample was formed in a lamellar phase. We obtained these conditions by depositing monoolein drops in methanol at 5°C and 100% RH. The methanol evaporated over a few minutes, and images were taken approximately 1 hour later. The films produced were approximately 10μm thick and we believe they contained, at most, only trace quantities of solvent. However, we made no measurement of solvent concentration.

3. Results

3.1 L_α to H_{II} Epitaxy

Figure 1 shows a typical example of the diffraction from the parent lamellar phase and the hexagonal phase which is grown from it. The image of the lamellar phase showed a good deal of mosaic spread. As the monoolein hydrated at 5°C this spread reduced considerably, until it achieved the levels exhibited in the H_{II} picture, which took an hour.

We have now observed this epitaxy between L_α and H_{II} phases in four type II systems. The presence of Bragg spots off the meridian and the high intensity of the meridional spots indicates that although the H_{II} tubes were co-planar with the lamellae, they had no preferred orientation in the plane. All the spots are linearly extended by a small amount in the transverse direction. We interpret this to mean that over macroscopic distances bundles of tubes are bent round whilst the (10) planes show small departures of approximately ±3° from being perfectly horizontal. Because of these effects we are unable to make an estimate of the average length of a bundle of rods, but we can set a lower limit of about 500Å.

3.2 L_α to Ia3d Epitaxy

We were unable in any of our attempts to produce a macroscopic cubic monodomain, i.e. with volumes greater than 1000x100x10μm^3. We believe this is

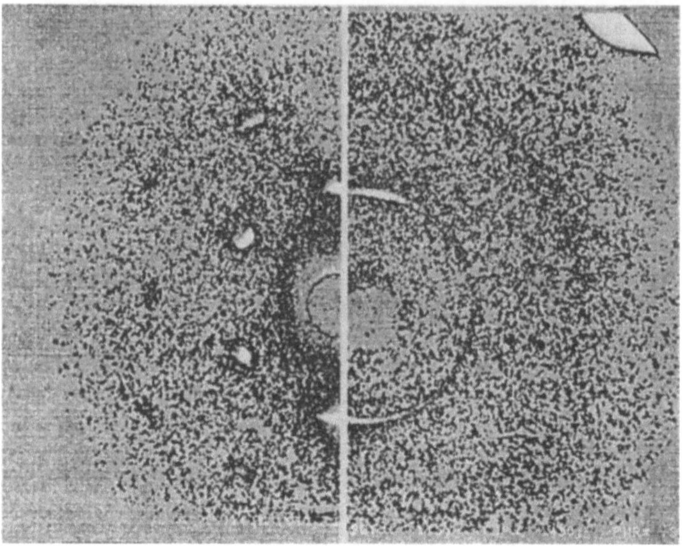

Figure 1. The diffraction patterns of the H_{II} and L_α phases.

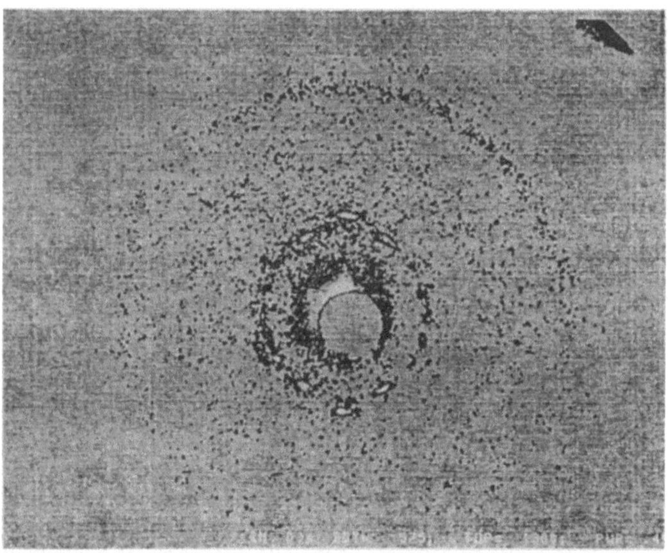

Figure 2. The diffraction pattern for the Ia3d cubic phase, L_α Bragg peaks aligned as in figure 1.

because the epitaxy is set by the lamellar planes, so that although a **111** axis is set parallel to the planes, many axis orientations in the plane are possible. However, in all of the runs in which we approached the phase transition slowly we were able to obtain patterns where a set of coupled reflections dominated the image as shown in figure 2.

The data in this pattern was obtained heating from 5°C to 25°C at a rate of approximately 0.2°Cs^{-1}. Faster rates lead to the production of many small domains of apparently random orientation. The real space epitaxy has the (220) plane parallel to the lamellae. We continued to obtain epitaxial relationships in which the **111** direction was parallel to the lamellae in subsequent runs, but sometimes the cubic lattice was rotated by 30° around **111**, thereby making the (211) plane parallel to the lamellae. Indeed the evidence we have obtained to date, suggests that this may be the epitaxial relationship between H_{II} and Ia3d, which corresponds to the case observed in type I systems [2].

By monitoring the X-Ray patterns over some tens of hours we have observed that the domains gradually move whilst retaining their **111** orientation. We believe that this occurs because the system is gradually reducing the number of domain walls, and in so doing domains end up being rotated around the meridional axis.

4. Concluding Remarks

We have yet to perfect techniques for creating very large monodomain inverse cubic phases, but even so we have obtained results which suggest that epitaxy between phases is set by the mutual coincidence of planes of high bilayer density.

Acknowledgements

We wish to thank Professors S.M. Gruner and E.F. Eikenberry for the software used to run our detector. This work was funded by the SERC (grant no. GR/F44052) and the Royal Society. RHT wishes to acknowledge the support of the Royal Society in the form of a University Research Fellowship.

References

1. Kékicheff, P. and Cabane B. (1988) *Acta Cryst.* **B44**, 395-406
2. Rançon, Y. and Charvolin, J. (1988) *J. Phys. Chem.* **92**, 2646-2651. Clerc, M., Levelut, A.M. and Sadoc, J.F. (1990) *Coll. de Physique* **51**, 97-104.
3. Templer, R.H., Gruner, S.M. and Eikenberry, E.F. (1988) *Advances in Electronics and Electron Physics* **74**, 275-283.

Films of Amphiphiles and Minimal Surfaces

J. Charvolin and J.F. Sadoc

Institut Laue–Langevin, 156X, F-38042 Grenoble Cedex and
Laboratoire de Physique des Solides, Bât. 510, Université
Paris-Sud, F-91405 Orsay, France

Abstract. A film of amphiphiles is made of two identical facing interfaces and its middle surface is a surface of symmetry. This symmetry leads the structures built by the film to be considered as organizations of geometrical surfaces partitioning space into two identical subspaces. Such surfaces may belong to different topological classes, according to the number and extension of the subspaces. We shall consider here only the case where the film builds a periodic structure partitioning space into two infinite subspaces, as for instance in the so-called "bicontinuous" cubic structures. It has been proposed in this case that the film is supported by infinite periodic minimal surfaces, or IPMS, which have the same topology and symmetries than the structures of the physical systems. We shall describe three of them, the P, D and G surfaces which can be considered as typical, in topological (connectivity, curvature) and crystallographic (symmetry) terms; we shall also examine their relevance as descriptors of real structures.

1. Introduction

We search for periodic organizations of films made of two identical facing amphiphile-water interfaces, the film may be either a film of amphiphiles limited by two layers of water or a film of water limited by two monolayers of amphiphiles. Such films are symmetric with respect to their middle surface which separates space in two identical subspaces, as shown in fig. 1. The two subspaces are identical in the sense that the physico-chemical potentials vary the same way along the normal to the middle surface on either sides of it but, as the amphiphilic and aqueous layers are liquid, there is no orientational correlation parallel to this surface. This implies that the symmetry element is not only a mirror but also, eventually, a mirror plus a rotation. This symmetry is the fundamental point of the development presented here and we shall focus our attention onto particular configurations of the middle surface which preserve it.

This symmetry constraint can be expressed in very simple terms, considering that elements of surfaces parallel to the middle surface and at

 Springer Proceedings in Physics, Vol. 66 The Structure and Conformation of Amphiphilic Membranes
Editors: R. Lipowsky · D. Richter · K. Kremer

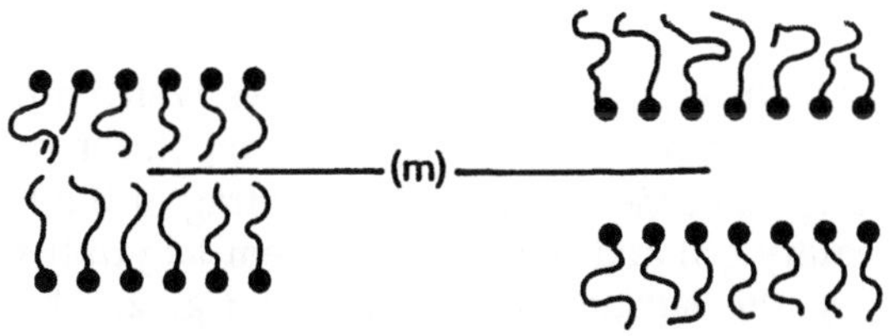

1) Films made of two facing interfaces, they are symmetric with respect to their middle surface (m).

equal distances on either sides of it must have the same area. If the middle surface is locally defined by its two principal curvatures $1/R_1$ and $1/R_2$, the element of parallel surface at a distance l on one side has an area which goes as $(R_1 + l)(R_2 + l)$, whereas that of the element on the other side goes as $(R_1 - l)(R_2 - l)$. The equality of these two area imposes that $R_1 + R_2 = 0$, i.e. the mean curvature $H = 1/R_1 + 1/R_2$ of the middle surface must be zero at every point. This is a local condition which defines what mathematicians call a minimal surface. In our Euclidean space there are only two possible surfaces: the flat plane with $R_1 = R_2 = 0$ and the saddle with $R_1 = -R_2$. Although these surfaces have the same mean curvature, they are obviously different; this is expressed in the fact that their Gaussian curvatures, $K = 1/(R_1R_2)$, have different values: zero for the flat plane and a negative one for the saddle. We focus our attention onto the presently much discussed periodic configurations of saddles.

2. Periodic configurations of saddle surfaces

The general idea underlying the building of such surfaces is rather simple. First, the saddles must be assembled in such a way that their minimal property is preserved when moving from one saddle to another, without any singularity along their line of contact: two saddles on either sides of this line must therefore be related by a simple symmetry operation. Second, as periodic surfaces are wanted, the borders must belong to symmetry elements of the periodic lattice. This problem was considered one century ago, for the first time, in the case of surfaces with cubic symmetries built with saddles supported by skew quadrilaterals with straight edges [1,2]; it was extended to non cubic symmetries and polygonal saddles with curved edges recently [3]. We shall describe three of these surfaces, with cubic symmetries and built with quadrilateral saddles, which can be considered as typical cases and which can be thought to be related to structures built by amphiphilic molecules (see also [4,5]).

2.1 Schwarz's P surface

The contour of the saddle is a quadrilateral with straight edges which are diagonals of faces of two cubes having one common face, the repetition of such a saddle by symmetry along its border leads to the building of the P surface. The saddle, one elementary cell and the two congruent labyrinths of connectivity 6 separated by the surface are shown in fig. 2. The orthoscheme triangle of the surface has vertex angles of $\pi/2$, $\pi/4$ and $\pi/6$, the elementary cell contains 96 of them and the organization has space group Im3m. The orthoscheme triangle is obviously not Euclidean, as expected for a triangle drawn on a hyperbolic surface and, if its hypotenuse is straight, its small edges are curved. It must be noted that the Gaussian curvature is not uniform on the surface, it is zero at points with 3-fold symmetry and negative at points with 2-fold symmetry, this is to be related to the very general fact that there is no hyperbolic surface with constant negative Gaussian curvature without singularity in our 3-D Euclidean space. Finally, the first elementary cell shown in fig. 2 is not the only possible one and we show also an equivalent cell, obtained from the first by a translation of vector (1/4,1/4,1/4) along the diagonal of the cube, which is better adapted to the forthcoming discussion. This cell encloses a piece of surface which is an irregular dodecagon of 96 orthoschemes, with short and long edges, and whose vertices are joined in six groups of two.

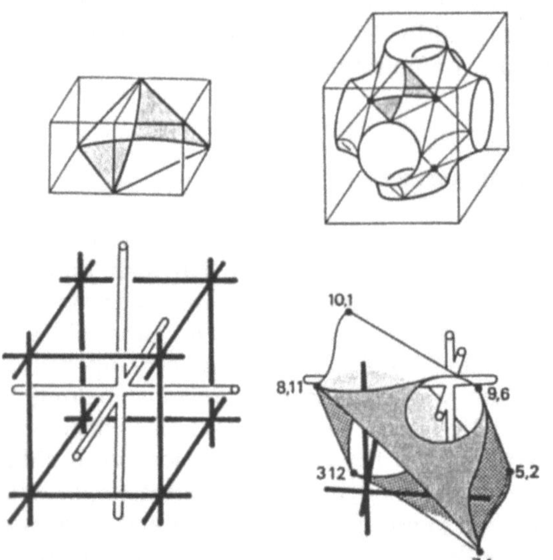

2) Schwarz' s P surface: the quadrilateral saddle with four orthoscheme triangles, an assembly of complete and incomplete saddles building an elementary cell, the two labyrinths separated by the surface and another elementary cell in which the surface appears as a dodecagonal sheet having vertices assembled in six groups of two.

There are some experimental evidences suggesting that films of amphiphiles might be organized with the same topology and symmetry than the P surface [6,7].

2.2 Schwarz's D or F surface

The contour of the saddle is a quadrilateral with straight edges which are diagonals of faces of one cube, the repetition of such a saddle by symmetry along its border leads to the building of the D surface. The saddle, a part of the elementary cell and the two congruent labyrinths of connectivity 4 separated by the surface are shown in fig. 3. The orthoscheme triangle has the same vertex angles than that of the P surface, the elementary cell contains also 96 triangles and the organization has space group Pn3m. However, although the angles are the same, the triangles are different in the sense that they do not have the same straight and curved edges: the hypotenuse is now curved and the small edges straight. Finally, a complete elementary cell is shown in fig. 3. As above, this cell encloses a piece of surface which is an irregular dodecagon of 96 orthoschemes, with short and long edges, and whose vertices are joined in six groups of two but in a different manner than for the P surface.

There are strong experimental evidences showing that films of amphiphiles may be organized with the same topology and symmetry than the D surface [8,9,6].

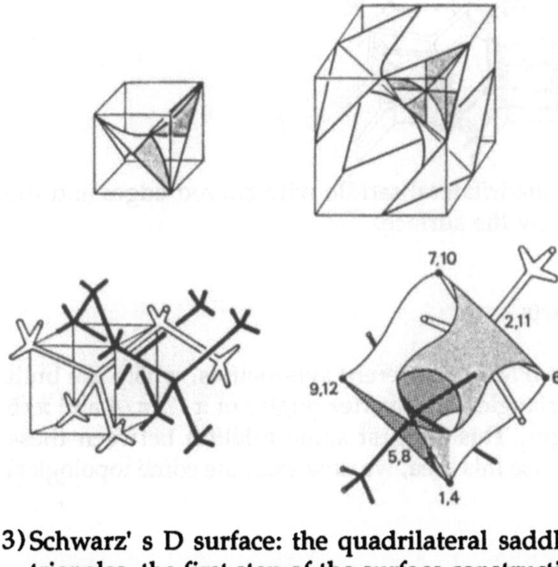

3) Schwarz' s D surface: the quadrilateral saddle with eight orthoscheme triangles, the first step of the surface construction assembling six saddles, the two labyrinths separated by the surface and an elementary cell in which the surface appears as a dodecagonal sheet having vertices assembled in six groups of two (in a different manner than that of fig. 2).

2.3 Schoen's G surface

The contour of the saddle is now a quadrilateral with curved edges, the repetition of such a saddle by symmetry along its border leads to the building of the G surface which has no straight line on it, it is therefore not easy to draw and we just present the two labyrinths of connectivity 3 separated by it in fig. 4, photographs of plastic models can be found in [3,4]. In this case the two labyrinths are not congruent but oppositely congruent, i.e. they have opposite chiralities. The orthoscheme triangle of the surface has still the same vertex angles than those of the P and D surfaces, but its three edges are now curved [3], there are still 96 orthoscheme triangles in an elementary cell and the organization has space group Ia3d. Here also the 96 orthoschemes of an elementary cell can be arranged in an irregular dodecagon, with short and long edges, but with six vertices joined three by three in two groups, the six others being left alone.

Such a complicated structure of labyrinths was indeed proposed for a system of amphiphiles more than twenty years ago, from the analysis of X-ray scattering data, at a time when physicists and physical chemists were not concerned by IPMS [10]. It has been since identified many times in various systems [6].

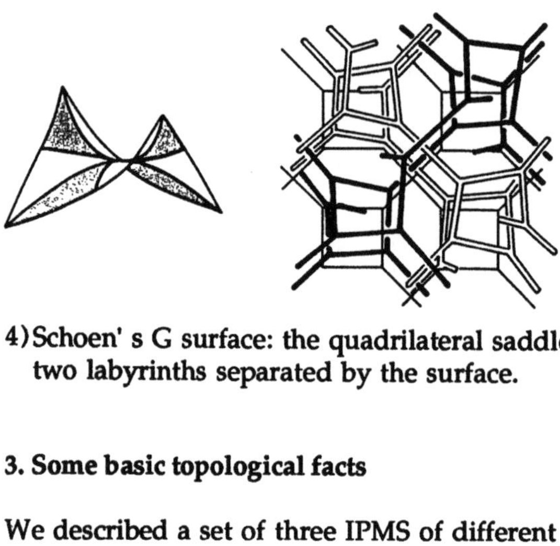

4) Schoen' s G surface: the quadrilateral saddle with curved edges and the two labyrinths separated by the surface.

3. Some basic topological facts

We described a set of three IPMS of different symmetries, which are built from similar orthoscheme triangles with vertex angles of $\pi/2$, $\pi/4$ and $\pi/6$ but differently curved edges. This suggest some relation between these surfaces and, in order to precise this idea, we now examine some topological properties.

3.1 Topological genus

The notion of a genus for the elementary cell of an IPMS is simple to introduce if the 3-D periodicity of the surface is expressed in the form of three cyclic conditions, corresponding to identifications of the six faces of the cell two by two. If this is done in each of the three cases, identifying the extremities of the rods of one labyrinth rather than those of the whole surface for simplicity, it can be seen that the three IPMS considered here have the same topological genus g = 3.

3.2 Gaussian curvature

The integral of the Gaussian curvature K on a surface is directly related to its topological genus g by the Gauss-Bonnet formula $\int Kds = 4\pi(1-g)$ [11]. In our case, as the elementary cells of the three IPMS considered here are equivalent to tori with genus 3, the pieces of surface they contain have the same integrated Gaussian curvature -8π.

3.3 Polygonal cell

A classical topological law states that a torus with genus g is built by identifications of the edges of a polygon with 4g edges two by two [11], a torus with genus 3 is therefore obtained with a dodecagon and, in the cases of the three surfaces studied here, the pieces of surface contained in their elementary cells must have a dodecagonal shape.

4. Crystallography in the hyperbolic plane

The P, D and G infinite periodic minimal surfaces have therefore the same topological genus 3, their orthoscheme triangles have the same vertex angles $\pi/2$, $\pi/4$ and $\pi/6$, the integrals of the Gaussian curvature in their elementary cells have the same value -8π, the pieces of surface in their elementary cells must be dodecagonal; moreover, if surfaces P and D are considered, the straight and curved edges of the orthoscheme triangles are interchanged. Thus the three surfaces, although of different symmetries, have many common topological and geometrical properties. The first may be indeed understood from the second. This can be done examining the properties of the surface having the same Gaussian curvature than the three IPMS, but distributed over the whole suface uniformly, which is called a hyperbolic plane. It does not exist in our 3-D Euclidean space, as already said, but a conformal representation of this surface in a 2-D Euclidean plane was worked out, which is commonly known as the representation of Poincaré [11] and which we are going to use [13].

4.1 Poincaré's representation

The representation of Poincaré of the hyperbolic plane with the orthoscheme triangle having vertex angles $\pi/2$, $\pi/4$ and $\pi/6$ is shown in

Fig.5

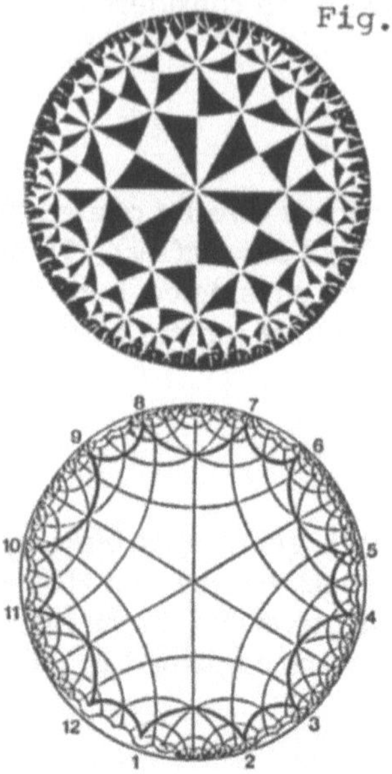

Fig.6

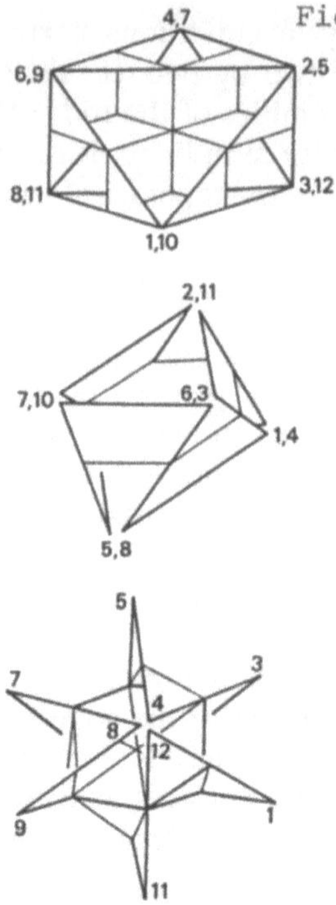

5) The Poincaré' s representation of the hyperbolic plane tiled with an orthoscheme triangle having vertex angles of $\pi/2$, $\pi/4$ and $\pi/6$ and its dodecagonal translation cell.

6) Topological equivalents showing the identifications of the vertices of the dodecagonal cell of the hyperbolic plane leading to the cells of surfaces P, D and G .

fig. 5 [12]. A closer look to the figure shows that the representation can be tiled with hexagons which meet four by four at every vertex, a tiling which is impossible on a flat plane, which admits three hexagons per vertex only, but is possible on a saddle. Of course, this is obtained adopting a particular metrics which manifests itself in the fact that the length of a constant segment is not itself constant in the representation.

4.2 Symmetry groups and dodecagonal cell

The orthoscheme triangle is the fundamental region of the symmetry group of the hyperbolic plane, repetitions of reflections along its edges cover the whole surface, without overlap. Among all the products of reflections possible certain correspond to translations on the surface and constitute its tranlation subgroups. If we limit ourselves to translations which can be assimilated with identifications corresponding to the building of a torus with topological genus 3, the fundamental region of the hyperbolic plane for this subgroup of translation is the dodecagon shown in fig. 5. This dodecagon is a set of 96 orthoscheme triangles, arranged in exactly the same manner than those in the dodecagonal piece of surface contained in the cells of the three IPMS, and the possible writing of certain translations as identities corresponds to the three groupings of vertices observed on the IPMS. These are shown in fig. 6 with topological equivalents in our 3-D Euclidean space.

5. Relationship between P, D and G surfaces

We have analyzed the relationship between these three IPMS adopting a topological point of view, indeed, we have shown that they are equivalent topologically, but other points of view are possible. For instance an analytical one, expressed in the so-called Bonnet's transformation [3,4], or a crystallographic one, inspired from the methods recently developed for building models of quasi-crystals [15].

In the first point of view the coordinates of the P, D, and G surfaces are analytically expressed as functions of an angular parameter θ in the so-called Weierstrass' representation. When this parameter is varied continuously from 0 to $\pi/2$, a set of surfaces is obtained, among which surfaces P, D, and G appear for 0, $\pi/2$ and 38,015° respectively and are the only ones without self intersection. This is an example of a Bonnet's transformation, at constant Gaussian curvature, by bending without stretching.

In the second point of view it is expected that the three surfaces might be obtained as 3-D cuts along three different directions of one unique surface built in a space of higher dimensionality. It was recently shown that, starting from a hypercubic surface in the 4-D Euclidean space, the P and D surfaces might be obtained as two orthogonal 3-D cuts of this surface but not the G surface, most likely a cubic surface built in a space of higher dimensionality is needed for obtaining them in one step.

6. Physical terms

We discussed the topology and symmetries of three basic periodic organizations of mathematical surfaces partitioning space into two identical subspaces, we also quoted a few examples of structures built by films of amphiphiles with the same topology and symmetries, we must therefore consider the question of the possible physical terms leading to such organizations. All known physical examples are met in regions of the phase diagrams where the concentration of amphiphiles is higher than

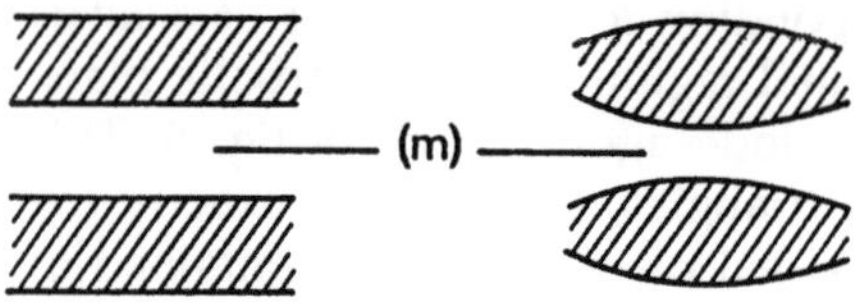

7) Frustration: constant interfacial distances and zero interfacial curvature are compatible in our 3-D space but not constant distances and non-zero curvatures.

0.5; there the distances between interfaces are of the same order of magnitude than their thickness, or smaller, and the chains of two films are in direct contact in their tail region. We have proposed to analyze these systems considering that their films are stabilized by forces normal to them, which maintain constant distances between the interfaces and forces parallel to them, which, as they exert their actions in different media, control the interfacial curvature (as in a bimetallic switch). When the latter is zero the obvious organization of the films is that of the periodical stacking of the lamellar structure; when the curvature is different from zero the constant distances between the interfaces cannot be preserved in this simple stacking, because of the symmetry of the film, as shown in fig. 7. In this case there is a conflict between normal and parallel forces, or "frustration", which can be solved by a change of configuration only. However in our flat Euclidean space, there is no configuration to relax such a frustration totally and we showed that it can but be relaxed locally in configurations with defects of rotation or disclinations [16]. A systematic investigation showed that these configurations have indeed the same topologies and symmetries as the structures observed for lyotropic liquid crystals, particularly for the cubic ones described in this article. These structures are to be understood as periodic organizations of frustrated fluid films driven by interaction forces within and between interfaces [17].

REFERENCES

1. H. A. Schwarz, *Gesammelte Mathematische Abhandlungen*, Springer, Berlin (1890).
2. A. Schœnflies, *Comptes Rendus Acad. Sc.* 112 478 (1882).
3. A. H. Schoen, *NASA Technical Note* D-5541 (1970).
4. S. T. Hyde and S. Anderson, Z. *Kristallographie* 168 221 (1984).
5. A. L. Mackay, *Nature* 314 604 (1985).
6. P. Mariani, V. Luzzati and H. Delacroix, *Jour. Mol. Biol.* 204 165 (1988).
7. P. Kékicheff and B. Cabane, *J. de Physique* 48 1571 (1987).
8 A. Tardieu, *Thesis* Université Paris Sud, Orsay (1972).

9. W. Longley and J. Mac Intosh, *Nature* 303 612 (1983).
10. V. Luzzati and P. A. Spegt, *Nature* 215 701 (1967).
11. D. Hilbert and S. Cohn-Vossen, *Geometry and the imagination*, Chelsea Publishing Company, New York (1983).
12. H. S. M. Coxeter, *Introduction to Geometry*, J. Wiley, New York (1961).
13. J. F. Sadoc and J. Charvolin, *Acta Cryst*. A 45 10 (1989).
14. P. A. Firby and C. F. Gardiner, *Surface topology*, Ellis Horwood Ltd (1982).
15. R. Mosseri and J. F. Sadoc, *J. de Physique* 51 C7-257 (1990).
16. J. F. Sadoc and J. Charvolin, *J. de Physique* 47 683 (1986). J. Charvolin and J. F. Sadoc, *J. de Physique* 48 1559 (1987).
17. J. N. Israelachvili, *Intermolecular and surfaces forces*, Academic Press, New York (1985).

Phase Transitions in Cubic Amphiphilic Crystals

M. Clerc and J.F. Sadoc

Laboratoire de Physique des Solides, Bât. 510,
Université Paris-Sud, F-91405 Orsay, France

Abstract -There is a wide variety of topological configurations in amphiphilic phases with phase transitions involving a cubic bicontinuous structure and another mesophases with simpler organization. We propose some topological arguments to explain these transitions and give some possible associated fluctuations .

1. Introduction

Amphiphilic molecules in solution in water can be organized in periodic fluid films when the concentration of amphiphilic molecules is large enough. The geometry of the different phases can be reduced to the one of the water medium, that can be described as a surface; in the case of the cubic phase, the involved surface is an I.P.M.S.. In this case, transitions between two fluid phases correspond to a change of the topology of this surface: simple assumptions about their shape and mutual interactions permit us to predict theoretically the different structures observed and their succession with concentration /1/.

The description of structures using a surface is only one point of view; however there is another approach where the amphiphilic subspace is characterized by rods. For example, a hexagonal phase consists of a 2D hexagonal network of infinite parallel channels of fluid amphiphilic molecules, often called rods, embedded in water. In the second description, each rod is denoted by a line. In the first one, the water is denoted by a honeycomb surface and described as a water film supported by this surface.

The different possible cubic space groups and related structures are, in many cases, established/2/. An important feature of the structures is the connectivity of the two media (water and surfactant). Cubic mesophases are observed at very well defined positions in phase diagrams and are usually characterized by a narrow range of composition. They appear as intermediate phases between other phases which have a wider range of stability /3/. Our purpose is to underline how the connectivity of a cubic mesophase depends on that of the neighbouring mesophases.

2. Epitaxial relations at mesophase transitions.

Recent X. ray studies /4,5/, using $C_{12}EO_6$ as an amphiphilic molecule, for the transition between a hexagonal and a cubic phase reveal clear epitaxial relations; the phase transition between these two structures does not involve the total destruction of

Springer Proceedings in Physics, Vol. 66 The Structure and Conformation of Amphiphilic Membranes
Editors: R. Lipowsky · D. Richter · K. Kremer

the initial structure. The two adjacent mesophases are always strongly related: A cubic mesophase grows with a fixed orientation with respect to a hexagonal one; a lamellar mesophase develops with the lamellar planes perpendicular to a [211] direction of the cubic mesophase. These relations are useful in understanding how the two interwoven labyrinths of the cubic phase can grow from the infinite amphiphilic rods of the hexagonal one. Such epitaxial relations are also observed at the transition between the lamellar and cubic phases.

When one considers the structure of the cubic phase, adjacent to the hexagonal one in the phase diagram, one sees a complete change of topology: the honeycomb surface is replaced by the Schoen Gyroïd IPMS and the set of parallel lines by two interwoven 3D networks of segments connected three by three. The growth of the final structure uses elements of the initial one. So one can imagine to analyse the change of topology implied by the transition with the help of local fluctuations from the initial structure.

3. Fluctuations in mesophases.

Even if the Bragg spots always remain clearly visible, characteristic diffuse scattering sometimes reveals strong fluctuation effects. In the case of the cubic mesophase the diffuse scattering is concentrated in lines between the Bragg spots. These lines characterize planes of the reciprocal space which are periodically stacked with four equivalent orientations, defined by their normals along the four [111] directions. This scattering has been interpreted /5/ as well ordered linear objects along the [111] directions but with weak coupling in the orthogonal direction allowing fluctuations.

For the hexagonal mesophase, two principal directions of diffuse scattering are observed. There is extension of the Bragg spots along the mesophase director and within the reciprocal plane containing the 2D lattice of Bragg spots, diffuse scattering is observed between the six first order reflections. So fluctuations are not density fluctuations. From a dynamic point of view, diffuse scattering is related to trnsverse undulations of the cylinders. The scattering in the other direction is due to deformations of the two-dimensional hexagonal packing.

4. Geometry of Transitions into the cubic mesophase

4.1 The cubic bicontinuous structure.

The existence of two equivalent surfactant networks separated by a film of water, associated with a 3D-periodic structure preserving the bicontinuity, imposes the shape of an IPMS on the centre of the water layer. The space group Ia3d imposes the choice of Schoen's Gyroïd surface. The two subspaces are filled with two 3D-networks of connected surfactant "rods". These two networks are two enantiomers (one D and the other L). All the rods have equal length and meet three by three at an angle of 120° at planar nodes. Each 3D-network contain closed polygons that are all congruent decagons.

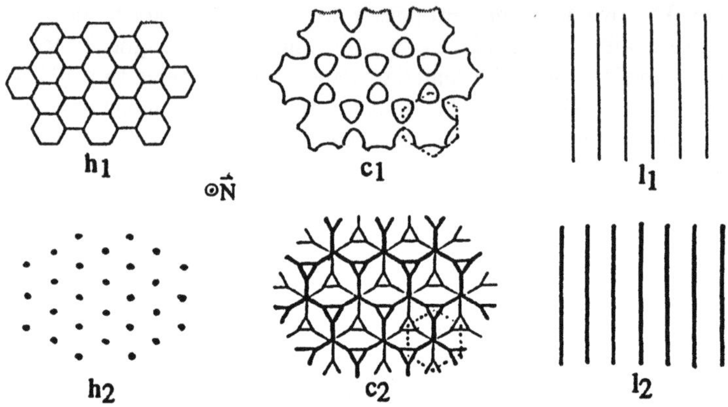

Figure 1: Projection of the three structures, along a [111] direction of the cubic structure.
h_1): The honeycomb water surface. h_2): The hexagonal packing of the infinite surfactant cylinders. Each cell of the honeycomb surface contains one cylinder.
c_1): The Gyroïd IPMS. c_2): The two 3D-networks of surfactant rods. Inside the dotted lines, the projection of a superposition of 3-connected nodes of both 3D-networks. A triangle is the projection of a threefold infinite helix.
l_1): The water planes. l_2): The surfactant planes. These planes are the dense planes that are continuously preserved at the two transitions.

In figure 1, we represent the Gyroïd and the two 3D-networks between the corresponding structures of the hexagonal and lamellar mesophases. For instance, the corresponding "water surfaces" are the stacking of parallel planes in the lamellar case and a "honeycomb surface" in the hexagonal case. So, the honeycomb surface is defined by the set of points which separate equally cylinder axes. Scales and relative orientations are deduced from the epitaxial relations at the transitions.

4.2 Hexagonal-Cubic transition.

How do the infinite parallel cylinders of the hexagonal mesophase establish connections between them to create the two 3D-networks of the cubic structure? The cubic mesophase develops preferentially along the cylinders direction that transforms into a [111] cubic direction. So we propose a local fluctuation within the hexagonal mesophase that transforms a cylinder as a whole. At the transition, the cylinders transform following two different ways (fig 2):

-two thirds of the cylinders become infinite threefold helices, each helix being entirely contained in one of the 3D-networks (all righthanded helices in one of them and the lefthanded in the other one). Using the surface description they show the position of tunnels in the structure.

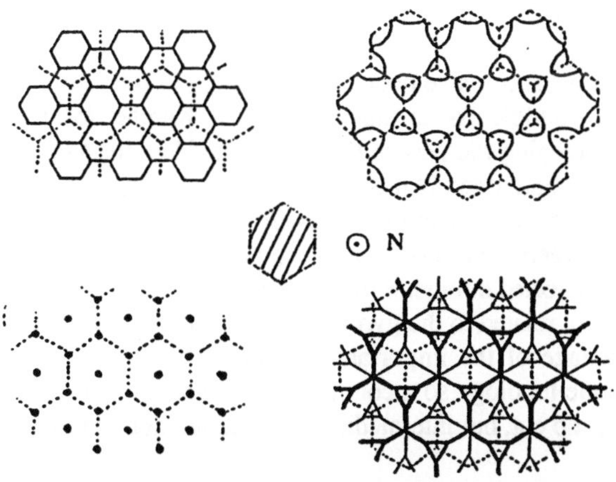

Figure 2: The hexagonal and the cubic (view along the (111) direction) structures. They are shown using the surface description (top) and the rods description (bottom). Dashed lines join those cylinders which are changed into helices (appearing as small triangles on the cubic projection-bottom right), the other cylinders are cut and changed locally into successive three coordinated nodes.

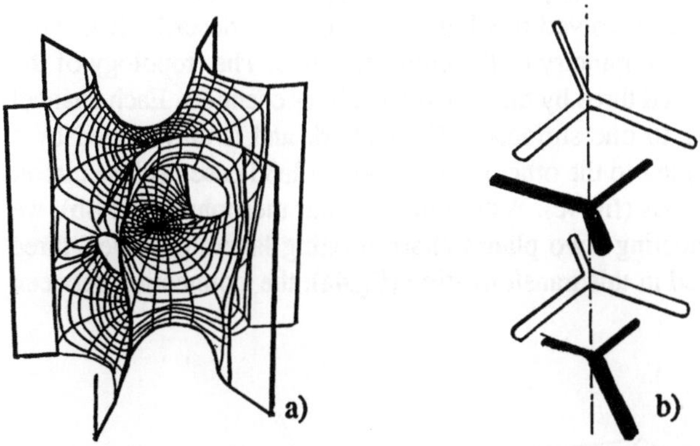

Figure 3: The "monkey saddle tower" (a) is a minimal surface,whose topology (inside a hexagonal cylinder) is similar to that of the Gyroïd (inside the same cylinder). It can be obtained from three plane sheets folded in order to have, far from a common axis, six half planes with a hexagonal symmetry.Then tunnels, with a three fold symmetry connect the three sheets, leading locally to "monkey saddles". Axes of the tunnels (b) form the succession of three coordinated nodes.

-the last third change into a succession of nodes, in turn contained in both networks. Theses nodes connect three coplanar rods of the networks.They are normal to the [111] direction and periodically stacked along this direction. Using the surface description these nodes are separated by locally parallel parts of the Gyroid surface.

This topological transformation of the cylinders is sufficient to entirely build the cubic structure. Looking at the correlated transformation for the honeycomb surface while a cylinder changes into the stack of nodes, one finds that a hexagonal channel becomes a "tower" surface, drawn in figure 3. This "tower" connects in turn three by three the six neighbouring rods at the successive nodes. The topology of this "tower" surface is taken from the Monkey Saddle Tower introduced by H. Karcher /6/. By simply "cutting" the Gyroïd and keeping the part of the surface inside a hexagonal cylinder parallele to the [111] direction, one obtains also the same topology and so we conclude that the whole Gyroïd is a juxtaposition of "towers", with an appropriate shift along the [111] direction between them. For example, a helix is the final state of a cylinder being successively connected through three "towers".

Even if the cubic structure is not related to the gyroïd, but to other IPMS with the same genus, we can take advantage of the important property relating the three possible IPMS: they have the same topology so a continuous deformation allows to go fom one to the others. There is no cut, but intermediate surfaces are not minimal.

4.3 Cubic/lamellar transition.

This transition corresponds to the transformation of the two 3D-networks into a periodic stacking of lamellae, oriented normal to a [211] direction. This orientation is related to the conservation of the dense planes at the mesophase transitions.

In order to understand the involved topological change, we describe it locally, independently of the detailed symmetry of the cubic structure. The topology of the IPMS is that of pipes connected three by three, and so defines channels. Each channel contains one rod belonging to one surfactant 3D-network and is surrounded by a closed circuit of rods belonging to the other network. In the Ia3d case, a close circuit is a non-planar decagon of rods (fig 4c). Within the lamellar mesophase (fig.4b), we now introduce a pipe connecting two planes characterizing layers of water, three surfactant layers are involved in this transformation (fig.4a): the connection between

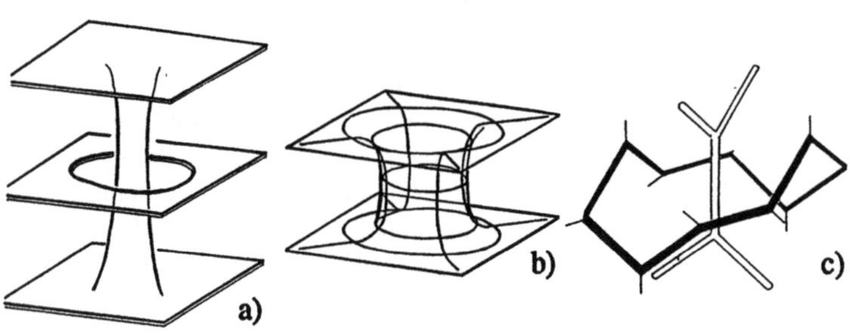

Figure 4: a) The connection between two next nearest neighbour layers. Here the surfaces denote the surfactant: in the cubic structure these surfaces are changed into networks of rods. The connection should be considered as the rod in (c) which is surrounded by the twisted decagonal ring.
b) The surface defined by two layers of water connected together by a tunnel.
c)A decagonal ring in one 3D-network of rods, surrounding a rod of the other network.

the top and bottom lamellae creates a rod ,which goes through an intermediate lamella. Then this last lamella contains a closed circuit around the channel and one recovers the same local configuration as in the cubic mesophase. Thusit is possible to construct the whole cubic structure by introducing water channels in the lamellar structure (for a complete description see /5/). The water channel mechanism for this transition appears to be a natural way to create a bicontinuous structure starting from the lamellar one. It has been observed /5/ that fluctuations in the lamellar phase near the transition into the cubic phase appear as local hexagonal domains. The oscillation of lamellae appearing in this model are in good agreement with this observation.

5. Conclusion

We have described two kinds of effects associated with the phase transitions:

- the fluctuations in a mesophase, like for example, the linear disorder in the cubic. These effects are deduced from the diffuse scattering interpretation and seem to correspond to dynamical fluctuations. No critical divergences of these fluctuations are observed, as is expected for first order transitions.

- the topological changes and their reduction to local connections between objects like rods or lamellae. These effects derive from the structural models of the mesophases and the epitaxial relations. The transformation of the whole topology can be decomposed into some elementary and local topological changes.

REFERENCES

/1/ J. Charvolin and J.F. Sadoc. "Morphology of stratified fluids" in "geometry in condensed matter physics" - J.F. Sadoc editor. W.S.P. (1990)

/2/ K. Fontell. Colloïd and polymer science 1990,268,264-285.

P. Mariani, V. Luzzati and H. Delacroix. J. Mol. Biol. 1988,204,165-189.

/3/ J. Seddon, J. L. Hogan, N. A. Warrender and E. Pebay-Peyroula. Progress in colloïd and polymer science 1990,81,189-197.

/4/ Y. Rançon and J. Charvolin. /a/ Journal de physique 1987,48,1067-1073. /b/ Journal of Physical Chemistry 1988,92,2646.

Y. Rançon. Thesis. Université Paris Sud, Orsay 1987.

/5/ M. Clerc, A.M. Levelut and J.F. Sadoc to appear in J. Phys. France (1991)

/6/ H. Karcher. Manuscripta mathematica 1988,62,83-114.

Elasticity and Excitations of Minimal Crystals

R. Bruinsma

Physics Department, University of California,
Los Angeles, CA 90024, USA

The elastic properties of the unusual crystals encountered in surfactant-rich solutions are investigated. Triply-periodic minimal surfaces provide a convenient frame-work for the understanding of such materials but, as is shown, degeneracy leads to vanishing elastic coefficients in the framework of the classical Helfrich energy. This degeneracy is lifted by higher-order corrections and by finite temperature effects. We show that, as a result, thermodynamic stability can be achieved at low levels of dilution but that with increasing dilution the P surface inevitably melts. The degeneracy also leads to an unusual collective excitation spectrum which has a smectic-like undulation dispersion, except at very long wavelengths where it becomes sound-like. The elastic moduli are found to have the same dependence on temperature and concentration as those of tethered stacked membranes and the shear moduli have a temperature and material independent ratio.

I. INTRODUCTION

Crystals are constructed by the endless addition of the same building block: the unit-cell. In solid-state physics, the canonical unit cell is a near rigid structure made from a small number of atoms kept firmly in position by strong potentials. But, crystals are also encountered in complex liquids such as oil/water/surfactant mixtures[1], block co-polymers[2] and in biological systems[3], where the unit-cell is of a fundamentally different nature.

As an example, surfactant molecules dissolved in water can form a range of structures, such as micelles, cylinders and stacked sheets, which are made out of surfactant bilayers.[4] In certain concentration ranges, X-ray diffraction demonstrates diffraction patterns consistent with cubic or diamond crystal structure. These crystals are believed to consist of a connected, triply-

Springer Proceedings in Physics, Vol. 66 The Structure and Conformation of Amphiphilic Membranes
Editors: R. Lipowsky · D. Richter · K. Kremer

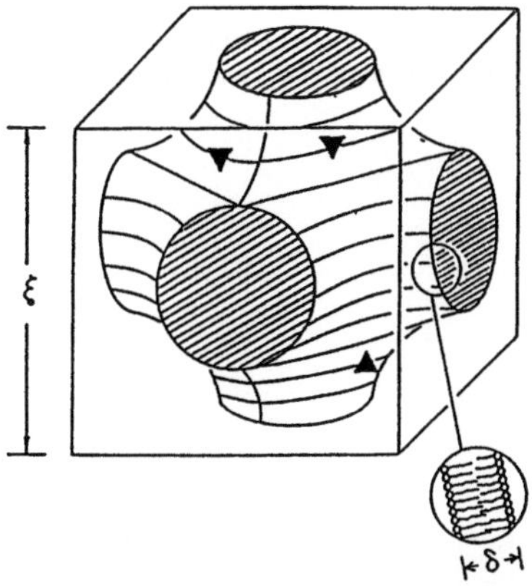

Fig. 1 Unit cell of the P surface. The lattice constant is ξ(~200A°) and the membrane thickness is δ (~ 30A°). The triangles mark the flat-points.

periodic bilayer extending across the sample.[5] The unit cells are very large, with lattice constants ~ 100-200A, containing very large numbers of molecules which are allowed to diffuse inside the bilayer.

These unit cells do not reflect ordering at the atomic level. Instead, the origin of the crystal structure is believed to be explained by the Helfrich bending-energy Hamiltonian[6] which gives the energy cost of bending a membrane-like surface. If thermal fluctuation effects are not too strong, then the minimization of this Hamiltonian leads to the prediction that the location of the bilayer should be described as a "minimal" surface[7]: a surface of zero average curvature. Indeed, among the (large) catalog of minimal surfaces one finds triply periodic structures with cubic and diamond symmetry - such as the simple cubic Schwartz P surface of Fig.1 - as well as many of the other cyrstal structures familiar from solid-state physics.[8] Quantitative tests - such as the relation between unit cell size and surfactant concentration[9] - have provided further support for the "minimal-surface" model. It must be realized though that for unit cell sizes in the range 100-200A and for film-thicknesses $\delta \approx$ 20-30A, significant corrections due to the finite membrane thickness must be present.

If we nevertheless accept the "minimal-surface" model, we could hope to calculate thermodynamic response functions such as the elastic moduli using the

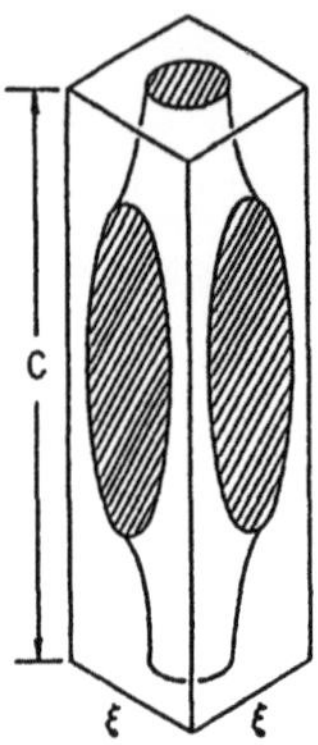

Fig. 2
Tetragonally deformed P surface. For c/ξ less than a critical value somewhat greater than one, the surface remains minimal and is stable against collapse.

Helfrich Hamiltonian. This program immediately runs into difficulties. The P surface is only a, particularly symmetric, member of a continuous (two-parameter) family of minimal surfaces.[10,11] Numerical studies by Maggs and Leibler showed that, for instance, tetragonal deformation (see Fig.2) does not violate the minimal surface constraint.[12] As a result, the crystal-structure would drift over this family (barring external constraints). In particular, the elastic coefficients can be shown to be zero within the Helfrich Hamiltonian which is inconsistent with thermodynamic stability as well as the requirement for long-range positional order in the crystal structure.

This ground-state degeneracy could be removed in three ways. The finite thickness of the membrane will produce, as mentioned, corrections to the Helfrich Hamiltonian.[13,14] These higher-order terms need not respect the degeneracy of the Helfrich Hamiltonian and may stabilize the P surface, as already pointed out by Helfrich himself.[15] Next, as is well known from magnetism, ground-state degeneracy can also be removed by thermal fluctuations. A P surface structure has, as we shall see, a complex spectrum of normal modes which at non-zero temperature are thermally populated. The associated entropy may stabilize certain crystal structures much like temperature tends to stabilize the BCC phase in solid-state physics. A closer analog is perhaps found by noting that if we stretch or compress a polymer we reduce its configuration space and thereby increase its free energy. This leads to elastic behaviour characterized by elastic coefficients which increase with temperature. If we think of a membrane as a d=2 polymer then we could expect similar "rubber-elasticity" for the P surface. Finally, long-range interactions such as electrostatic and van der Waals interactions could also lift the degeneracy. However, in the presence of strong screening, we can neglect electrostatic interaction while it is easy to see that attractive interactions - like the van der Waals interaction - would break the degeneracy in the wrong way: it would de-stabilize the P surface. We will neglect this last possibility in the following.

It is the aim now to investigate the elastic properties and the low-energy excitations of the P surfaces. The low-energy spectrum controls the response to external perturbations such as elastic deformations and changes in temperature (specific heat) as well as the X-ray structure factor and the phase-diagram. In addition they are, as we shall see, important to understand the thermodynamic stability of the P surface and the problem of the melting of the P surface. The high-energy spectrum of membranes has been theoretically well explored[16] but, except for the case of stacked membranes,[17] the low-energy spectrum of ordered structures is less well understood, although a global phase-diagram has been proposed[18] (which included the P surface). It was also suggested[19] that the P surface upon melting could produce the isotropic L_3 phase. Experiments on the P surface structures, such as small-angle X-ray structural studies[1] and NMR[20], have centered on static properties so there is little known about the low-energy spectrum.

The nature of the low-lying excitations of the P surface can be discussed in terms of the solution of a quantum-mechanical problem involving a Schrödinger-like equation defined on a periodic surface imbedded in three-dimensional space. Borrowing methods from solid-state physics it is possible to classify the mode spectrum. We find that the low-energy spectrum is rich and complex (Fig. 3). At the lowest wave-vectors there is a conventional 3-fold q^2 acoustic-mode spectrum. However, for $|\vec{q}| \gtrsim q_{co}$ with $q_{co} = \left(\bar{\bar{\kappa}}/\kappa\xi^4\right)^{1/2}$ This crosses over to a 3-fold q^4 "undulation" spectrum. Here, κ is the Helfrich bending energy, $\bar{\bar{\kappa}} \sim \kappa\delta^2$ with δ (~30A) the membrane thickness, and ξ (~100A) is the unit cell size. Next, we can identify a 2-fold libration band of optical modes and a 1-fold "tight-binding" local-mode band. For wavevectors

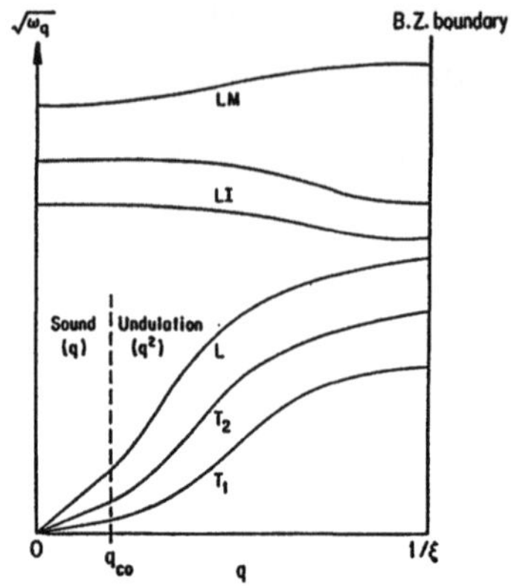

Fig. 3 Collective mode dispersion (qualitative). The lowest 3 fold acoustic branch (L, T_1, T_2) has a sound-like dispersion for $q \lesssim \delta/\xi^2$ and an undulation dispersion for $q \gtrsim \delta/\xi^2$. Next, there is a 2 fold libration (LI) band and finally a 1 fold local-mode (LM) band.

large compared to $1/\xi$ the usual single-membrane undulations[16] are of course still present.

It is easy to show that the P surface is elastically unstable within the context of the Helfrich Hamiltonian, due to the above-mentioned degeneracy, but that the dominant correction term to the Helfrich Hamiltonian is exactly right to stabilize the P surface among the family of minimal surfaces. Thermally excited "collective" modes as well as single-membrane short-wavelength fluctuations both lead to strong temperature dependence in the elastic moduli. The bulk-modulus B and shear-moduli $\mu_{1,2}$ have a dependence on temperature T and lattice constant ξ given by

$$B = \alpha_1 \bar{\bar{\kappa}}/\xi^5 + \alpha_2 k_B T/\xi^3 \qquad 1.1$$

and

$$\mu_{1,2} = C_{1,2}\left\{\beta_1 \bar{\bar{\kappa}}/\xi^5 - \beta_2 k_B T/\xi^3\right\} \qquad 1.2$$

with $\alpha_{1,2}$ and $\beta_{1,2}$ constants. The constants $C_{1,2}$ are geometrical invariants. Equation 1.2 thus predicts that the μ_1/μ_2 ratio should be a temperature and material independent invariant. As the lattice constant increases, the shear-moduli decrease. The maximum allowed lattice constant is

$$\xi_m \sim \left(\bar{\bar{\kappa}}/k_B T\right)^{1/2} \qquad 1.3$$

Since the surfactant concentration C_s is proportional to δ/ξ, Eq. 1.3 defines a boundary in the phase-diagram such that for $C_s \lesssim \delta/\xi_m$, the P surface is unstable.

REFERENCES

1. V. Luzatti in Biological Membranes edited by D. Chapman, Vol.1, p.71 (Academic Press, New York, 1968); P. Mariani, V. Luzatti and H.J. Delacroix, J. Mol. Biol. 204, 165 (1988); K. Fontell, Liquid Crystals and Plastic Crystals, edited by G.W. Gray and P. Winson, Vol. 2, 80 (Ellis Harwood, Chicester, 1974); K. Larsson, Z. Phys. Chem. 56, 173 (1973); P. Kelucheff and B. Cabane, J. Phys. 48, 1571 (1987).
2. E.L. Thomas, D.B. Alward, D.J. Kinning, D.C. Martin, D.L. Handlin, L.J. Fetters, Macromolecules, 19, 2197 (1986); H. Hasegawa, K. Tanaka, K. Yamasalii, T. Hashimoto, Macromolecules 20, 1651 (1987); E.L. Thomas, J.R. Reffner and J. Bellare, J. Phys. (Paris), C7-363 (1990).
3. Y. Bouligand, J. Phys. (Paris, C7-35 (1990).
4. See for instance: Physics of Amphiphilic Layers, edited by J. Meunier, D. Langevin and N. Boccard (Springer, Berlin, 1987).
5. See for instance: A. Tardieu, Thesis, orsay (1972); D. Anderson, Thesis, University of Minnesota (1986); J. Charvolin and J.F. Sadoc, J. Phys. (Paris) 48, 1559 (1987); S.T. Hyde, J. Phys. Chem. 93, 1458 (1989).
6. P.B. Canham, J. Theor. Biol. 26, 61 (1970); W. Helfrich, Z. Naturforsch. 28c, 693 (1973); H.J. Deuling and W. Helfrich, J. Physique 37, 1335 (1976); E.A. Evans, Biophys. J 30, 265 (1980); J.T. Jenkins, J. Math. Biology 4, 149 (1977); M.A. Peterson, Mol. Cryst. Liq. Cryst. 127, 257 (1985) and J. Math. Phys. 26, 711 (1985).
7. See for instance: J.C.C. Nitsche, Lectures on Minimal Surfaces, Vol.1 (Cambridge University Press, 1989); D.M. Anderson, H.T. Davis, J.C.C. Nitsche, Adv. Chem. Phys. 77, p.337 (Interscience, New York, 1990); A.T. Fomenko, The Plateau Problem I & II, (Gordon Breach, New York, 1989).
8. For a table of minimal surfaces and their symmetry see: E. Koch and W. Fischer, Acta Cryst. A46, 33 (1990) and references therein.
9. P. Barois, D. Erdam, and S.T. Hyde, J. Phys. (Paris), C7-25 (1990).

10. S. Lidin and K. Hyde, J. Phys. (Paris), 48, 1585 (1987). The degeneracy is also known for the D,H and CLP Schwartz surfaces and the G, S' S" Schoen surfaces.

11. H. Karcher, Manus Math. 62, 83 (1988).

12. D. A. Hoffman, J. Phys. C7, 197, (1990).

13. M. Mitov, Comptes Rendus de l'Academie Bulgare des Sciences 31, 513 (1970).

14. W. Helfrich, Liquid Crystals 6, 1647 (1989).

15. W. Helfrich and H. Rennschuh, J. Phys. (Paris), C7-189 (1990)

16. For a review: Statistical Mechanics of Membranes and Surfaces, (Jerusalem Winter School), edited by D.N. Nelson, T. Piran and S. Weinberg (World Scientific, Singapore, 1989).

17. See: C.R. Safinya et al., Phys. Rev. Lett. 57, 2718 (1986), and references therein.

18. D.A. Huse and S.J. Leibler, J. Phys. (Paris) 49, 605 (1988).

19. G. Porte, J. Marignan, P. Bassereau, and R. May, J. Phys. (Paris) 49, 511 (1988); G. Porte, J. Appel, P. Bassereau, and J. Marignan, J. Phys. (Paris) 50, 1335 (1990).

20. D.M. Anderson, J. Phys. (Paris), 51, C7-1 (1990).

Inverse Micellar Cubic Phases of Lipids

J.M. Seddon and E.A. Bartle*

Department of Chemistry, The University,
Southampton SO2 5NH, UK
*Address from July 1991: Department of Chemistry,
Imperial College, London SW7 2AY, UK

Abstract. We describe the observation of a novel inverse face-centred cubic phase, of space group Fd3m, in two different systems of hydrated binary lipid mixtures. This cubic phase occurs between the inverse hexagonal (H_{II}) phase and the inverse micellar solution (L_2), with increasing concentration of the lipid component with the less strongly hydrophilic headgroup, and appears to be the first example of a lyotropic cubic phase composed entirely of closed inverse micellar aggregates.

1. Introduction

There are now many examples known of lipids which adopt inverse bicontinuous cubic phases [1,2]. We have previously found that saturated phospholipids such as phosphatidylethanolamines can, in certain cases, adopt as many as three different inverse bicontinuous cubic phases in water, of probable space groups Ia3d (No. 230), Im3m (No. 229) and Pn3m (No. 224). We found that these cubic phases could be induced to appear by reducing the chainlength, or by increasing the hydrophilicity of the headgroup of the phospholipid molecule. All of these cubic phases are located in the phase diagrams between the lamellar (L_α) and the inverse hexagonal (H_{II}) phases, and are based on underlying periodic minimal surfaces (reviewed in reference [3]). For such cubic phases, the average interfacial mean curvature H is negative (curvature towards the water) and adopts a value intermediate between those of the L_α and H_{II} phases.

2. Inverse Micellar Cubic Phases

Although cubic phases based on packings of normal topology micelles are now quite well established, it is frequently stated that corresponding cubic phases based on packings of inverse micelles do not exist. However, it has previously been suggested that the I_2''' cubic phase observed in certain ternary surfactant / alkanol / water systems [4] might have a structure based on a simple fcc packing of inverse micelles [5]. Such a face-centred cubic phase of inverse micelles, with spherical aqueous cores, has also been claimed to be observed in hydrated double-chain lipids having peptide-containing polar headgroups [6]. In this latter case the structure appears to be incompatible with the stated composition of the phase. In the absence of any detailed structural evidence, it remains uncertain whether or not cubic phases based on simple packings of inverse micelles do in fact exist. Such structures might be expected in systems which prefer to have a mean interfacial curvature which is more negative than that of the H_{II} phase. In fact, for binary lipid systems the H_{II} phase tends to remain even upon complete dehydration.

However, there are a number of systems containing mixtures of lipids where a cubic phase does appear to form on the "far side" of the inverse hexagonal phase, away from the lamellar phase. One example is the ternary system oleic acid / monoolein / water, which has been identified as being of spacegroup Fd3m [1]. Another example is the acid-soap system oleic acid / sodium oleate / brine (0.25 M NaCl), where a viscous isotropic phase appears between H_{II} and an inverse micellar L_2 solution [7]. We have identified this as a cubic phase of spacegroup Fd3m [8]. A third example is the ternary system dioleoylphosphatidylcholine / dioleoylglycerol / water, which also adopts the same Fd3m cubic phase [9].

3. Structure of the Fd3m Cubic Phase

The Fd3m cubic phase was first identified in a lipid extract from *Pseudomonas fluorescens*, and a structure based on a lipid monolayer together with a packing of inverse micelles was proposed, based on a crystallographic analysis involving a novel pattern recognition approach to phase the reflections [1,10]. However, we suggested that a possible alternative structure for the Fd3m cubic phase was one based solely on a packing of two types of inverse micelle [8, 9]. Such a structure had been recently predicted by a topological / geometrical study of the possible structures formed by periodic systems of frustrated fluid films [11]. The *Pseudomonas fluorescens* cubic phase was then found to be contaminated with a small amount of H_{II} phase, and a new analysis of the diffraction data from the Fd3m cubic phases of all four of the above-mentioned systems indicates that the structure in each of the four lipid systems is in fact identical. The low-resolution electron density maps show that the structure does indeed consist solely of a packing of quasi-spherical inverse micelles, 8 larger and 16 smaller ones per unit cell [12].

A schematic representation of the structure of the Fd3m inverse micellar cubic phase is shown in figure 1. The 8 larger inverse micelles are located at positions (a) in the unit cell, ie (0,0,0), (3/4,1/4,3/4) etc. They thus lie on a diamond lattice. The 16 smaller inverse micelles occupy positions (d), ie (5/8,5/8,5/8), (3/8,7/8,1/8), (7/8,1/8,3/8), (1/8,3/8,7/8) etc, arranged as tetrahedral clusters in the four vacant octants of the unit cell. An alternative view of the structure is that each larger micelle is surrounded by 12 of the smaller micelles, and tetrahedrally by 4 of the larger ones.

The structure is closely analogous [11] to that of the 17 Å cubic crystalline clathrate hydrates, where the water molecules are arranged into 16 dodecahedral and 8 hexakaidecahedral cages per unit cell. In the lipid Fd3m cubic phase, the polyhedral surfaces are essentially mapped out by the methyl ends of the hydrocarbon chains.

The packing of the inverse micelles is also the same [11] as the $MgCu_2$ structure (Laves phase C_{15}) found in certain binary metal alloys of composition AB_2. For close packing of an AB_2 system of spheres in spacegroup Fd3m with the $MgCu_2$ structure, the ratio of the radii of A and B should be equal to 1.225, leading to a

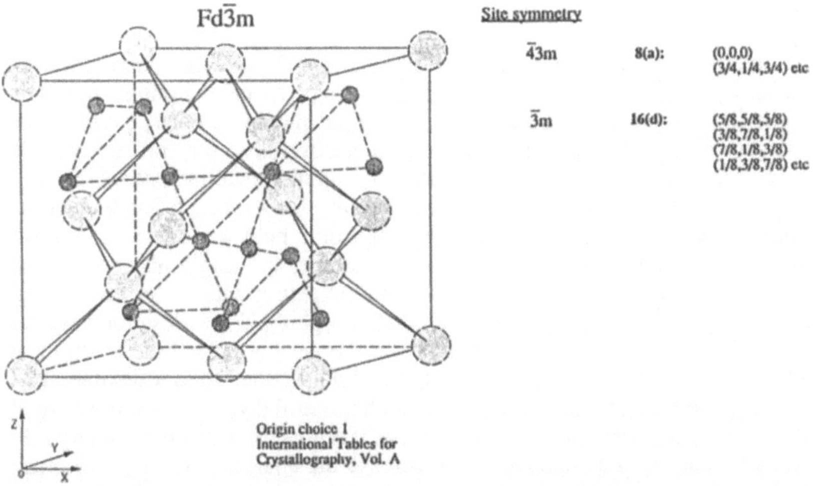

Figure 1. Schematic representation of the Fd3m inverse cubic phase. The structure consists of 8 larger inverse micelles (polar cores shaded light) and 16 smaller inverse micelles (polar cores shaded dark). The remaining volume is filled by the fluid hydrocarbon chains of the lipids. The connecting lines and rods are drawn merely as a guide for the eye.

packing fraction of 0.71, which is halfway between those of simple bcc (0.68) and fcc (0.74) packings. It is interesting to note that colloidal crystalls formed from dispersions of binary mixtures of monodisperse polystyrene latex spheres of slightly different diameters have also been observed to adopt this MgCu2 cubic structure [13].

4. Topology of the Fd3m Cubic Phase

There is very strong indirect evidence that the Fd3m cubic phase in all of the lipid systems studied so far has an inverse topology. Its location between the inverse hexagonal (H_{II}) phase and the L_2 inverse micellar solution is in itself strong evidence; furthermore, it can coexist with a large excess water phase, which is common for inverse phases, but unknown (or at least very unusual) for normal topology structures.

Pulsed field gradient NMR measurements of amphiphile and water self-diffusion should be able to confirm both the inverse topology and the discrete nature of the aggregates in the Fd3m cubic phase.

It remains to be seen whether a version of Fd3m with normal topology will also be discovered. It is a curious observation that the other well established micellar cubic phase, of spacegroup Pm3n (No. 223), has so far only been observed in normal topology versions.

5. Mechanism of Formation

The data obtained so far indicate that formation of the Fd3m cubic phase does not occur for binary lipid / water systems, but requires the presence of a minimum of two amphiphilic lipid components. Furthermore, one lipid should be polar, but the other should be very weakly polar. In fact, in all cases discovered so far, this second lipid component is too weakly amphiphilic to form any liquid-crystalline phases in water on its own, but does in principle have the possibility for hydrogen-bonding to the more polar lipid. This could enhance its solubility, help to locate its weakly hydrophilic headgroup at the polar/ nonpolar interface, and might also reduce the hydration of the more polar lipid by effectively competing with water for binding sites on the headgroup of the latter. The fact that phospholipid / cholesterol mixtures do not form the Fd3m cubic phase suggests that the weakly polar lipid component must not have a too rigid hydrophobic part, but should rather have a flexible alkyl chain (it should be borne in mind however that cholesterol does have a more limited solubility in phospholipids than do diglycerides). It seems clear that the primary role of the weakly polar lipid component is to increase the tendency for negative mean interfacial curvature of the lipid layer. At very large mole fractions, this tendency becomes so great that the cylindrical curvature of the H_{II} phase ($H = -1/2R$) is not sufficient, and the aggregates break up into quasi-spherical inverse micelles ($H = -1/R$). The surprising thing is that they then form two quite distinct types of micelle, of different sizes. It is very likely that the composition is inhomogeneous between the two types, with the more polar lipid being preferentially located in the larger inverse micelles, and the less polar component being preferentially in the smaller ones. Indeed, this may partly explain why Fd3m does not appear to form in binary lipid / water systems such as phosphatidylethanolamine / water, where we have previously shown that the inverse hexagonal H_{II} phase remains at all hydrations, even down to the dry state, where the tendency for negative mean curvature should be strongest. For such binary systems, the free energy cost of maintaining two different sizes of inverse micelle, yet without the possibility of inhomogeneous compositions, would certainly be very great.

6. Effect of Chainlength

In order to assess the effect of chainlength on the formation of the Fd3m cubic phase, we have investigated the phase behaviour of a fully hydrated homologous series of saturated phosphatidylcholine / digyceride mixtures [E. A. Bartle and J. M. Seddon, unpublished results]. A highly significant finding is that for chainlengths between C_{22} and C_{16}, the system forms the Fd3m cubic phase upon heating; however, reducing the chainlength to C_{15} or below suppresses the cubic phase and replaces it with the inverse hexagonal H_{II} phase. This effect of chainlength is precisely the opposite behaviour to that which we have previously observed for inverse bicontinuous cubic phases, where reducing chainlength *induces* the cubic phase to appear. In this latter case, the likely explanation is that the chain packing problem of filling the hydrophobic interstices of the H_{II} phase becomes more severe as the chainlength becomes shorter, and the inverse bicontinuous cubic phase offers a partial relief from this, although at the expense of having a less optimal interfacial mean curvature. The opposite behaviour observed for the Fd3m cubic phase is fully consistent with its inverse micellar structure. There is clearly a greater frustration in the packing of the hydrocarbon chains than in the H_{II} phase, in the sense of the degree of variation of the average chainlength which is required to fill the hydrophobic region of the phases uniformly. Thus shortening the chainlength has the effect of favouring the H_{II} phase, where the chain packing problem is less severe.

Acknowledgements

We thank the Science and Engineering Research Council for a CASE award to EAB, and for research grant GR/C/95428 to JMS.

References

1. Mariani, P., Luzzati, V. and Delacroix, H. (1988) J. Mol. Biol. **204**, 165-189

2. Lindblom, G. and Rilfors, L. (1989) Biochim. Biophys. Acta **988**, 221-256

3. Seddon, J. M. (1990) Biochim. Biophys. Acta **1031**, 1-69

4. Ekwall, P. (1975) Adv. Liq. Cryst. **1**, 1-142

5. Helfrich, W. (1981) in Physics of Defects (Balian et al, eds.) pp. 715-755, North-Holland, Amsterdam.

6. Murakami, Y., Kikuchi, J., Takaki, T. and Uchimura, K. (1987) Bull. Chem. Soc. Jpn. **60**, 1469-1479

7. Stenius, P., Palonen, H., Ström, G. and Ödberg, L. (1984) in Surfactants in Solution, Vol. 1 (Mittal, K. L. and Lindman, B., eds.) pp. 153-174, Plenum, New York.

8. Seddon, J. M., Bartle, E. A. and Mingins, J. (1990) J. Phys.: Condens. Matter **2**, SA285-SA290

9. Seddon, J. M. (1990) Biochemistry **29**, 7997-8002

10. Mariani, P., Rivas, E., Luzzati, V. and Delacroix, H. (1990) Biochemistry **29**, 6799-6810

11. Charvolin, J. and Sadoc, J. F. (1988) J. Physique **49**, 521-526

12. Luzzati, V., Vargas, R., Gulik, A., Mariani, P., Seddon, J. M. and Rivas, E. (1991) Biochemistry, in press

13. Hachisu, S. and Yoshimura, S. (1987) in Physics of Complex and Supermolecular Fluids (Safran, S. A. and Clark, N. A., eds.), pp 221-240, Wiley-Interscience, New York.

Swollen Lyotropic Cubic Phases in Fully Hydrated Mixtures of Monoolein, Dioleoylphosphatidylcholine, and Dioleoylphosphatidylethanolamine

R.H. Templer, K.H. Madan, N.A. Warrender, and J.M. Seddon*

Department of Chemistry, Imperial College, London SW7 2AY, UK
*Present address: Reverse Osmosis Plants, Water Supply Directorate, Ministry of Works, Power and Water, Bahrain

Abstract. In this work we have recorded the phase behaviour of mixtures of monoolein (MO), dioleoylphosphatidylcholine (DOPC) and dioleoylphosphatidylethanolamine (DOPE) in excess water, in the temperature range 20-100°C. In this temperature range we have detected the presence of four phase topologies: fluid lamellar, inverse hexagonal, bicontinuous cubic Pn3m and bicontinuous cubic Im3m. By measuring how the cubic phases swell with composition we have attempted to determine the relative magnitudes of the curvature elastic parameters for each component as well as the part each plays in swelling and/or destabilizing the mesophase.

1. Introduction

By using continuum elastic theories to model the curvature free energy of the bilayer one can subsume all of the intermolecular interactions acting in the plane of the bilayer into a few elastic terms: k_M, the elastic splay modulus; k_G, the elastic saddle splay modulus, and c_0, the spontaneous curvature [1]. This approach represents an attractive route for the development of our understanding of the monolayer lateral stresses which we believe drive *bilayer* phase transitions, since both the magnitude and tranverse distribution of chain pressure, interfacial tension and headgroup interactions set the values of k_M, k_G, and c_0.

In this work we have begun to explore the qualitative effects and comparative magnitudes of the above parameters in the pseudo-ternary system MO/DOPC/DOPE. We chose this combination of components because, in excess water, each by itself exhibits a different combination of mean and gaussian curvatures within their predominant phases (respectively cubic Pn3m, L_α and H_{II}). Measured at the middle of the bilayer, MO has zero mean curvature and negative gaussian curvature, DOPC has zero mean and gaussian curvature, and DOPE has negative mean curvature and zero gaussian curvature (although strictly speaking for DOPE we cannot determine the curvature at the hydrocarbon/hydrocarbon interface). By determining the phase behaviour in excess water we allow the curvature component of the total free energy to be fully expressed, and hence, as long as other free energy terms do not vary greatly, we can use continuum elastic theories to interpret the behaviour of this system in bilayer cubic phases.

To this end we have used a development of Helfrich's original theory due to Ljunggren and Eriksson [2]. These authors have shown that in order to obtain an

Springer Proceedings in Physics, Vol. 66 The Structure and Conformation of Amphiphilic Membranes
Editors: R. Lipowsky · D. Richter · K. Kremer

overall minimum for the bending free energy of an infinitely periodic minimal surface (IPMS) that is also thermodynamically stable to size variation, the bending energy must be expanded to at least second order in the surface curvature. We have taken their general expression for the bilayer surface tension and derived an expression for the lattice parameter, a, of an IPMS cubic phase which is expressed in terms of three topological variables and four physical parameters [3].

$$a=\sqrt{\frac{-8\pi\chi_u f}{A_0}}\;\frac{\xi}{\sqrt{k_G/k_M-4c_0\xi}}\,. \tag{1}$$

The four physical parameters, k_G, k_M, c_0 and ξ are all *monolayer* quantities, with ξ being the monolayer thickness. This is especially important as far as c_0 is concerned, since it is this term that reflects the degree of physical frustration being built up across the bilayer [4]. Note that for inverse topologies all the physical parameters, except for c_0, are positive. As for the topological variables, χ_u is the Euler characteristic of the unit cell, A_0a^2 is the IPMS surface area in the unit cell [5] and f is equal to $\langle K^2\rangle/\langle K\rangle^2$ over the unit cell [2].

2. Results

In figure 1 we present our preliminary phase diagrams for 20 and 100°C. We have determined the phase behaviour for 50 mixtures, and used these data to interpolate the phase boundaries.

It is evident that at 20°C the lamellar phase dominates the phase diagram, whereas at 100°C the curved phases dominate. This trend is characterised by the H_{II} phase eating into both the lamellar and cubic phase boundaries. This reduces the region of lamellar phase stability, whilst simply shifting the cubic region towards the DOPC corner.

Over most of the L_α phase region the lamellar spacing apparently varies by no more than 2%, from the value for pure DOPC (63.8Å at 20°C). It is only near the boundary with the H_{II} phase that we have observed a sizeable shrinkage in lattice size. For example, in the case of 75mol% DOPE and 25mol% DOPC we obtain a lamellar spacing of 59.5Å at 25°C. This is in sharp contrast to the distinct and often dramatic changes in the lattice dimensions of the hexagonal and cubic phases.

For the H_{II} phase we find that the lattice is shrunk close to the cubic phase boundary. For example a mixture containing 75mol% of MO and 25mol% DOPE has a lattice parameter of 66.0Å at 25°C compared to 74.3Å for pure DOPE. Conversely close to the L_α phase the addition of DOPC swells the lattice. For example with 25mol% DOPC and 75mol% DOPE we obtain a lattice parameter of 77.3Å at 55°C compared to 68.5Å with pure DOPE.

The most dramatic changes in lattice size are observed in the cubic phase region. Pure MO has a lattice parameter of 98Å at 25°C. With 33.1mol% of DOPC and 9.1mol% DOPE we obtain a Pn3m unit cell at 25°C with sides 207Å in length. This is in coexistence with an Im3m phase which, at 269Å, is even larger. The cubic lattice is swollen both by the addition of DOPC and DOPE, but we observe distinct differences in their effect on the cubic phase. DOPE appears to swell the lattice

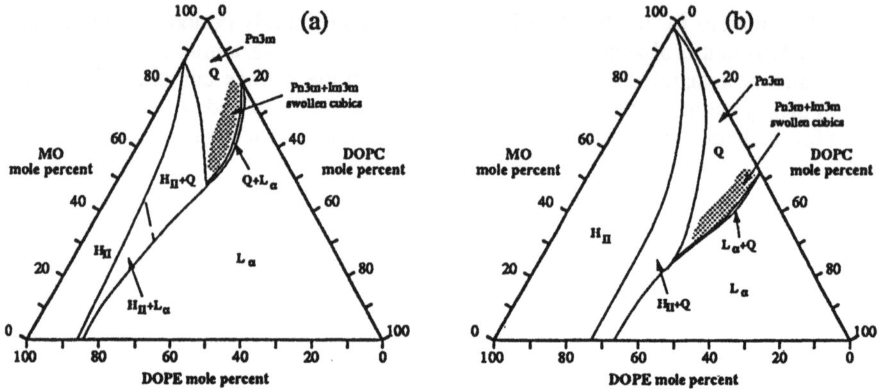

Figure 1. The pseudo-ternary phase diagram of fully hydrated DOPC/MO/DOPE mixtures at (a) 20°C and (b) 100°C.

less rapidly, whilst also destabilizing the cubic phase in favour of an H_{II} topology. With the addition of DOPC we observe a more rapid swelling of the mesophase, with no apparent decrease in the phase stability. This means that the lattice parameter increases more slowly as we travel anti-clockwise from pure MO along the phase boundaries toward the MO/DOPC axis than if we travel toward pure DOPC along the axis itself.

3. Discussion

Using data from the highly swollen cubic phases (where $1/H>2\xi$) and Ljungren and Eriksson's continuum elastic model [2] it should be possible to determine the values of the elastic parameters for each lipid component, so long as the parameters combine linearly. It is our intention to do this in the near future, but even without this we can set comparative limits on these parameters, which we do below for our data at 20°C.

Since the addition of MO to DOPE shrinks the H_{II} lattice, this has the surprising implication that $c_0(\mathrm{MO})<c_0(\mathrm{DOPE})$. Furthermore, since DOPC swells the H_{II} phase, and exhibits no desire for curvature on its own this implies that we have $c_0(\mathrm{DOPE})<c_0(\mathrm{DOPC})$. Using eq. 1 we can see that dilation of the cubic phase will occur as $k_G/k_M-4c_0\xi$ becomes small. With the assumption that ξ does not vary as we change the size of a swollen cubic it can be seen that one reason that the addition of DOPC to MO leads to swollen cubics is the reduction in the net magnitude of c_0. Of course a decrease in k_G/k_M will have the same effect, but an increase of k_M with respect to k_G will end up by promoting the stability of the H_{II} phase with respect to the cubic [6]. This it would appear is what is occuring when DOPE is used to swell the cubic. This tells us that we can inflate the cubic phase stably if we keep k_G/k_M above some threshold level, whilst simultaneously reducing the magnitude of c_0. Another way of expressing this is to say that large k_G stabilizes cubics, whilst large k_M stabilizes H_{II}.

Acknowledgements

This work was funded by the SERC (grant no. GR/F44052) and the Royal Society. RHT wishes to acknowledge the support of the Royal Society in the form of a University Research Fellowship. KHM wishes to acknowledge the support of the Ministry of Works, Power and Water of Bahrain. We thank Professor J.C. Eriksson for presenting us with a preprint of his and S. Ljunggren's article, which has been used as the basis for our interpretation of the experimental results.

References

1. Helfrich, W. (1973) *Z.Naturforsch.* **28c**, 693-703.
2. Ljunggren, S. and Eriksson, J.C. (submitted) *Langmuir*
3. Templer, R.H., Madan, K.H., Warrender, N.A. and Seddon, J.M. (in preparation)
4. Charvolin, J. and Sadoc, J.F. (1987) *J. Physique* **48**, 1559-1569.
5. Anderson, D., Wennerstrom, H. and Olsson, U. (1989) *J. Phys. Chem.* **93**, 4243-4253.
6. Kirk, G.L., Gruner, S.M. and Stein, D.E. (1984) *Biochem.* **23**, 1093-1102.

Surfactant Phases with Bilayer Structures and Their Rheological Properties

C. Thunig, G. Platz, and H. Hoffmann

Institut für Physikalische Chemie I, Universitätsstr. 30,
W-8580 Bayreuth, Fed.Rep.of Germany

The ternary phase diagram Dodecyldimethylaminoxide/Hexanol/Water is reported for low volume fractions of surfactant. Besides the well known L_1, L_α and L_3 phases, three more single phase regions are observed in the phase diagram and called L_1*, L_3* and $L_{\alpha l}$. The L_1* and the L_3*-phases are optically isotropic, slightly turbid and the $L_{\alpha l}$-phase is birefringent and highly viscous. With increasing cosurfactant concentration, the sequence of the phases is L_1, L_1*, $L_{\alpha l}$, L_3*, $L_{\alpha h}$ and L_3. The last five phases have all a bilayer type structure. The L_1* and the L_3* phase consist of single and multilamellar vesicles. The L_3*-phase disappears at higher surfactant concentration and the two L_α- phases border then on each other but seem to be separated by a two phase region. At low volume fractions of about one % both L_α-phases show iridescent colours which are due to interference of white light from the ordered bilayers. The colours of the $L_{\alpha h}$-phase are much brighter than the colours of the $L_{\alpha l}$- phase. The bilayers of the two L_α phases have probably a different thickness. The L_1*, $L_{\alpha l}$ and the L_3*-phases are viscoelastic phases with very long structural relaxation times. For all three phases both, the storage and the loss moduli are about the same and independent of the oscillating frequency over several decades. The $L_{\alpha h}$ and the L_3-phase have a low viscosity. For the L_3-phase the zero shear viscosity is explained on the basis of a simple network model in which η_0 is the product of the shear modulus G_0 and the orientation time τ_s of the bilayers which have a persistence length of the size of the interlamellar spacing of the $L_{\alpha h}$-phase.

1.Introduction

The shape of micellar aggregates and the microstructures which occur in different mesophases are controlled by the spontaneous curvature of the micellar interface [1]. For uncharged single chain surfactants, this curvature can be continously varied by the addition of cosurfactants like intermediate chain alcohols from a strongly convex curved to a planar and finally to a concave interface. As a result of this change in curvature, single chain surfactant solutions undergo several phase transitions with increasing cosurfactant concentration[2]. The sequence of the phases which have been observed and studied in detail are the L_1, the L_α and the L_3 or sponge phase [3,4,5,6,7]. Recently we have observed two more phases in dilute zwitter

Springer Proceedings in Physics, Vol. 66 The Structure and Conformation of Amphiphilic Membranes
Editors: R. Lipowsky · D. Richter · K. Kremer

ionic surfactant systems. They were called L_1* and L_3* in order to distinguish these phases from the normal L_1 and L_3 phase[2,8]. The appearance of these phases depends on the chain length of the surfactant and the cosurfactant which are used for the measurements. In this investigation we present further evidence for the existence and the properties of these phases. We shall show a detailed phase diagram in which the previously known and the new phases occur. We will also present some information on the microstructure of the various phases and in particular we will characterize the rheological properties of the phases. During this investigation we became aware of the existence of two different L_α-phases and a large region in the phase diagram in which the two L_α-phases seem to coexist. All of the reported phases have a bilayer type structure on a local scale but on a longe range the bilayer have a different morphology.

2. Phase Diagrams and Microscopic Structures

In Fig. 1 a section of the phase diagram for the ternary system Dodecyldimethylaminoxide, hexanol, H_2O is shown[9].

In a previous investigation we had reported some data on the phase diagram of Tetradecyldimethylaminoxids. For this system the equilibration time of some of the phases takes a very long time (several weeks) and we thought a shorter chainlength surfactant would lead to a more rapid equilibration. While this is true,we still found areas in the phase diagram in which long times are necessary in order to reach equilibrium. The phase diagram was established by observing the mixtures in calibrated test tubes under temperature control for several weeks. The samples were placed between crossed polarisors in order to check for birefringence. Two phase regions were usually observed between the single phase regions. In many cases, but not in all, the phases separated macroscopically and clear phase boundaries could be distinguished between the phases. The whole multiphase region is of triangu-

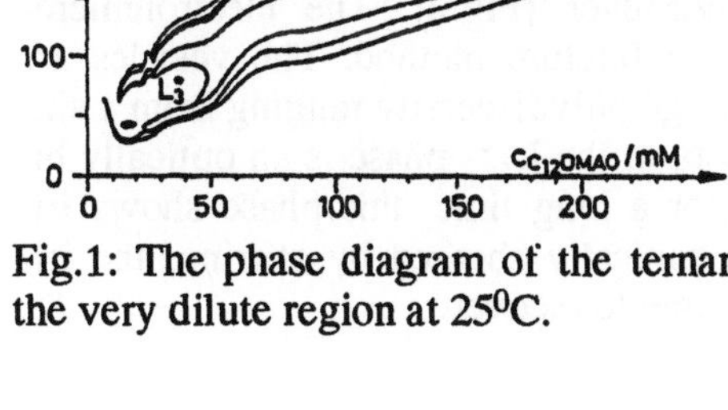

Fig.1: The phase diagram of the ternary system $C_{12}DMAO/C_6OH/H_2O$ in the very dilute region at 25°C.

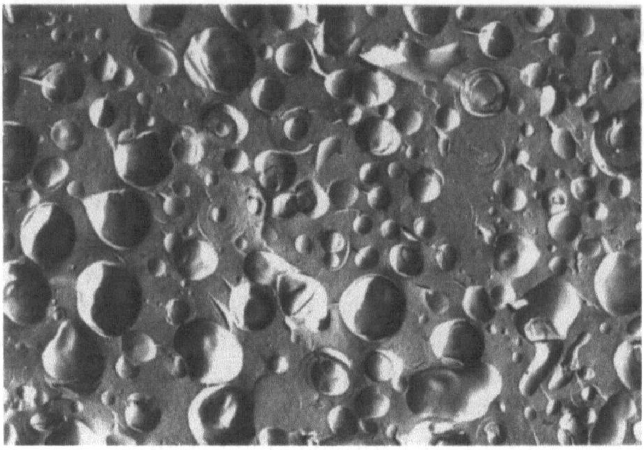

Fig.2: Electronmicrograph of the L_1^* phase of a 50 mmol surfactant solution (magnification 7650 fold)

lar shape. This fundamental shape is determined from the fact that the micellar structure in the controlling phases have a constant cosurfactant/surfactant ratio which is independent in the total concentration of the surfactant. The most complicated behavior is found for small surfactant concentrations. Because of the low concentration and the small density difference between the various phases, macroscopic phase separation can last a long time. To establish exact phase boundaries in this situation, conductivity measurements can be very helpful in which small amounts of salt are added to systems in order to assure a constant background conductivity. Abrupt changes of the conductivity occur at the L_1/L_1^* transition [10]. The L_1^*-phase is an optically isotropic but slightly turbid phase with a very strong flow birefringence. Previously, we had assumed that the phase consists of micellar rods which are in a condensed state. In such a situation all the water and also the salt should be in a continous aqueous phase and the mobility of the ions should hardly be effected by the presence of the small volume fraction of the rods as is the case in the L_1-phase. The drop in the conductivity is a most direct indication that part of the charge carriers do not contribute to the conductivity. Electronmicrographs of the L_1^*-phase clearly show vesicles (Fig.2).

The electronmicrographs were made in Jena at the Inst. for Electronmicroskopie of Prof.Dr.H.W.Meyer (Fig.2). The electronmicrographs were obtained with the freeze fracture method. The vesicles are single and multiwalled. They have a large polydispersity ranging from a size of a few 100 Å to several µm. Adjacent to the L_1^*- phase is an optically birefringent phase ($L_{\alpha l}$). On standing for a long time, this phase shows iridescent colours. If the phase is mechanically sheared by stirring, the iridescence disappears and takes a long time to recover.

Fig.3: Electronmicroscopy of a L_3 phase for a 50 mmol surfactant solution (magnification 27300 fold)

Contrary to the iridescent colours which can be observed in the $L_{\alpha h}$-phase which appears at a high cosurfactant/surfactant ratio, the colours are less bright in the $L_{\alpha l}$ phase [11]. For surfactant concentration which are higher than 50 mM a two phase region exists in which the two L_α-phases coexist.

The two phases do not separate easily on a macroscopic level. This is probably due to the rheological properties of the phases. For surfactant concentrations below 50 mM, part of the two phase $L_{\alpha h}$ /$L_{\alpha l}$ region is taken up by a phase which shows no birefringence and which we previously have called L_3* in recognition of the fact that it occured above a L_α-phase. Finally for the highest cosurfactant/surfactant ratio the usual L_3-phase or sponge sphase appears which has been observed in many systems by now. It should be noted that this phase begins at the solubility line of the cosurfactant in water. From results on the dynamic behavior of this phase we had previously concluded that the phase consists of discrete platelike structures and the diameter of the plates is decreasing linearly with the surfactant concentration[4]. Electronmicrographs show however clearly that the phase consists of a bicontinous structure (Fig.3).

For a 40 mM surfactant solution the phases have the following sequence: L_1, L_1*, $L_{\alpha l}$, L_3*, $L_{\alpha h}$, L_3. We thus note 6 distinct phases when the cosurfactant/surfactant ratio is increased continuously from 1:1 to 3,5:1.

3. Rheological results

In Fig.4 a plot of the viscosity of a 40 mM C_{12}DMAO solution is shown against the hexanol concentration. Up to the phase boundary of the 2 phase region L_1, L_1* the viscosity remains very low. The small increase is due to

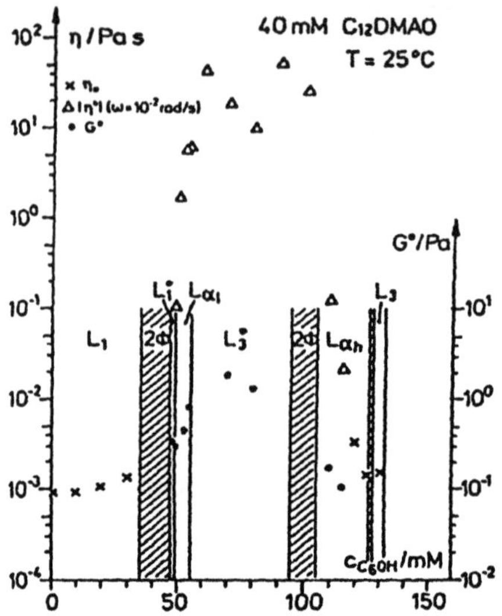

Fig. 4 : The magnitude of the complex viscosity $|\eta^*|$ ($\omega=10^{-2}$rad/sec), the zero shear viscosity η_0 and the storage modulus G_0 as a function of the hexanol concentration for a 40 mmol C_{12}DMAO solution at 25°C.

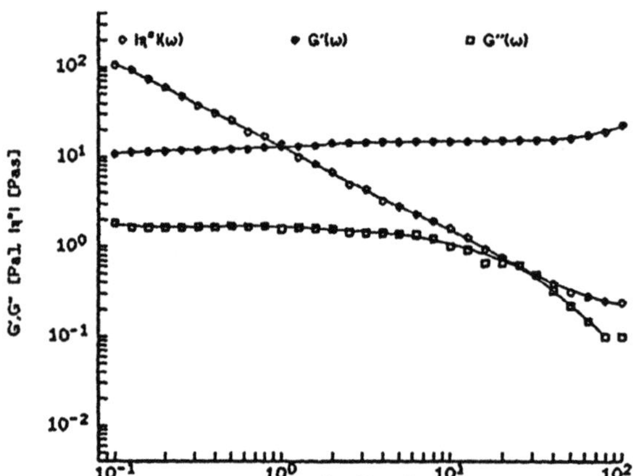

Fig.5: The complex viscosity $|\eta^*|$, the loss modulus G" and the storage modulus G' of a 100mmol C_{14}DMAO and 100 mmol C_6OH as a function of the angular velocity ω at 25°C.

the growth of globular micelles to rodlike micelles. Within the narrow L_1*-region the viscosity increases several orders of magnitude.Actually, the viscosity in this region depends on the oscillating frequency for the measurements. A typical curve is shown in Fig.5 where the complex viscosity is plotted against the frequency in a double log plot. The rheological results of the L_1*-phase look like the results of systems with a yield stress value. For such systems the applied shear stress must be higher than the yield value in order to cause flow. For a lower shear stress, the systems behave like a solid with an indefinitely long time constant. The L_1* phases do not have a yield stress value but have a finite long structural relaxation time. Small bubbles which are dispersed in the phase, rise within a day or two, even though they seem stationary on a small time scale. The viscosity which was plotted in Fig.4 is the complex viscosity at ω=O,O1 Hz. The dynamic viscosity decreases linearly with the frequency over several decades of the frequency(Fig 5).

Such a behavior is typical for a network system for which the product of $\omega * \tau > 1$. The slope of the double log plot is however somewhat smaller than one. A further characteristic feature of the rheological results of the L_1* phase is that the storage and the loss modulus are about the same and depend relatively little on the frequency. The storage modulus is usually somewhat lower than the loss modulus. This means that the system responds elastically but shear waves will always be strongly damped. It can be concluded from the rheological data that the L_1*-phase must consist of particles or structures which are linked together to a superstructure.

If the aggregates in the phase would be isolated from each other the phase should behave like a Newtonion liquid. Since it is known from the electronmicrographs that the phase consists of vesicles, it must be concluded that the vesicles are in contact and form a network in the phase. The network does however not behave like an entanglement network which can be stretched strongly without a noticable decrease of G_o. The storage modulus of the L_1*-phase begins already to decrease at a deformation of 10%. Microscopically this means that the contacts in the network are broken.

The elasticity of the system does not seem to arise from entropy, but seems to come from energy change. The storage modulus of the L_1*-phase depends very strongly on the total concentration of the surfactant. The rheological data for the $L_{\alpha l}$ and the L_3*- phase look very similar as the results for the L_1* phase (Fig.6 and Fig7).

In particular situations G' and G" have the same values over four decades in frequency. The similarity of the rheological data for the L_1*, the $L_{\alpha l}$ and the L_3*-phase is a strong indication that the structures in the phases are very similar on a local scale. We thus can conclude that the L_3* phase consists also of multilamellar vesicles which are still small enough in diameter that the systems shows no birefringence. Since for a higher cosurfactant/surfactant ratio we find a second L_α- phase we can also consider the L_3*-phase to be the L_1*-phase for the $L_{\alpha h}$-phase. The $L_{\alpha h}$- phase has a much lower viscosity and a shorter structural relaxation time as the $L_{\alpha l}$- phase. It seems appropriate to look upon the upper $L_{\alpha h}$ phase in the phase diagram as

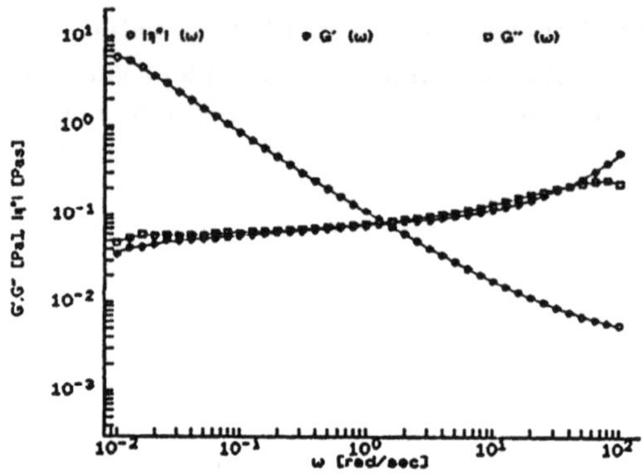

Fig.6: The complex viscosity |η*|, the loss modulus G" and the storage modulus G' of a 40mmol C_{12}DMAO and 55 mmol C_6OH as a function of the angular velocity ω at 25^{0}C.

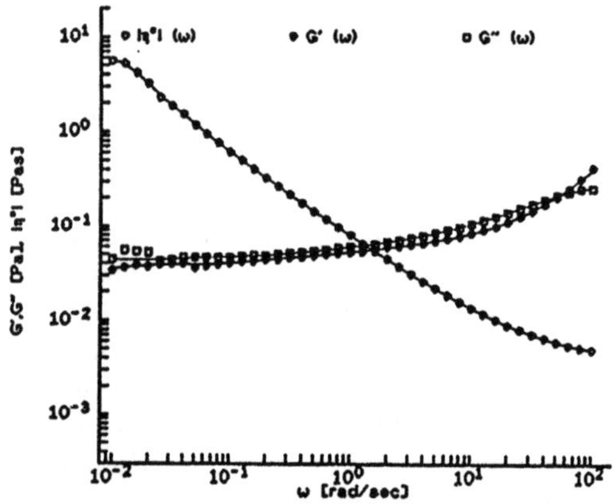

Fig.7: The complex viscosity |η*|, the loss modulus G" and the storage modulus G' of a 40mmol C_{12}DMAO and 60 mmol C_6OH as a function of the angular velocity ω at 25^{0}C.

the L_α-phase of the cosurfactant while the lower L_α-phase is the lamellar phase of the surfactant. It is likely that the bilayer of the two phases have a different thickness and also different bending constants. The L_3- phase has a viscosity in the range of the solvent viscosity. It increases approximately linearly with the volume fraction of the surfactant. For small shear rates the phase behaves like a Newtonion fluid. It undergoes however a structural reorganisation for shear rates above a critical shear rate and is transformed into the L_α-phase which shows the iridescent colours when the volume fraction is around 1% [12]. Such a shear induced L_α-phase can remain many days in this state when the shear is stopped. We have observed samples for which it takes a week for the shear induced state to disappear. This is very remarkable in view of the fact that the phase contains so much hexanol which labilizes all dynamic processes. It is also very remarkable with respect to the short structural relaxation times in the L_3-phase in the range of msec to μsec. Obviously this means that the L_3-phase can relax with this short time constants when it is deformed by electric or shear fields in the linear viscoelastic range. When it is deformed strongly, it switches however to a metastable state which behaves completely differently.

For a given cosurfactant, the macroscopic properties of the L_3-phase including the viscosities, are remarkably insensitive to the chainlength of the surfactant. The absolute values of the viscosities and the concentration dependence were explained on the basis of a model which assumed discrete disks for the L_3-phases whose diameter is about the same as the interlamellar spacing of the L_α-phase [4]. In this model the zero shear viscositiy can be looked at as the viscosity of a viscoelastic solution and should be given by $\eta_0 = G_o * \tau$, where τ is the structural or rotational time constant of the disks and G_o the shear modulus which is given by the number density of the disks ($G_o = \nu kT$). Both, the modulus and the time constant vary with the third power of the volume fraction. Since the structural relaxation time varies inversely, the two effect compensate and the viscosity varies very little with the volume fraction. This model can still be used for the interpretation of the viscosity data. Since it is clear however by the Electronmicrograph that the L_3-phase is indeed a bicontinous phase, the meaning of the structural relaxation time has to be modified somewhat. It can be looked upon as the structural relaxation time of a folded bilayer with a persistence length l_p which corresponds to the diameter in the disk-like model. A more detailed interpretation of the dynamic properties of the L_3-phase was given by M. Cates during this conference [13].

Acknowledgement:

The authors of this article would like to thank very much Prof. H.W. Meyer and W.Richter from the Friedrich-Schiller-Universität in Jena for the preparation of the electronmicrographs.

References :

1] Wang,Z.G, Safran S. A. (1990), Europhys. Lett., 11 (5), 425-430

[2] Hoffmann H., (1990), Progr. Colloid Polym. Sci. 83, 16-28

[3] Porte G., Appell J., Bassereau P., Marignan J., (1990), J. Phys. France, 50, 1335-1347

[4] Miller C. A., Gradzielski M., Hoffmann H., Krämer U., Thunig C., (1990), Colloid Polym. Sci. , 268, 1066-1072

[5] Balinov B., Olsson U., Söderman O., (1991), J. Phys. Chem., 5931-5936

[6] Cates M. E., Roux D., (1990), J. Phy. Condens. Matter, 2, SA 339

[7] Cates M. E., Roux D., Andelman D., Milner S. T., Safran S. A., (1988), Europhys. Lett., 5 (8), 733-739

[8] Miller C. A., Gradzielski M., Hoffmann H., Krämer U., Thunig C., (1991), Progr. Colloid Polym. Sci., Proceedings of the ECIS meeting in Catanzarro, Sept. 199O, 84, 1

[9] A detailed paper is in preparation

[10] Unpublished results

[11] Thunig C., Hoffmann H., Platz G., (1989), Progr. Colloid Polym. Sci. , 79, 297-307

[12] Hoffmann, H., Hofmann, S.,Rauscher A.,Kalus, J. (1991), Progr. Colloid Polym. Sci.,84,1

[13] Cates M. E., contribution during this conference

Sponge Phases

M.E. Cates

Cavendish Laboratory, Madingley Road, Cambridge CB3 0HE, UK

Abstract. The L_3 phase of surfactant solutions appears to consist of two interpenetrating solvent domains separated by a continuous bilayer film. The hamiltonian of the system is symmetrical with respect to interchange of the two solvent domains. This imparts a special Ising-like symmetry over and above the usual translational symmetry of the liquid state. The symmetry may be spontaneously broken at a phase transition whose proper description requires the introduction of two order parameters: one describing the special symmetry and one describing the fluctuating density of surfactant. A theory based on these ideas can explain the unique small angle light scattering signal observed in these materials. Our understanding of the *dynamics* of sponge phases is in a less mature state. Some preliminary ideas concerning the dynamic structure factor $S(q,t)$, which can be calculated within the same model, are briefly discussed.

1. Introduction

In many bilayer-forming surfactant systems, the lyotropic smectic phase (L_α) is stable under dilution to a layer spacing of thousands of nanometres. In these cases, the state which is ultimately reached on further dilution is not a micellar phase, but a dilute phase of randomly connected bilayers or "sponge phase" (L_3). There is now decisive evidence, from conductivity and other studies in several different systems [1-3] that the L_3 phase is bilayer-continuous. This rules out alternative models of L_3 which invoke a structure of large discs [4]. The pore size of the sponge is typically in the range of 10-100 nm.

Apart from conductivity, there is another special property that arises from having a continuous web of bilayer percolating through the system. This comes from the fact that a continuous surface, without line defects consisting of either seams or edges [5], partitions space into exactly two disjoint regions. These two regions can be thought of as defining two different species of solvent, and can be arbitrarily labelled the "Inside" (I) and the "Outside" (O) of the dividing surface.

With this definition, two distinct types of spongelike structure can be envisaged: a *symmetric* sponge, in which the statistics of I and O regions are equivalent, and an *asymmetric* sponge in which they differ. By varying the relative amounts of I and O, a continuous family of asymmetric sponge phase can be constructed that connects smoothly to a phase of disconnected vesicles. However, the symmetric state cannot be so connected without first breaking the symmetry at a phase transition [5,6]. The I/O symmetry is Ising like: its breakdown can be compared to the Curie transition of a ferromagnet.

The I/O symmetry is an exact property of the coarse-grained Hamiltonian of the system, which emerges when the self-assembly of molecules into continuous sheets is considered. It is obvious that in practise the bilayer structure is not perfect but is (for example) punctured by holes in certain places. A very important result of Huse and Leibler [7] is that defects of this kind do not destroy the Ising like I/O symmetry unless the defect density exceeds a critical value (perhaps corresponding to some sort of percolation transition on the infinite random membrane). Thus, although the simple partitioning of space into two parts cannot be performed in the presence of holes, the underlying symmetry, and resulting phase transitions, still remain.

The transition from symmetric to asymmetric sponge is one of the most interesting features of sponge phases. Since the exact symmetry of the underlying hamiltonian is present throughout the sponge region of the phase diagram, and not just at some special point (as would arise for example with a gas-liquid critical point) one can expect unusually rich critical behaviour. For example, in a binary system of bilayer-forming surfactant there can be not just an isolated critical point on the T, ϕ plane (temperature, volume fraction plane) but a continuous line of such points at which the symmetry is broken. This has now been observed, not in a T, ϕ plane but in a two dimensional cut through a higher-dimensional parameter space for a multicomponent system [8,9]. The chances of this happening by accident are almost zero: the observation provides strong evidence that the structure of the L_3 phase indeed consists of a continuous random bilayer that divides space.

2. Static Light Scattering

The static light scattering signal from the L_3 phase has a very strange dependence on scattered wave vector q. It does not take a simple Ornstein Zernicke form but instead follows an apparent $1/q$ dependence in a certain range. This unusual property can be understood in terms of the underlying I/O symmetry of the system [10,9]. Our description is based on the following Landau Ginzburg Hamiltonian that describes the fluctuations of the system close to the transition at which I/O symmetry is spontaneously broken:

$$\Delta\Phi = \mu\rho + \lambda\rho^2 + \nu\rho\eta^2 + \beta\eta^2 + \gamma(\nabla\eta^2) + \tau(\nabla\rho)^2 + \zeta\nabla\rho\nabla\eta \tag{1}$$

Here $\Delta\Phi$ is the thermodynamic potential; μ the chemical potential; $\lambda, \nu, \beta, \gamma, \tau, \zeta$ are phenomenological constants; and ρ and η two fluctuating order parameters. The first of these represents the density of surfactant molecules, whereas the second it an order parameter which represents the local amount of asymmetry between I and O regions. The expansion (1) is symmetric in η since the Hamiltonian does not distinguish I from O.

From Eq.(1), the light scattering signal can be calculated analytically so long as the last two terms are dropped [10]. The signal comes entirely from the ρ fluctuations since the I/O order parameter η has no contrast in the signal (because I and O regions are made of identical solvent). The result is

$$I(q) = A[B + 2\arctan(q\xi/2)/q\xi] \tag{2}$$

a function which has been found to fit very well the data on several systems [10]. Note that $I(q) \sim q^{-1}$ as $\xi \to \infty$ which is precisely the novel q-dependence mentioned above. The fact that (2) is confirmed experimentally provides very strong evidence for a structure in which a bilayer divides space. To obtain this form, one must have a hidden Ising-like transition in an invisible order parameter (η) to which the surfactant density is quadratically coupled. It is hard to see how this could arise in (say) a phase of discs.

In fact, the neglect of terms in $\nabla\rho$ in (1) is not always justified and various approximate calculations have been devised to deal with them [9,11]. In particular, these terms become important close to a tricritical point, which seems to be relevant for the system SDS/dodecane/pentanol bilayers in water [9]. That system shows scattering which is intermediate between (2) and ordinary Ornstein-Zernicke behaviour, a feature which has been analyzed in terms of tricriticality. There is every reason to expect tricritical behaviour in sponge phases, due to the exceptionally high symmetry of system [10].

3. Dynamics

The existence of a percolating bilayer film in the sponge phase might lead one to expect a strongly viscoelastic character for the phase. However, this does not seem to be the case experimentally; in particular, Snabre and Porte have reported a simple newtonian behaviour for aqueous symmetric sponge phases [12]. It is very interesting that the viscosity of the phase is not simply that of the solvent, but extrapolates to several times the solvent viscosity as the limit of $\phi \to 0$ is taken. This suggests that the bilayers distort streamlines locally and enhance the dissipation rate by a purely geometrical factor [12]. It had previously been suggested that the viscosity would be enhanced by a slow process involving the fusion and rupture of membranes [13] which, one might expect, should be a limiting factor in the flow of the material. However, the data suggest that either such processes are in fact rapid, or that flow can be accomodated by fluid motion within the bilayer, without any need for changes in film topology. This issue remains open theoretically. With the benefit of hindsight, it should be no surprise that a structure containing a continuous fluid film (here a bilayer) can be relatively inviscid: this has long been known for bicontinuous microemulsions where the film is a monolayer separating domains of oil and water. It is possible that the topological rearrangement time shows up in other sorts of dynamics experiment, such as temperature jump [14].

4. Flow Induced Phase Transitions

It was suggested first theoretically by Cates and Milner [15] that the sponge phase L_3 could undergo a shear-induced phase transition to the lamellar phase L_α. The theory has since been developed further to include possible hexagonal phases [16], and to incorportate the effect of having two fluctuating order parameters as described in Sect.2 above [17]. This transition was invoked to explain the extreme flow birefringence of the L_3 phase [18-20] which has a strongly nonlinear dependence on flow rate and which seems to show hysteresis. Thus a sample which is gently shaken in a test tube can become birefringent in a manner which persists for seconds or, in favorable cases, much longer than this. Preliminary reports are that the transition is indeed observed under carefully controlled (Couette) flow conditions [19,20], and full results from a very detailed study should be available soon [21]. The interpretation can be complicated since the flow-rate dependence of the birefringence even in a pure lamellar phase is nontrivial [21].

5. Dynamic Light Scattering

It is obviously of interest to predict the dynamical structure factor $I(q,t)$ for a sponge phase, which can be measured by photon correlation spectroscopy. As for the static case (see Sect.2 above), the observable correlations are those in the local surfactant density ρ. The unusual static scattering behavior arises since these are coupled to a second fluctuating order parameter η which can develop very long-range correlations near the critical line at which the I/O symmetry is spontaneously broken.

The same remarks clearly apply to the dynamic case, but the position is more complicated. The surfactant order parameter is a conserved hydrodynamic variable (it can only relax by a diffusive process), whereas the I/O order parameter might be conserved (in a system of perfectly unpunctured bilayers) but in practise is probably not conserved at long wavelengths due to the presence of defects in the film. Another "pseudoconserved" quantity, which could be relevant in some experiments, is the Euler character or density of handles in the bilayer, which might only relax on some long time scale associated with topological rearrangement. The interplay of these various hydrodynamic modes is discussed qualitatively in Ref.[13].

In fact, quite a detailed treatment is possible [22] and can be based on a generalization of the two order-parameter theory of Sect.2, which successfully describes the static limit. The dynamical description is based on two coupled Langevin equations in which the free energy (1) plays the role of a driving force. In this picture, many different regimes can arise, depending on the relative magnitudes of the gradient terms in (1) and on the relevant Onsager coefficients. However, a clear preliminary result is that, because ρ is coupled to the *square* of the I/O order parameter η, there can be a substantial q-range for which the time-dependent correlator $I(q,t) = \langle\rho(q,t)\rho(-q,0)\rangle$ has a strongly nonexponential lineshape, even if the decay of η, the primary order parameter, is exponential.

This departure from single exponential decay is the dynamical analogue of the departure from static Ornstein Zernicke scattering represented in Eq.(2), and has precisely the same physical origin. The details of this nonexponential dependence remain the subject of our current research [22]. In the mean time, it seems likely thatexperimental studies of the dynamic structure factor $I(q,t)$ may play an important role in advancing our understanding of sponge phases, and we suggest that such studies would be very timely at present.

Acknowledgements: This work was funded in part under EC Grant SCI 0288 C. Much of the work reviewed above was done in collaboration as detailed in the references. I am grateful to all my colleagues for their assistance.

References

[1] For a review, see: D. Roux and M. E. Cates, Springer Proc. Phys., **52**, 19 (1990) (Dynamics and Patterns in Complex Fluids, Eds. A. Onuki and K. Kawasaki, Springer, Heidelberg.)
[2] D. Gazeau, A. M. Bellocq, D. Roux and T. Zemb, Europhys. Lett. **9**, 447 (1989).
[3] G. Porte, J. Appell, P. Bassereau, J. Marignan, J. Physique **50**, 1335 (1989).
[4] C. A. Miller, M. Gradzielski, H. Hoffmann, U. Kraemer and C. Thunig, Colloid Polym. Sci.,in press.
[5] M. E. Cates, D. Roux, D. Andelman, S. T. Milner and S. A. Safran, Europhys. Lett. **5**, 733 (1988)
[6] D. Huse and S. Leibler, J. Physique **49**, 605 (1988).
[7] D. Huse and S. Leibler, Phys. Rev. Lett. **66**, 437 (1991).
[8] C. Coulon, D. Roux, and A. M. Bellocq, Phys. Rev. Lett. **66**, 709 (1991).
[9] C. Coulon, D. Roux and M. E. Cates, J. Phys. Chem., to be published.
[10] D. Roux, M. E. Cates, U. Olfsson, R. C. Ball, F. Nallet and A. M. Bellocq, Europhys. Lett. **11**, 229 (1990).
[11] R. Granek, unpublished results.
[12] P. Snabre and G. Porte, Europhys. Lett. **13**, 641 (1990).
[13] S. T. Milner. M. E. Cates and D. Roux, J. Physique **51**, 2629 (1990).
[14] G. Porte, private communication.
[15] M. E. Cates and S. T. Milner, Phys. Rev. Lett. **62**, 1856 (1989).
[16] C. M. Marques and M. E. Cates, J. Physique **51**, 1733 (1990).
[17] M. E. Cates, Physica A, in press.

[18] D. J. Mitchell, G. J. T. Tiddy, L. Waring, T. Bostock, and M. P. McDonald, J. Chem. Soc. Faraday Trans. I, **79**, 1975 (1983); J. C. Lang and R. D. Morgan, J. Chem. Phys. **73**, 5849 (1980); P. G. Nilsson and B. Lindman, J. Phys. Chem. **88**, 4764 (1984); C. A. Miller, O. Ghosh, Langmuir **2**, 321 (1986).
[19] H. Hoffmann, S. Hofmann, A. Rauscher and J. Kalus, Prog. in Colloid Polym. Sci., in press.
[20] D. Roux and O. Diat, private communication.
[21] O. Diat and D. Roux, to be published.
[22] R. Granek and M. E. Cates, work in progress.

Structural Inversion Processes in Three-Component Ionic Microemulsion Studied by Small Angle Neutron Scattering

S.H. Chen[1], *S.L. Chang*[1], *R. Strey*[2], *and P. Thiyagarajan*[3]

[1]Department of Nuclear Engineering and Center of Materials Science and Engineering, 24–211, Massachusetts Institute of Technology, Cambridge, MA 02139, USA

[2]Max-Planck-Institut für Biophysikalische Chemie, Postfach 2841, W-3400 Göttingen, Fed.Rep.of Germany

[3]Intense Pulsed Neutron Source, Argonne National Laboratory, Argonne, IL 60439, USA

Abstract. The three–component microemulsion system, consisting of AOT/water(.6wt% NaCl) /decane, shows a typical $\bar{2}\phi$–1ϕ–$\underline{2}\phi$ progression at moderate surfactant weight fraction (γ=8%) when temperature of the system is raised through the hydrophile–lipophile balanced temperature. At low oil–water weight fraction (α=10%), it exhibits a structural inversion from L_3–phase (water–in–oil microemulsions) at low temperatures to L_1–phase (oil–in–water microemulsions) at high temperature through L_α–phase (lamellar microemulsions) in the intermediate temperatures within the one–phase channel. At equal oil–water volume fraction (α=40%), on the other hand, the inversion takes place through an intermediate disordered bicontinuous structure. We investigated these different inversion pathways for different α values using SANS intensity distributions. The thickness and polydispersity of surfactant bilayers in the L_3 and L_α phases and surfactant monolayers in the bicontinuous phase are measured. In particular, the structural evolution as a function of α from the bicontinuous structure to the L_3 structure has been explored.

1. Introduction

We have been investigating the microstructure of three–component mixtures of AOT (an ionic surfactant sodium–bis–ethylhexylsulfosuccinate), water (with .6wt% of NaCl added), and decane in regions of phase diagram where the structure can be described as bicontinuous [1–4]. In a bicontinuous microemulsion, multiply connected randomly oriented monolayer of surfactant sheets separate water– and oil–rich subdomains. Both the water and the oil domains are continuous, as evidenced by electrical conductivity and NMR self–diffusion experiments [1,5]. For some nonionic systems, freeze fracture electron micrographs do show water and oil domains forming an interpenetrating network [6]. The bicontinuous microemulsions are normally found in the region of phase diagram where the surfactant concentration is low and the water and oil volume fractions are comparable. However, when the water–oil volume fractions become highly asymmetric, the bicontinuous microemulsion is found to transform to another type of

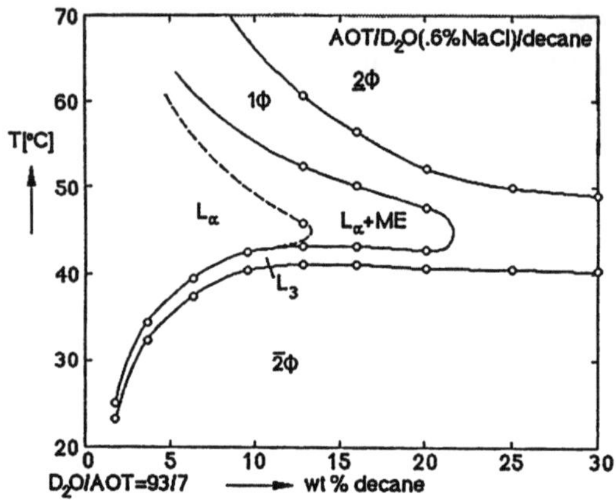

Fig.1 Phase diagram of AOT/D_2O(.6wt% NaCl)/decane microemulsion system in the T–α plane at γ=8%.

random structure known as a L_3–microemulsion. In the L_3–phase a multiply connected randomly oriented surfactant bilayer, incorporating the oil in–between, divides space into two equivalent water continuous subdomains [7,8]. In this paper, we report a preliminary result of the study of continuous evolution of the bicontinuous microemulsions to the L_3 microemulsions when the oil weight fraction is varied from α=30% down to α=10% (α=wt.oil/(wt. oil + wt. water)). We also summarized the result of a separate investigation of the structural inversion process: L_3–L_α–L_1, at α=10% as the temperature is raised [9]. Fig.1 shows the phase diagram in T–α plane. At α=30% the one–phase channel spans a temperature interval of 40°C to 50°C, on the other hand, at α=10% the one–phase channel splits into a lower channel, occupied by the L_3–phase, and an upper channel, occupied by the L_1–phase, interrupted in–between by a L_α–phase extending from 42°C to 50°C.

2. Data Analysis

We shall analyze SANS intensity distributions using two models. The first one focus attention on large Q portion of the intensity distributions where the data are sensitive only to the local structure of surfactant sheets. the second one is applicable to the description of the global structure of the oil–water domains. This means that we are attempting to model the scattering intensity at the entire Q range.

2.1 The Asymptotic Model

When Q is large compared to 2π/(correlation length of the microstructure) in a microemulsion, the scattering intensity distribution will reflect the Fourier transform of the shortest distance scale microstructure. For the interface contrast case of bicontinuous microemulsions and also for L_3 microemulsions, one would then be probing the thickness of the monolayers or the bilayers in the respective cases.

The scattering intensity for a system of homogeneous, randomly oriented monolayer or bilayer of an average thickness $\bar{t}$ suspended in a solvent (this can be a mixture of D_2O and deuterated decane, in the monolayer case, or just D_2O, in the bilayer case), in the Q range where the interference effect is negligible, can be written as [9]:

$$I(Q) = 4\pi\,\varphi\,(\Delta\rho)^2\,Q^{-4}\,\left(\frac{Z+1}{Z}\right)\frac{1}{\bar{t}}\,[1-(\cos\psi)^{Z-1}\cos(Z\psi)], \tag{1}$$

where φ is the volume fraction of the mono– (or bi–) layers, $\Delta\rho$ the contrast between the layer and the solvent, and Z is related to the polydispersity parameter of the layer thickness by $\Delta t/\bar{t}=(Z+1)^{-1/2}$. The angle ψ is given by $\psi=\tan^{-1}[Q\bar{t}/(Z+1)]$. The asymptotic intensity I(Q) given by expression eq.(1) shows a crossover from a Q^{-2} to Q^{-4} behavior occurring at $Q\bar{t}\approx10$. This model has been shown to fit SANS intensities for the interface contrast case of bicontinuous microemulsions, having α=40% and γ=12%, giving an average monolayer thickness $\bar{t}=10.7\pm0.3$Å, which is slightly shorter than the fully stretched AOT tail length of 12.5Å, and the polydispersity $\Delta t/\bar{t}$=56% [9].

2.2 The Generalized Teubner–Strey Model

This model applicable only to the bulk contrast case, is a generalization of the original T–S model [10]. It can be written as:

$$I(Q) = (\Delta\rho)^2\,\varphi_1\,\varphi_2\,\frac{8\pi/\xi}{a^2-2bQ^2+Q^4}, \tag{2}$$

where φ_1 and φ_2 are respectively the water and oil volume fractions. In practice, the AOT molecules are partitioned between water and oil, with the head group incorporated into the water and tails into the oil. The model is derived from an assumption that the microstructure of bicontinuous microemulsions can be characterized by two length scales, $d=2\pi/k$ and ξ, through a Debye correlation function of the form: $\Gamma_{TS}(r)=j_0(kr)\exp(-r/\xi)$. The two constants appearing in eq.(2) are given in terms of the two lengths as: $a=(k^2+1/\xi^2)$, and $b=k^2-1/\xi^2$. Originally the model was introduced for the case b>0 only. We shall generalize the model to include the case $b\lesssim0$ also. Note that this latter case corresponds to the ratio $d/2\pi\xi>1$.

3. Results

As stated in the introduction, the L_3 microemulsions consist of oil–swollen surfactant bilayers dispersed in water. We shall first show this evidence by comparing a SANS intensity distribution of a L_3–microemulsion, in the bulk contrast, with that of a

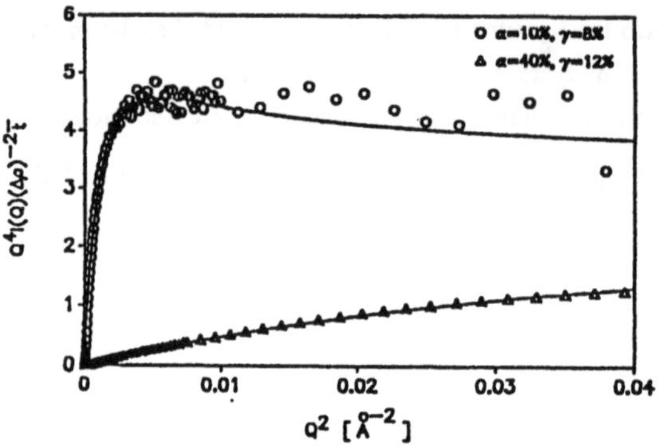

Fig.2 Plots of the dimensionless intensity $Q^4I(Q)(\Delta\rho)^{-2}\bar{t}$ vs. Q^2. Open circles represent data from a sample with a bulk contrast in the L_3–phase, having α=10% and γ=8%, at 38°C. Open triangles represent those from monolayer of surfactant sheets in a sample with an interface contrast, having α=40% and γ=12%, at 41.5°C. Solid lines are fitted curves using the asymptotic model.

disordered bicontinuous microemulsion, in the interface contrast. Fig.2 shows plots of the L_3–microemulsion at α=10%, γ=8% and T=38°C, and the bicontinuous microemulsion at α=40%, γ=12% and T=41.5°C. According to the asymptotic model, eq.(1), the dimensionless intensity should be a function of variables $Q\bar{t}$, Z and the volume fraction φ of the surfactant sheet (including oil in the L_3 case). The plots should have an initial linear slope at small Q, reflecting Q^{-2} behavior and gradually approaches a constant value, reflecting Q^{-4} behavior of the intensity. Solid lines in the figure are the best fits by the asymptotic model. The model parameters we obtain are: $\bar{t}$=69.9Å, $\Delta t/\bar{t}$=53%, for the L_3–microemulsion and $\bar{t}$=10.9Å, $\Delta t/\bar{t}$=56%, for the bicontinuous microemulsion. $\bar{t}$=69.9Å for the bilayer thickness is reasonable if we assume that the packing densities of surfactant tails and oil molecules are about the same in the oil–swollen bilayer. The ratio of the two volume fractions is $\varphi_{ST}:\varphi_0$=1:2, which is consistent with the bare thickness of the surfactant bilayer being 22Å and that of oil 48Å, considering the skewing effect of the polydispersity of 53%.

Turning next to the structural inversion across the one–phase channel at α=10%, γ=8%. We fit the data, in an absolute scale, using the generalized T–S model. The model parameters are given in Table I. The parameter d is the interdomain distance of two neighboring water–rich regions in the L_3 and L_α cases, and it is two times the average diameter of the oil droplets in the case of L_1–microemulsion (see ref.9 for a detailed explanation). The parameter $d/2\pi\xi$, a measure of the polydispersity of the

Table I

Generalized T–S model parameters for AOT/D_2O(.6wt% NaCl)/H–decane sample

γ	α	Temp($^\circ$C)	d(Å)	ξ(Å)	2b ($\times10^{-4}$)	$\frac{d}{2\pi\xi}$
8	10	38	522	62.7	–2.18	1.32
8	10	46	359	67.6	+1.75	0.85
8	10	65	243	71.7	+9.50	0.54
8	20	43	458	69.0	–0.66	1.06
8	30	43	504	96.6	+0.49	0.83

domain sizes in the case of bicontinuous microemulsions [2], can be taken to be a measure of disorder of the microstructure in this low α case. We then see a trend that the disorder decreases as the temperature increases. However, the most important point to make here is that the coefficient b changes sign from L_α to L_3. This is the consequence of the disappearance of the peak in the intensity distribution.

Lastly, we present results of a T–S model analysis of a series of three microemulsions with γ=8% but with α=10%, 20% and 30%. Fig. 3 gives the I(Q) vs. Q plots in log–log scale. The T–S parameters are listed in Table I as well. Two points should be noted. The polydispersity parameter $d/2\pi\xi$, is generally between 0.5 and 1.0, for the disordered bicontinuous microemulsions investigated by us before [3]. Firstly, for the sample at α=30%, $d/2\pi\xi$=0.83, which is in this range. But for α=10% and 20%, the values of the parameter are out of this range. Secondly, the parameter b is seen to be positive for α=30% but is negative for α=10% and 20%. We have thus a suggestive indication that within the lower channel the transition from bicontinuous microemulsion to L_3–microemulsion may be occurring at about α=20% or somewhat less than this value.

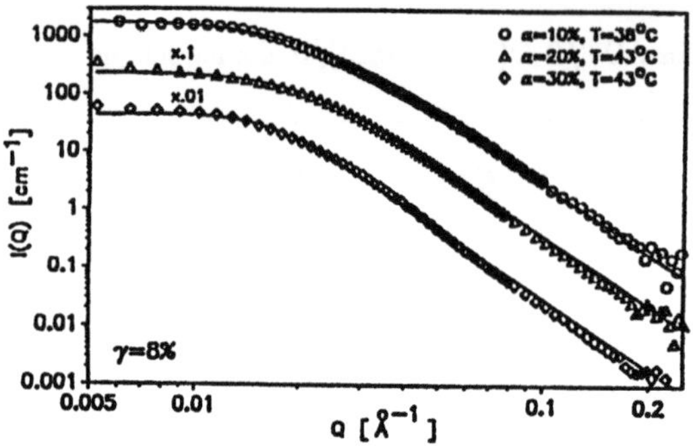

Fig.3 SANS intensity distributions I(Q), for bulk contrast samples having γ=8%, and three α–values, 10%, 20% and 30%. Solid lines are the calculated results using the generalized T–S model.

This research is supported by grant from the Division of Materials Science of the Department of Energy, DE–FG01–90ER45429.

References

1. S.H. Chen, S.L. Chang, and R. Strey, J. Chem. Phys. 93, 1907(1990).
2. S.H. Chen, S.L. Chang, and R. Strey, Progr. Colloid Polym. Sci. 81, 30(1990).
3. S.H. Chen, S.L. Chang, and R. Strey, "Simulation of Bicontinuous Microemulsions: Comparison of Simulated Real Space Microstructure with Scattering Experiments", to appear in J. Appl. Crystal. (Oct. 1991).
4. S.H. Chen, S.L. Chang, R. Strey, J. Samseth, and K. Mortensen, "Structural Evolution of Bicontinuous Microemulsions", to appear in J. Phys. Chem. (Oct. 1991).
5. U. Olsson, K. Shinoda, and B. Lindmann, J. Phys. Chem. 90, 4083(1986).
6. W. Jahn and R. Strey, J. Chem. Phys 92, 2294(1988).
7. D. Gazeau, A.M. Bellocq, T. Zemb, and D. Roux, Europhys. Lett. 9, 447(1989); G. Porte, J. Appell, P. Bassereau, J. Marignan, J. Phys. France 50, 1335(1989).
8. R. Strey, R. Schomacker, D. Roux, F. Nallet, and U. Olsson, J. Chem. Soc. Faraday Trans. 86, 2253(1990).
9. S.H. Chen, S.L. Chang, R. Strey, and P. Thigagarajan, "SANS Investigation of Structural Inversion in a Three–component Ionic Microemulsions", to appear in J. Phys.: Condensed Matter (1991).
10. M. Teubner and R. Strey, J. Chem. Phys. 87, 3195(1987).

The L_3 Phase Microstructure in the AOT–Brine System Has a Low Average Coordination Number

U. Olsson, B. Balinov, and O. Söderman

Division of Physical Chemistry 1, Chemical Center,
Lund University, P.O. Box 124, S-22100 Lund, Sweden

Abstract. We present data on surfactant and water self-diffusion in the L_3 and V_2 phases of the AOT-water-NaCl system. By comparing the data from the two phases and from a quantitative analysis of the L_3 phase data we arrive at the conclusion that the L_3 phase microstructure has a low (possibly 3-4) average coordination number.

1. Introduction

In certain cases, surfactants self-assembly into a bilayer which forms a multiply connected dividing surface, continuous in three dimensions [1]. This microstructure exists in a disordered form in the liquid L_3 phases or in a crystallized or ordered form in bilayer continuous cubic phases.

In some systems both the L_3 and cubic phases are present in the phase diagram. Here, a crystallization of the microstructure occurs at higher concentrations when interactions get sufficiently strong. One such system, which offers an interesting opportunity to compare the properties of the two phases, is the AOT-water-NaCl system [2], which was choosen for the present study. The L_3 phase of this system has recently been investigated by a number of different techniques [3-5].

In the cubic phases the bilayer topology, which can be quantified in terms of the coordination number of the two solvent labyrinths, can be determined from the space group. In the liquid L_3 phase only an average coordination number can be defined. In what follows, we present data on the self-diffusion of surfactant and water molecules in the L_3 and V_2 (cubic) phases. By comparing the results from the two phases and from a quantitative analysis of the L_3 phase data we draw conclusions regarding the average coordination number of the L_3 phase microstructure.

2. Results and Discussion

Data on molecular self-diffusion, conveniently obtained with the FTPGSE-NMR technique, provide unique and valuable information regarding microstructure in aggregated surfactant systems [6]. In a bilayer continuous microstructure, solvent molecules diffuse within the labyrinths while macroscopic transport of surfactant molecules occurs through lateral diffusion within the two monolayers. The

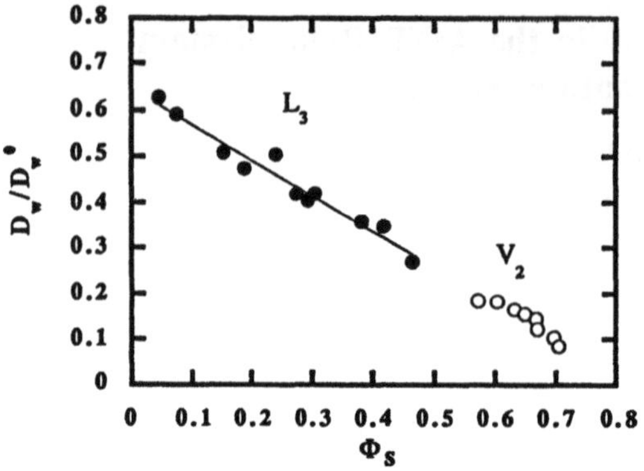

Figure 1. The variation of the reduced water self-diffusion coefficient with Φ_s in the L_3 (filled symbols) and the V_2 (open symbols) phases. The solid line corresponds to $D_w/D_w^0=0.66 - 0.77\Phi_s$ and is a linear fit to the data from the L_3 phase (see text). Data are taken from ref. [3].

geometrical obstruction effects on the solvent and surfactant self-diffusion was calculated by Anderson and Wennerström for some cubic phase model structures [7]. An important result of these calculations is that obstruction effects are topology dependent, *e.g.* the obstruction increases, leading to a lower diffusion constant, with increasing coordination number. This result was recently experimentally confirmed in a self-diffusion study of a sequence of cubic phases occuring on a dilution line in a ternary system [8].

In the present system, the water and surfactant self-diffusion coefficients, D_w and D_s, were measured at various concentrations in the L_3 and cubic (V_2) phases. The data are presented in Figures 1 (water) and 2 (surfactant) as a function of the surfactant volume fraction, Φ_s. The water diffusion data are presented relative to the diffusion constant in brine of the same salinity.

An important observation is the smooth variation of D_w and D_s across the L_3 to V_2 transition; D_w and D_s appears to be unaffected by the crystallization of the microstructure. This result demonstrates that the microstructure of the two phases are topologically related. Moreover, it indicates that diffusion properties derived for ordered structures also apply for disordered structures.

The cubic phase belongs to the spacegroup Ia3d [9] which implies a coordination number of three. Assuming similar topology dependent obstruction effects for a disordered microstructure, the smooth variation of the diffusion constants indicate that the L_3 phase microstructure has a low (≈3) average coordination number.

From the calculations of ref. [7] we have:

$$D_w/D_w^0 = a' - b'\Phi_b \quad (1)$$
$$D_s/D_{s1} = 2/3 - b''\Phi_b \;. \quad (2)$$

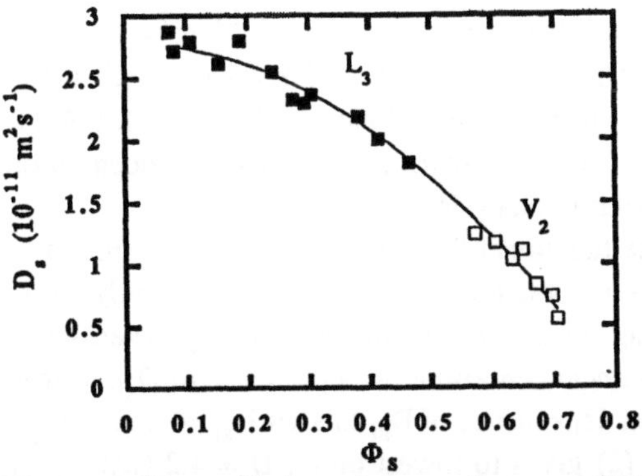

Figure 2. The variation of the surfactant self-diffusion coefficient with Φ_s in the L_3 (filled symbols) and the V_2 (open symbols) phases. The solid line corresponds to D_s=4.2 (2/3 - (b"+ 2c/3)Φ_s^2) 10^{-11} m^2s^{-1} (see text). Data are taken from ref. [3].

Here, D_{sl} is the surfactant lateral diffusion constant, Φ_b is the bilayer volume fraction and a', b' and b" are constants. a' is basically topology independent (a'-values in the range 0.65-0.67 were reported for various families of H-surfaces) while the constants b' and b" depend more strongly on the topology. Eqs (1) and (2) are accurate for $\Phi_b \leq 0.4$.

Extrapolating the water diffusion data to infinite dilution we obtain D_w/D_w^0=0.66 in agreement with a bilayer continuous structure. In order to analyze the Φ_s dependence, however, we need to take into account hydration effects. These can be estimated from spin relaxation and quadrupolar line splittings of water ^2H and ^{17}O nuclei suggesting that roughly 15 water molecules per AOT molecule are dynamically perturbed by the water-surfactant interface [10]. Furthermore, the reorientation dynamics in the perturbed state is less than a factor of ten, probably a factor of three, slower than in bulk [10].

A simple model that takes into account the hydration effects on the self-diffusion is a two-site discrete exchange model with a hydration and a bulk site. Here we assume that n_h water molecules per AOT hydrate the water-surfactant interface and that these water molecules are free to diffuse latterally within the hydration layer, with a diffusion constant D_{hw}^0. The remaining (bulk) water is obstructed by a bilayer volume fraction which equals Φ_s plus the volume fraction of the hydration water. With fast exchange between the bulk and the hydration states, eq. (1) is rewritten as [3]:

$$D_w/D_w^0 = a' - [b' + n_h \frac{\rho_s M_w}{\rho_w M_s}(a' + b' - \frac{2}{3}(D_{hw}^0/D_w^0))] \Phi_s \qquad (3)$$

where ρ and M refer to density and molecular weight respectively. Setting n_h=15 and D_{hw}^0/D_w^0=1/3, in accordance with spin relaxation data, the slope in Figure 1 corresponds to b'=0.27 which is consistent with a coordination number of four.

Extrapolating D_s to infinite dilution we obtain 2.8 10^{-11} m^2s^{-1} which, according to eq. (2), corresponds to D_{sl}=4.2 10^{-11} m^2s^{-1}. This value is higher than the lateral diffusion constant 2.7 10^{-11} m^2s^{-1} measured previously at 60 wt.% AOT in the binary AOT-water L_α phase [11]. This indicates that D_{sl} is concentration dependent which complicates a quantitative analysis.

The D_s data follow roughly the functional form of eq.(2), however, from a fit (shown as the solid line in Figure 2) we obtain b"=1.1 which is significantly larger than 0.3-0.4 expected for a low average coordination number. It is easy to verify that the D_s data are not consistent with a linear concentration dependence of D_{sl}. Instead we will assume that the concentration dependence of D_{sl} follows $D_{sl} = 4.2\,(1 - c\Phi_s^2)\,10^{-11}\,m^2s^{-1}$ which inserted in eq. (2) gives to lowest order: $D_s = 4.2\,(2/3 - (b'' + 2c/3)\Phi_s^2)\,10^{-11}\,m^2s^{-1}$. From our two values of D_{sl} we obtain c=1.2. and with this value, the concentration dependence of D_s is consistent with b"=0.3. Hence, if our assumption regarding the concentration dependence is correct, the D_s data from the L_3 phase are quantitatively consistent with a bilayer continuous microstructure with a low (< 6) average coordination number.

To summarize, we conclude that the L_3 phase microstructure in the AOT-brine system has a low average coordination number.

References

1. D. M. Anderson, H. Wennerström and U. Olsson, *J. Phys. Chem.* **93**, 4243 (1989).
2. K. Fontell, in "*Colloidal Dispersions and Micellar Behavior;*" ACS Symposium Series No. 9; American Chemical Society: Washington, DC, 1975 p 270.
3. B. Balinov, U. Olsson and O. Söderman, *J. Phys. Chem.* **95**, 5931 (1991).
4. M. Skouri, J. Marignan and R. May, *P. Colloid Polym. Sci.*, (in press, 1991).
5. R. Strey, W. Jahn, M. Skouri, G. Porte, J. Marignan, and U. Olsson, (submitted).
6. B. Lindman, K. Shinoda, U. Olsson, D. Anderson, G. Karlström and H. Wennerström, *Colloids and Surfaces* **38**, 205 (1989).
7. D. M. Anderson and H. Wennerström, *J. Phys. Chem.* **94**, 8683 (1990).
8. P. Ström and D. M. Anderson, (submitted)
9. K. Fontell, *J. Colloid Interface Sci.* **43**, 156 (1973).
10. G. Carlström and B. Halle, *Langmuir* **4**, 1346 (1988).
11. G. Lindblom and H. Wennerström *Biophys.Chem.* **6**, 167 (1977).

Hydrodynamic Modes of a Viscoelastic Membrane or Interface

H. Pleiner[1] *and J.L. Harden*[2]

[1]Department of Physics, Universität Essen, W-4300 Essen 1, Fed. Rep. of Germany
[2]Department of Applied Physics, Nagoya University, Nagoya 464, Japan

Abstract. We discuss the general hydrodynamic mode structure of a thin, flat polymeric (viscoelastic) layer between two Newtonian fluids. Dispersion relations and dynamic structure factors are obtained by deriving generalized boundary conditions at the interface.

1. Introduction

The surface and interfacial properties of viscoelastic material has become the subject of growing scientific and technological interest. Especially the dynamical properties of surfaces and interfaces have come into experimental reach. Here we deal with the case of a thin viscoelastic layer serving as an (in equilibrium flat) interface between two (different) Newtonian fluids. This scenario models different relevant situations like e.g. a polymeric membrane or an oil film floating on top of a simple liquid, like e.g. the surface of a dilute polymer solution with its vapor, if the surface tension of the solvent is much larger than that of the polymer (and the polymer is strongly concentrated at the surface [1]), like e.g. a viscoelestic membrane separating different fluids. We start with the proper hydrodynamic description of the three phases involved (Newtonian$_1$ – viscoelastic – Newtonian$_2$), which are linked by the appropriate boundary conditions. Taking the limit of a very thin viscoelastic layer the situation can be reduced to an effective two phase system (Newtonian$_1$ – Newtonian$_2$) with generalized boundary conditions in between, which contain the relevant influence of the viscoelastic layer. In contrast to previous similar treatments e.g. [2] we do not postulate such generalized boundary conditions, but derive them from the hydrodynamic properties of the viscoelastic layer. Our approach is quite similar to [3], [4], although we use a more realistic description of the viscoelasticity. The method can easily be generalized to deal with membranes of different or additional internal structures and dissipation mechanisms, like e.g. liquid crystalline and polymer liquid crystalline or even biological membranes.

2. Boundary Conditions

Boundary conditions amending the hydrodynamic equations in the case of finite geometries are generally obtained by integration the differential equations over an infinitesimally small volume that contains the boundary. Assuming that the viscoelastic layer (membrane) itself is infinitesimally thin this procedure leads for the momentum conservation equation to

$$\sigma_{iz}^{(1)} - \sigma_{iz}^{(2)} + \nabla_j \sigma_{ij}^{m} = \hat{L} \, , \tag{1}$$

where $\sigma_{ij}^{(\cdot)}$ is the stress tensor of the Newtonian fluids 1, 2, and σ_{ij}^{m} is the stress tensor (times the thickness) of the membrane. Eq. (1) has to be taken at $z = 0$ (linearized theory) with the z-axis being the surface normal. We have added on the r.h.s. of eq. (1) the well-known Laplacean stress $\hat{L}$ due to the surface tension of the liquids [5]. For the flow we use no-slip conditions at the surface

$$v_i^{(1)} = v_i^{(2)} = v_i^{m} \quad \text{and} \quad \nabla_z v_z^{(1)} = \nabla_z v_z^{(2)} \, , \tag{2}$$

where the latter relation holds for incompressible Newtonian fluids (the membrane is not assumed to be incompressible in the following). Since we are neglecting the thermal degree of freedom, we do not need boundary conditions for the heat flow.

It is straitforward to express in eq. (1) the stress tensor of the Newtonian fluids by the velocity and pressure fields. The subtleties lie in the viscoelastic properties of the membrane stress tensor. Here we use the hydrodynamic description of viscoelasticity that treats explicitly the (short time) elastic degree of freedom as an additional macroscopic (slow) variable [6], which relaxes, however, in a finite time. This concept has already proven fruitful in the hydrodynamic description of liquid crystalline polymers [7], [8]. For concentrated polymer solutions the results obtained by this procedure are equivalent [9], [10] (under the present restrictions) to those of a model [11] of two strongly interacting "fluids": one Newtonian liquid and one Eulerian solid. For plain waves (with the wave vector k in x-direction and with the frequency ω) we find

$$\nabla_j \sigma_{ij}^{m} = C_{(i)} (k^2 / i\omega) v_i^{m} \, , \tag{3}$$

where the $C_{(i)}$ are frequency- (and for $i = z$ also k-) dependent and describe the elastic and dissipative response of the membrane to longitudinal (density) variations, to in-plane shear deformations, and to out-of-plane stretching and bending of the membrane, respectively,

$$C_{(x)} = \chi + i\omega\nu_1 + \frac{i\omega c_1}{i\omega + c_1/\tau_1} \, , \tag{4a}$$

$$C_{(y)} = i\omega\nu_2 + \frac{i\omega c_2}{i\omega + c_2/\tau_2} \, , \tag{4b}$$

$$C_{(z)} = c_s + c_b k^2 + i\omega(\nu_s + \nu_b k^2) \, . \tag{4c}$$

Here $c_{1,2}$ are the (short time) elastic constants of the membrane, $\tau_{1,2}$ the relaxation coefficients (i.e. the Maxwell times multiplicated by the elastic constants), χ the (liquid-like bulk) compressibility of the membrane, $\nu_{1,2}$ the (Newtonian) viscosities of the membrane, c_s, c_b the surface stretching and bending elastic constants of the membrane, and ν_s and ν_b their dissipative counterparts. The out-of-plane coefficients (c_s, ν_s) are generall different from the in-plane quantities (c_2, ν_2), since in three dimensions the membrane is effectively a uniaxial system and the material parameters are of the uniaxial form (cf. eq.(2.3) in [7]). The complete relation between the material parameters introduced here and the (directly measurable) isotropic bulk parameters will be given elsewhere. The Laplacean stress $\hat{L}$ simply renormalizes c_s in eq. (4c) by adding the surface tensions of the two Newtonian fluids.

3. Normal Modes and Structure Factors

It is straitforward to calculate the normal mode structure from eqs. (1)–(4) for surface/interface waves, which decay exponentially normal to the surface, $\sim \exp(-q\,|z\,|)$. It is well known [2]–[4], that there are three different types of modes, which are of in-plane-shear, in-plane-compression and transverse-undulation character, according to $C_{(y)}$, $C_{(x)}$, $C_{(z)}$, respectively. The first one is decoupled from the others and in addition it is hardly measurable, so we will concentrate on the latter two. However, even in the special case that they too are decoupled from each other (i.e. for the symmetric case, fluid 1 $\equiv$ fluid 2), they show a very complicated dispersion relation due to the complicated frequency and wave vector dependency of the $C_{(i)}$-expressions

$$2\,\omega^2\rho q = C_{(z)}k^3(q-k) \qquad \text{and} \qquad 2\,\omega^2\rho = C_{(x)}k^2(q-k)\ , \tag{5}$$

where for the complex and frequency dependent transverse wave vector $q(k,\omega)$ one finds $(q/k)^2 = 1 + i\omega/(\nu k^2)$, with ρ and ν the density and the viscosity of the Newtonian fluids; the membrane parameters enter via eq.(4b,c). Depending on k and on the material parameters the low frequency normal modes contained in eq.(5) can be propagating or diffusive with rather unusual wave vector dependencies. Details will be given elsewhere.

Another interesting special case we want to mention briefly is a free surface covered by the membrane. Here the modes are coupled

$$\begin{aligned}\omega^2(\omega - 2i\nu k^2)^2 + 4\nu^2\omega^2 q k^3 + \omega^2 k^2(kC_{(x)} - qC_{(z)})/\rho \\ - C_{(x)}C_{(z)}k^5(k-q)/\rho^2 = 0\ ,\end{aligned} \tag{6}$$

where again a good deal of the frequency dependence is hidden in q and the $C_{(i)}$.

In order to make contact with recent light scattering experiments we calculate the dynamical structure factor $S(k,\omega)$ for the case of a free surface by the usual method of linear fluctuating hydrodynamics. Here the hydrodynamic currents are amended by fluctuating parts with zero means and variances given by generalized fluctuation–dissipation relations. These are then used to calculate the correlation function of transverse surface fluctuations from the linearized hydrodynamic equations including the boundary conditions. We obtain

$$S(k,\omega) = 8k_BT\,(\nu/\rho)\,k^3\,\Lambda(k,\omega)\,|\,N(k,\omega)\,|^{-2}\ , \tag{7}$$

where $N(k,\omega)$ is the expression on the left hand side of eq.(6) and $\Lambda(k,\omega)$ a very complicated function of wave vector and frequency, which we do not write down here. The eigenfrequencies in eq.(6) give rise to the peaks in S, while Λ governs their strength. Further detailed studies to connect these results with experiments are necessary and under way.

Support of H.P. by the Deutsche Forschungsgemeinschaft is gratefully acknowledged.

References

[1] The opposite case of a homogeneous or even depleted polymer surface has been discussed in J.L. Harden, H. Pleiner, and P.A. Pincus, *J.Chem.Phys.* **94**, 5208 (1991).

[2] J. Lucassen and M.v.d. Tempel, *J.Coll.Int.Sci.* **41**, 491 (1972).

[3] L. Kramer, *J.Chem.Phys.* **55**, 2097 (1971).

[4] C.F. Tejero, M.J. Rodriguez and M. Baus, *Phys.Rev.* **A29**, 2179 (1984).

[5] V.G. Levich, *Physiochemical Hydrodynamics*, Prentice Hall, Englewood Cliffs p. 372 (1962).

[6] H.R. Brand, H. Pleiner, and W. Renz, *J.Phys.(Paris)* **51**, 1065 (1990).

[7] H. Pleiner and H.R. Brand, *Mol.Cryst.Liq.Cryst.* **199**, 407 (1991).

[8] H.R. Brand and H. Pleiner, *Makromol.Chem.Rap.Commun.* **11**, 607 (1990).

[9] H. Pleiner, J.L. Harden, and P.A. Pincus, *Euro.Phys.Lett.* **7**, 383 (1988).

[10] J.L. Harden, H. Pleiner, and P.A. Pincus, *Langmuir* **5**, 1436 (1989).

[11] P.-G. de Gennes and P.A. Pincus, *J.Chemie Phys.* **74**, 616 (1977).

The Undulation Mode of Freely Suspended Liquid Films

H. Pleiner and H.R. Brand

Department of Physics, Universität Essen,
W-4300 Essen 1, Fed. Rep. of Germany

Abstract. We argue that for real, freely suspended liquid films the undulation mode (transverse sound mode) has a linear dispersion relation of the usual sound-like structure, i.e. a damping contribution proportional to the wave vector squared. Thus, there is no need for introducing nonlinear fluctuation effects, which would be necessary, if the regular damping mechanism would be absent, and which would lead to singular contributions in the damping.

1. Introduction

The physical properties of liquid films have found growing interest in recent years. Very widespread are thin (at least bilayer) smectic liquid crystal films, bilayer membranes, monolayer interfaces, or soap films [1].

Here we discuss the transverse dynamics (undulation mode or transverse sound mode) of those real, liquid films in the hydrodynamic limit. In contrast to two recent publications [2],[3], we argue that there *are* internal dissipative processes within the liquid films, which provide the usual sound-like damping ($\mathrm{Im}\,\omega \sim k^2$) of the undulation mode with frequency ω and wave vector k, even in the absence of any surrounding medium. If there were no damping $\sim k^2$ present, the nonlinearities in the hydrodynamic equations would provide the coupling to and scattering at thermal fluctuations giving rise to an anomalous, renormalised dispersion relation.

On the other hand for mathematical, i.e. ideally two-dimensional surfaces the arguments in [2] and [3] apply and there is no damping in order k^2. But in addition, there is also no sound-like propagation ($\mathrm{Re}\,\omega \sim k$) in such mathematical surfaces due to the lack of surface tension.

2. Undulation Mode in Real Films

In the incompressible limit the transverse momentum conservation for thin flat films is expressed by the linearized equation (gravity neglected)

$$\rho \dot{v}_z - \nabla \Phi = \nu_2 \Delta v_z, \tag{1}$$

where any dependence on the position across the film can be neglected. For the same reason it is sufficient to use one displacement field **u** to describe the non-equilibrium position of the film. The elastic stress Φ in eq. (1) is due to the surface tension, which is proportional to the change of surface area [4], i.e. $\Phi = \gamma \nabla u_z$, where γ is the surface tension divided by the layer thickness. (In addition also bending gives a linear restoring force $\nabla\Phi = -K\,\Delta^2 u_z$.) The viscous contribution (ν_2) in eq.(1) describes dissipative processes within the film during shearing. It has been argued in [2] and [3] that those dissipative processes have to be absent, because for films shearing and rotating would be equivalent processes. Obviously, this is not true for real films. After a constant rotation the film is still in a perfect equilibrium state, where no elastic or dissipative restoring forces arise. The state obtained after shearing however, is a non-equilibrium state and dissipation occurs during the relaxation into any equilibrium state. Thus, shearing and rotation of a real film lead to fundamentally different states.

In the simplest approximation we are left with the kinematic relation $\dot{u}_z = v_z$, and eq. (1) leads to the usual sound-like dispersion relation. Thus nonlinear fluctuation effects are not a priori necessary to be considered for real films, only in the special case that ν_2 is extraordinarily small.

3. Undulation Mode in Mathematical Films

For mathematical films, which are infinitesimally thin (and thus also of laterally infinite extent in order to have a non-vanishing total physical mass), there is no compression and no stretching during shear and thus no compressional elasticity and no surface tension ($\gamma \equiv 0$). But such a system can have curvature elasticity ($K \neq 0$).

Since shearing and rotating are identical operations for mathematical films, it is clear that shear viscosity cannot occur as a dissipative process, because (homogeneous) rotations must not lead to any entropy production (a homogeneously rotated system is still in an equilibrium state). Thus, $\nu_2 \equiv 0$ and in the momentum conservation equation only dissipation due to bending is allowed – a process also present in real films (permeation [5] or Gibbs diffusion [6]), but usually neglected against ordinary shear viscosity. Therefore, in the case of mathematical films a dispersion relation of the form $\omega = \pm Ek^2 + iHk^4$ is obtained (in linear approximation), which is not a sound-like excitation.

Support by the Deutsche Forschungsgemeinschaft is gratefully acknowledged.

References

[1] Y. Couder, J.M. Chomaz and M. Rabaud, Physica **D37**, 384 (1989).
[2] E.I. Kats and V.V. Lebedev, J. Phys. France **46**, 2093 (1985).
[3] Y. Marathe and S. Ramaswamy, J. Phys. France **51**, 2143 (1990).
[4] V.G. Levich, *Physiochemical Hydrodynamics*, (Prentice Hall, Englewood Cliffs, 1962) p. 372.
[5] W. Helfrich, Phys.Rev.Lett. **23**, 372 (1969).
[6] J.M. Chomaz and B. Cathalau, Phys. Rev. **A41**, 2243 (1990).

Index of Contributors

Springer Proceedings in Physics

Managing Editor: H. K. V. Lotsch

1 *Fluctuations and Sensitivity in Nonequilibrium Systems*
Editors: W. Horsthemke and D. K. Kondepudi

2 *EXAFS and Near Edge Structure III*
Editors: K. O. Hodgson, B. Hedman, and J. E. Penner-Hahn

3 *Nonlinear Phenomena in Physics* Editor: F. Claro

4 *Time-Resolved Vibrational Spectroscopy*
Editors: A. Laubereau and M. Stockburger

5 *Physics of Finely Divided Matter*
Editors: N. Boccara and M. Daoud

6 *Aerogels* Editor: J. Fricke

7 *Nonlinear Optics: Materials and Devices*
Editors: C. Flytzanis and J. L. Oudar

8 *Optical Bistability III*
Editors: H. M. Gibbs, P. Mandel, N. Peyghambarian, and S. D. Smith

9 *Ion Formation from Organic Solids (IFOS III)*
Editor: A. Benninghoven

10 *Atomic Transport and Defects in Metals by Neutron Scattering*
Editors: C. Janot, W. Petry, D. Richter, and T. Springer

11 *Biophysical Effects of Steady Magnetic Fields*
Editors: G. Maret, J. Kiepenheuer, and N. Boccara

12 *Quantum Optics IV*
Editors: J. D. Harvey and D. F. Walls

13 *The Physics and Fabrication of Microstructures and Microdevices*
Editors: M. J. Kelly and C. Weisbuch

14 *Magnetic Properties of Low-Dimensional Systems*
Editors: L. M. Falicov and J. L. Morán-López

15 *Gas Flow and Chemical Lasers*
Editor: S. Rosenwaks

16 *Photons and Continuum States of Atoms and Molecules*
Editors: N. K. Rahman, C. Guidotti, and M. Allegrini

17 *Quantum Aspects of Molecular Motions in Solids*
Editors: A. Heidemann, A. Magerl, M. Prager, D. Richter, and T. Springer

18 *Electro-optic and Photorefractive Materials*
Editor: P. Günter

19 *Lasers and Synergetics*
Editors: R. Graham and A. Wunderlin

20 *Primary Processes in Photobiology*
Editor: T. Kobayashi

21 *Physics of Amphiphilic Layers*
Editors: J. Meunier, D. Langevin, and N. Boccara

22 *Semiconductor Interfaces: Formation and Properties*
Editors: G. Le Lay, J. Derrien, and N. Boccara

23 *Magnetic Excitations and Fluctuations II*
Editors: U. Balucani, S. W. Lovesey, M. G. Rasetti, and V. Tognetti

24 *Recent Topics in Theoretical Physics*
Editor: H. Takayama

25 *Excitons in Confined Systems*
Editors: R. Del Sole, A. D'Andrea, and A. Lapiccirella

26 *The Elementary Structure of Matter*
Editors: J.-M. Richard, E. Aslanides, and N. Boccara

27 *Competing Interactions and Microstructures: Statics and Dynamics*
Editors: R. LeSar, A. Bishop, and R. Heffner

28 *Anderson Localization*
Editors: T. Ando and H. Fukuyama

29 *Polymer Motion in Dense Systems*
Editors: D. Richter and T. Springer

30 *Short-Wavelength Lasers and Their Applications*
Editor: C. Yamanaka

31 *Quantum String Theory*
Editors: N. Kawamoto and T. Kugo

32 *Universalities in Condensed Matter*
Editors: R. Jullien, L. Peliti, R. Rammal, and N. Boccara

33 *Computer Simulation Studies in Condensed Matter Physics: Recent Developments*
Editors: D. P. Landau, K. K. Mon, and H.-B. Schüttler

34 *Amorphous and Crystalline Silicon Carbide and Related Materials*
Editors: G. L. Harris and C. Y.-W. Yang

35 *Polycrystalline Semiconductors: Grain Boundaries and Interfaces*
Editors: H. J. Möller, H. P. Strunk, and J. H. Werner

36 *Nonlinear Optics of Organics and Semiconductors*
Editors: T. Kobayashi

37 *Dynamics of Disordered Materials*
Editors: D. Richter, A. J. Dianoux, W. Petry, and J. Teixeira

38 *Electroluminescence*
Editors: S. Shionoya and H. Kobayashi

39 *Disorder and Nonlinearity*
Editors: A. R. Bishop, D. K. Campbell, and S. Pnevmatikos

40 *Static and Dynamic Properties of Liquids*
Editors: M. Davidović and A. K. Soper

41 *Quantum Optics V*
Editors: J. D. Harvey and D. F. Walls

42 *Molecular Basis of Polymer Networks*
Editors: A. Baumgärtner and C. E. Picot

43 *Amorphous and Crystalline Silicon Carbide II: Recent Developments*
Editors: M. M. Rahman, C. Y.-W. Yang, and G. L. Harris

Springer Proceedings in Physics

Managing Editor: H. K. V. Lotsch

44 *Optical Fiber Sensors*
Editors: H. J. Arditty, J. P. Dakin, and R. Th. Kersten

45 *Computer Simulation Studies in Condensed Matter Physics II: New Directions*
Editors: D. P. Landau, K. K. Mon, and H.-B. Schüttler

46 *Cellular Automata and Modeling of Complex Physical Systems*
Editors: P. Manneville, N. Boccara, G. Y. Vichniac, and R. Bidaux

47 *Number Theory and Physics*
Editors: J.-M. Luck, P. Moussa, and M. Waldschmidt

48 *Many-Atom Interactions in Solids*
Editors: R .M. Nieminen, M. J. Puska, and M. J. Manninen

49 *Ultrafast Phenomena in Spectroscopy*
Editors: E. Klose and B. Wilhelmi

50 *Magnetic Properties of Low-Dimensional Systems II: New Developments*
Editors: L. M. Falicov, F. Mejía-Lira, and J. L. Morán-López

51 *The Physics and Chemistry of Organic Superconductors*
Editors: G. Saito and S. Kagoshima

52 *Dynamics and Patterns in Complex Fluids: New Aspects of the Physics–Chemistry Interface*
Editors: A. Onuki and K. Kawasaki

53 *Computer Simulation Studies in Condensed Matter Physics III*
Editors: D. P. Landau, K. K. Mon, and H.-B. Schüttler

54 *Polycrystalline Semiconductors II*
Editors: J. H. Werner and H. P. Strunk

55 *Nonlinear Dynamics and Quantum Phenomena in Optical Systems*
Editors: R. Vilaseca and R. Corbalán

56 *Amorphous and Crystalline Silicon Carbide III, and Other Group IV-IV Materials*
Editors: G. L. Harris, M. G. Spencer, and C. Y.-W. Yang

57 *Evolutionary Trends in the Physical Sciences*
Editors: M. Suzuki and R. Kubo

58 *New Trends in Nuclear Collective Dynamics*
Editors: Y. Abe, H. Horiuchi, and K. Matsuyanagi

59 *Exotic Atoms in Condensed Matter*
Editors: G. Benedek and H. Schneuwly

60 *The Physics and Chemistry of Oxide Superconductors*
Editors: Y. Iye and H. Yasuoka

61 *Surface X-Ray and Neutron Scattering*
Editors: H. Zabel and I. K. Robinson

62 *Surface Science: Lectures on Basic Concepts and Applications*
Editors: F. A. Ponce and M. Cardona

63 *Coherent Raman Spectroscopy: Recent Advances*
Editors: G. Marowsky and V. V. Smirnov

64 *Superconducting Devices and Their Applications*
Editors: H. Koch and H. Lübbig

65 *Present and Future of High-Energy Physics*
Editors: K.-I. Aoki and M. Kobayashi

66 *The Structure and Conformation of Amphiphilic Membranes*
Editors: R. Lipowsky, D. Richter, and K. Kremer

67 *Nonlinearity with Disorder*
Editors: F. Abdullaev, A. R. Bishop, and S. Pnevmatikos

68 *Time-Resolved Vibrational Spectroscopy V*
Editor: H. Takahashi

69 *Evolution of Dynamical Structures in Complex Systems*
Editors: R. Friedrich and A. Wunderlin

Volumes 1–43 are listed on the back inside cover

MIX
Papier aus verantwortungsvollen Quellen
Paper from responsible sources
FSC® C105338

If you have any concerns about our products,
you can contact us on
ProductSafety@springernature.com

In case Publisher is established outside the EU,
the EU authorized representative is:
Springer Nature Customer Service Center GmbH
Europaplatz 3, 69115 Heidelberg, Germany

Printed by Libri Plureos GmbH
in Hamburg, Germany